Frontiers in Biotransformation
Vol. 1

Frontiers in Biotransformation Volume 1

Basis and Mechanisms of Regulation of Cytochrome P-450

Edited by

Klaus Ruckpaul and Horst Rein

Akademie-Verlag Berlin

1989

Gesamt-ISBN 3-05-500456-6
Vol. 1-ISBN 3-05-500457-4

Erschienen im Akademie-Verlag Berlin, Leipziger Straße 3—4, Berlin, DDR-1086

Lizenznummer: 202 · 100/486/88
Printed in the German Democratic Republic
Gesamtherstellung: VEB Druckhaus „Maxim Gorki", 7400 Altenburg
Lektor: Christiane Grunow
Gesamtgestaltung: Martina Bubner
LSV 1315
Bestellnummer: 7637949 (9090/1)
06400

Preface

An increasing number of papers in the field of biotransformation has led to a real need for review articles to keep abreast of the new information. This flow of papers implies that the frontier of our knowledge is moving forward rapidly, requiring a continuous re-evaluation of existing data and accepted principles. Consequently this series has been planned to publish overviews of the current state-of-the-art by experts actively working in the field. The editors do not aim to produce a textbook with well established data, rules and principles but rather to focus on the borderline of our knowledge by presenting innovative results, controversial ideas and aspects, and so to stimulate the progress of understanding.

The pivotal role of cytochrome P-450 in biotransformation is well established, but biotransformation involves more enzymes than only cytochromes P-450. Therefore, to provide the reader with an in-depth presentation of a well defined topic, biotransformation has been chosen as the more appropriate subject for this series rather than the narrower field of cytochrome P-450 research. Biotransformation as an essential biological process has developed into a research field with an interdisciplinary character in which the biochemist, biophysicist, pharmacologist, toxicologist, biologist and other bioscientists are all interested. The editors will keep this in mind. On the other hand the broad scope of the series will not interfere with the interest of readers with specialized interest because each volume will focus on a distinct topic such as regulation, induction or protein/lipid interaction.

Besides new results in basic research concerning theory, models, updating of thermodynamic and physicochemical data for better understanding of, for example, reaction mechanisms, structure/activity relationships and kinetically based interdependences, the editors intend also to follow carefully the progress that molecular biology has brought about, and will continue to, in classical biotransformation research and likewise to cover recent developments in which the role of cytochrome P-450 in integrated biological regulation processes has become evident.

But there are also close relationships between biotransformation and environmental pollution problems, for example, food additive toxicity, occupational diseases and toxicity of environmental pollutants. Clearly there is increasing interest in application of knowledge from biotransformation research. In addition to applications in toxicology, another growth area will undoubtedly be biotechnology based on progress in molecular biology and protein engineering. This series hopes to contribute to a successful linkage between basic research and such applications by providing up-to-date reviews of these topics.

Klaus Ruckpaul
Horst Rein

Contents

Introduction

K. Ruckpaul and H. Rein

Scientific knowledge — as a simplified rule — passes from the discovery of a new phenomenon (fact, subject, reaction) through its disintegrative analytical phase, to the integration of resolved detailed data and on into a more or less reconstituted system of mutual relationships. These general characteristics may similarly refer to cytochrome P-450 research which started about thirty years ago. Now it is one of the most intensively studied enzymes. In the last decade alone, more than 10,000 papers relating to cytochrome P-450 have been published. This abundance of information requires a comprehensive review to condense the data to the very essentials.

In the beginning, cytochrome P-450 investigations mainly dealt with physiological, pharmacological and toxicological aspects. These comprised studies on its occurrence in different species and tissues together with analyses of its localization within the cell. Finally, examinations of its implications in detoxifying and toxifying processes, including the responce to inducers and the quantitation of the degree of induction.

The analytical phase is characterized by the isolation and purification of the essential components of the different cytochrome P-450 systems, and the determination of their molecular weights, their spectral and other physicochemical properties. It is further characterized by studies on the broad substrate specificity of the catalytic activity and the analysis of the different pathways of oxygen activation. This period led to the recognition of the reaction cycle and the detection of the functional versatility and the structural multiplicity including sequencing of the primary structure of cytochrome P-450. The full-length (or nearly full-length) cDNA and/or amino acid sequences for about 80 gene products have currently been analysed. One of the high-lights in this period was the analysis of the 3-dimensional structure of cytochrome P-450cam in 1985.

The phase of a comprehensive treatise is characterized by an evaluation of accumulated analytical data, the integration of which permits deeper insights into the complex relationships which constitute the structural and more so the functional integrity of cytochrome P-450 systems. Molecular biology and its novel techniques have become of special importance by accelerating the process, thus enabling the alignment of different complete primary structures and the quantitation of overall homologies among these structures resulting in the statement that cytochromes P-450 exhibit a gene superfamily which — dependent on the inducer — can be divided into families and subfamilies. In addition, new experimental approaches have become available to investigate comparative and integrative considerations at different levels of integration.

Problems inherent to this stage include:

- which structural features of the multitude of compounds that interact with cytochromes P-450 determine their capability to control oxygenatic and oxidatic pathways of the enzyme?
- how is the catalytic activity of the complex enzyme system regulated?
- at what level of cytochrome P-450 biosynthesis is the induction process regulated in dependence on different inducers — and what consequences result with regard to active interference?
- to what extent is hormonal control of importance for the induction of sex specific cytochromes P-450, what is triggering that process and by what route is it accomplished?

The first volume of 'Frontiers in Biotransformation', following the considerations outlined above, is intended to review selected topics which all contribute to the concept of 'Regulation'.

Regulation is conceivable at different levels of integration. At the molecular level interactions can be amended by low molecular effectors (substrate, cosubstrate) and by high molecular effectors (reductase, cytochrome b_5). The sum of the interactions results in changes of either conformation, redox behaviour and/or turnover rates of cytochrome P-450. Due to the enzyme's membrane bound character the surrounding phospholipids provide a matrix which itself has been proposed as regulating enzymatic activity by affecting the e^--transfer through influencing the stability of the cytochrome P-450/reductase-complex and the organization of the enzyme system within the matrix. Finally, the physiologically most important regulation proceeds at the cellular level. Induction, inducibility as well as sex, species and tissue specificity determine via the isozymic distribution and the enzyme concentration the extent of toxification and detoxification.

A reliable comprehensive review naturally requires much accumulated data. Looking at the current state-of-the-art for various cytochromes P-450 it can be seen that the level of knowledge achieved to date differs greatly. As compared to the remarkable progress in the elucidation of structural and functional details in cytochrome P-450cam and the mammalian cytochromes P-450, the analytical period of microbial and plant cytochromes has just begun. This implies that a sufficient amount of data necessary for a comprehensive review of those cytochrome P-450 systems is still missing.

Due to its pharmacological and toxicological origin it is not surprising that the liver system and especially that of the endoplasmic reticulum has been most intensively studied. Consequently most papers published so far are concerned with this system. That emphasis is likewise reflected in the articles of volume 1. The concentration on the endoplasmic system, however, implies correspondingly less research on other systems. An extrapolation of mechanisms found valid for one distinct system has therefore to be carefully carried out taking into account the biochemical diversity of cytochrome P-450 dependent monooxygenatic systems so far characterized (soluble — membrane bound; endoplasmic — mitochondrial).

The present volume is introduced by an article which is mainly concerned with regulation of the liver exoplasmic cytochrome P-450 LM2 (cytochrome P-450 II B1 according to recently published recommendations for a unified nomenclature). Several considerations favour this chapter as an introductory one. (1) Studies on an isolated cytochrome P-450 isozyme which represents a simplified system are suited to the elucidation of principle mechanisms of action and regulation. (2) A remarkable amount of data on cytochrome P-450 II B1 has been accumulated which provides a reliable basis for such a review. (3) Cytochrome P-450 II B1, due to its involvement in biotransformation reactions of a huge number of exogenous substrates, drugs and environmental contaminants, is of high relevance with regard to pharmacological and toxicological implications.

In the following articles the scope outlined above is detailed, starting with a comprehensive contribution on model systems in chapter 2. In general, utilization of models has proved an appropriate means to convert complex biological systems into relevant subjects which can be experimentally examined. Therefore, metalloporphyrins are surveyed, the aim being to study the physicochemical, geometrical and electronic requirements for a catalytically-active metalloporphyrin and to derive conclusions on the chemical reactivity and functional diversity of the heme group in the enzyme. Moreover, the studies on models of cytochrome P-450 also represent a possible line which may initiate future developments of engineered synthetic enzymes having optimized functions.

A comprehensive review which is aimed at elucidating regulation phenomena has to consider the structural suppositions which are necessary for the catalytic function. Therefore, structural multiplicity which at least in part originates from the existence of isozymes is an essential for understanding the functional versatility and — based on these structure/function-relationships — to understand further the molecular basis of regulation. The data accumulated so far are summarized in chapter 3.

Cytochrome P-450-catalyzed reactions depend on the enzyme's capability to activate and split molecular oxygen by insertion of electrons. This unique function brings about functional versatility in that the insertion of a different number of electrons leads to differential degrees of oxygen activation, generating either an oxygenatic or an oxidatic pathway. Due to their reactivity, uncoupled activated oxygen species are of toxicological importance. Besides direct toxicological effects on cytochrome P-450, the formation of activated oxygen species represents an important side effect of monooxygenatic reactions, the regulation of which is discussed in chapter 4.

A more complex mechanism of cytochrome P-450 regulation proceeds at the cellular level. It is hormonal in nature and involved in sexually differentiated steroid drug metabolism. Great efforts in the last 5 years have revealed an immense progress in insight, not only into the molecular basis of the sex-specific isozymes but also into the regulation of their biosynthesis. The detailed knowledge about that differentiation may be assumed to have practical pharmacological consequences on sex-differentiated biotransformation recativities in the near future (chapter 5).

Molecular biology and especially genetic engineering will allow not only new experimental approaches to the analysis of structure/function relationships by site directed mutagenesis but also the construction of arteficial (e.g. chimeric) cytochromes P-450. By way of these new techniques the molecular basis of genetic relatedness of structural multiplicity and the mechanisms of its regulation have become available. Further these investigations include studies on the phylogenetic relations of evolutionarily distant cytochromes P-450. Based on the results derived therefrom, their application to clinical problems may become relevant.

The articles of this volume, being mainly concerned with endoplasmic cytochrome P-450, present comprehensive information on the principles of regulation. Therefore they contribute knowledge and strategies for integrational treatment of cytochrome P-450 systems outside this selection. Following these lines, data will be elaborated which may prove the general validity of the outlined regulation principles as challenging task for future progress.

Chapter 1

Regulation Mechanisms of the Activity of the Hepatic Endoplasmic Cytochrome P-450

K. Ruckpaul, H. Rein, and J. Blanck

1. Introduction

Hardly any other field of enzymology has developed as dynamically as that of cytochrome P-450 dependent monooxygenases. The reason for the extent and intensity of this progress is the key function of this enzyme system in the metabolism of endogenous substrates (steroids, fatty acids) as well as in the biotransformation of xenobiotics (drugs, environmental contaminants). However, cytochrome P-450 research is following the same general approach used to investigate other complex biological systems, based on an enhanced understanding of molecular mechanisms and structures.

Studies on cytochrome P-450 dependent monooxygenase systems over the last decade have resulted in the elucidation of biosynthetic pathways of various important classes of biologically active compounds (steroids, prostaglandins and vitamin D) and their regulation. In view of the increasing exposure to pollutants in the highly developed industrialized countries, chemical carcinogenesis has become of increasing importance. For many carcinogens the incidence of cancer is closely linked to the extent of activation and so the toxicity of xenobiotics is regulated by the activity of cytochrome P-450. The elucidation of this relationship has led to the study of toxicology at the molecular and cellular level. Recent results using molecular biology have allowed this enzyme system to be applied for diagnostic, therapeutic and biotechnological purposes.

In contrast to this remarkable progress, that molecular biology has already introduced into cytochrome P-450 research, analysis of the regulation mechanisms of its activity has shown less spectacular development. The biological importance of cytochrome P-450 and its ubiquity, however, demands an understanding of such mechanisms in detail. The purpose of the present survey is to bring together relevant results, to assess them critically and to deduce new insights.

Cytochrome P-450 dependent monooxygenases belong to the class of oxygen-activating enzymes which are known to transfer electrons to molecular oxygen thereby catalyzing biological oxidations. Biological oxidations in general proceed via two different mechanisms. One mechanism in which hydration and hydrolysis are involved is catalyzed by dehydrogenases. In this case the oxygen atom is derived from a water molecule. The other mechanism which involves the activation of molecular oxygen, is catalyzed by a class of enzymes which can be divided into oxidases and oxygenases. Oxidases transfer electrons from a substrate to oxygen, thereby oxidizing the substrate. Depending on the num-

Abbreviations used: PC: phosphatidylcholine; PE: phosphatidylethanolamine; PS: phosphatidylserine; PI: phosphatidylinositol; PA: phosphatidic acid; DLPC: dilauroyl PC; DMPC: dimyristoyl PC; DPPC: dipalmitoyl PC; DOPC: dioleoyl PC; PAH: polycyclic aromatic hydrocarbons; AHH: aryl hydrocarbon hydroxylase; PCB: polychlorinated biphenyl; PBB: polybrominated biphenyl; TCDD: 2,3,7,8-tetrachlorodibenzo-p-dioxin; PCN: pregnenolone-16α-carbonitrile; BNF: β-naphthoflavone; 3-MC: 3-methylcholanthrene.

ber of electrons which are transferred to oxygen, either superoxide anion, hydrogen peroxide or water is formed. Oxygenases, on the other hand, transfer oxygen to a substrate after reductive splitting of molecular oxygen (HAYAISHI, 1974). The electrons for this activation process are provided by the NADPH (NADH)-dependent cytochrome P-450 reductases (Fig. 1).

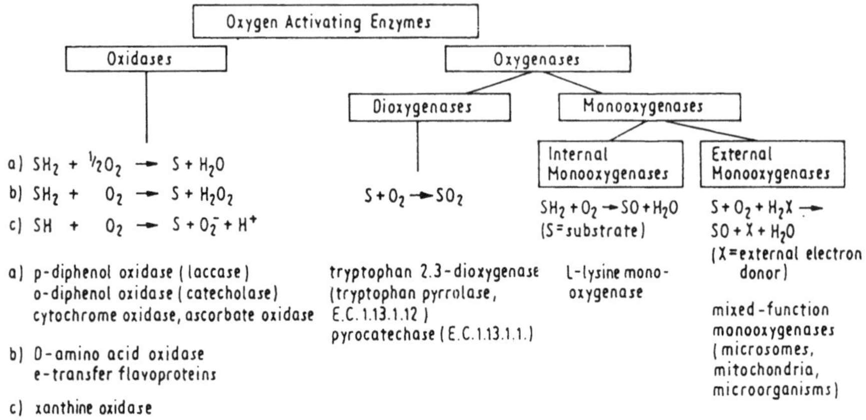

Fig. 1. Oxygen activating enzymes.

The basic mechanism of monooxygenases consists in the reductive cleavage of molecular oxygen and the splitting of bonds in the substrate for the insertion of oxygen, e.g. a C-H bond. From a thermodynamic point of view oxygenation reactions in general should proceed spontaneously, because the dissociation energy for O_2 (about 460 kJ/mol) and that of the C-H bond of the substrate (about 420 kJ/mol) is more than compensated by formation of the product bonds. The overall reaction (1) in the living cell can be formally described in terms of several intermediate steps (2)—(5) (JUNG and RISTAU, 1978)

$$RH + O_2 + NADPH + H^+ \xrightarrow{\Delta G} ROH + H_2O + NADP^+ \qquad (1)$$

$$NADPH + H^+ \rightarrow H_2 + NADP^+ \qquad \Delta G_1 = 19.3\ \text{kJ/mol} \qquad (2)$$

$$O_{2(g)} \rightarrow 2O_{(g)} \qquad \Delta G_2 = 460.5\ \text{kJ/mol} \qquad (3)$$

$$H_{2(g)} + O_{(g)} \rightarrow H_2O_{(l)} \qquad \Delta G_3 = -470.1\ \text{kJ/mol} \qquad (4)$$

$$RH_{(l)} + O_{(g)} \rightarrow ROH_{(l)} \qquad \Delta G_4 = -393.9\ \text{kJ/mol} \qquad (5)$$

$$\Delta G = -384.2\ \text{kJ/mol}$$

RH = cyclohexane.

Reactions with molecular oxygen, however, are kinetically restricted. The low reactivity of the oxygen biradical is due to the peculiar electronic configuration of the molecule. The triplet ground state is characterized by the following electron configuration: $(\sigma_s)^2$, $(\sigma_s^*)^2$, $(\sigma_p)^2$, $(\pi_p)^4$, $(\pi_p^*)^1$, $(\pi_p^*)^1$. The two unpaired electrons of the antibonding π_p^*-orbitals cause the paramagnetic spin state of the oxygen molecule which makes its reduction a spin-forbidden process with a high activation energy, because all stable organic compounds as reactants do not have unpaired electrons (Malmström, 1982). This kinetic barrier can be overcome by enzymatic activation of molecular oxygen.

As indicated by the thermodynamic data, an activation energy of at least 420—460 kJ/mol is expected for the overall reaction. The cytochrome P-450 catalyzed reaction, however, can proceed with an activation energy of only 40—70 kJ/mol (Jung and Ristau, 1978). This decreased activation energy is due to the cytochrome P-450 catalyzed reductive activation of the enzyme-bound molecular oxygen which results in a lengthening of the bond between the oxygen atoms from 1.21 Å in molecular oxygen via 1.28 Å in O_2^- to 1.49 Å in O_2^{2-} which finally leads to a cleavage of the oxygen molecule.

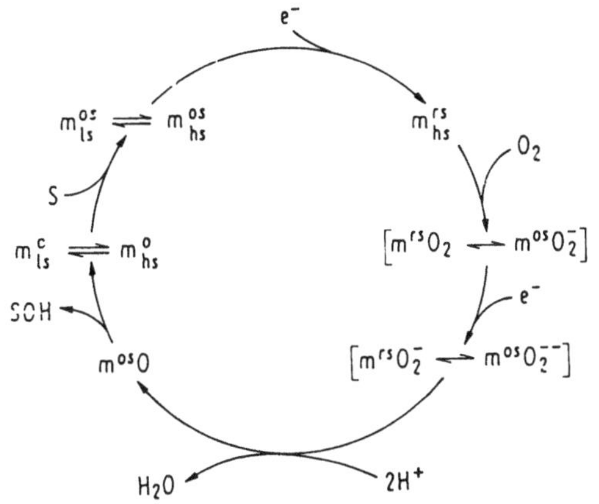

Fig. 2. Reaction cycle of spin shift regulated cytochromes P-450.

The cytochrome P-450 LM2 catalyzed multi-step reaction cycle (Fig. 2) first leads to the formation of an enzyme/substrate complex which is accompanied by shifting a low-spin/high-spin conformation equilibrium to the high-spin conformer. This partial reaction provides the first step with regulatory function in the whole reaction cycle. By insertion of one electron the complex formed by substrate binding is reduced to the ferrous heme iron complex which in a further step binds oxygen (ferrous dioxygen/cytochrome P-450/substrate complex). By intramolecular electron transfer the dioxygen coordinated ferrous heme iron is converted to the ferric form and superoxide anion. The oxygen species after insertion of the second electron is formally equivalent to a ferric iron coordinated peroxide complex, but the actual electron distribution is not known. The iron in the peroxide complex is in the formal oxidation state of +5 (perferryl) but could be highly resonance stabilized (White and Coon, 1980).

The binding energy between the oxygen atoms in ferric iron coordinated peroxide is too high to be split. Consequently in the next activation step the distance between the two oxygen atoms is further increased by acylation and release of water. The cleavage of the heme iron-bound dioxygen molecule has been proposed to proceed via one of two possible substrate-dependent pathways: the heterolytic and the homolytic cleavage (BLAKE and COON, 1981). In both pathways acylation of the iron-bound peroxide is included (WHITE et al., 1980). In a final step the hydroxylated product is liberated and the ferric form of the enzyme is ready to enter the next reaction cycle.

Due to the membrane-bound character of mammalian cytochrome P-450 systems and to its multicomponent nature, a reasonable basis for discussing the mechanisms by which activity of cytochrome P-450 is regulated would be the different molecular, membrane and cellular levels of integration.

For elucidating the basic principles of regulation, the microsomal polysubstrate monooxygenase system of the liver (EC 1.14.14.1), which consists of several forms of cytochrome P-450 and associated NADPH-cytochrome P-450 reductase provides an appropriate tool. The discussion in this chapter (except section 3) focuses mainly on the phenobarbital-inducible cytochrome P-450 LM2 because of its physiological importance, based on its broad substrate specificity and its well characterized physicochemical properties. This means that other endoplasmic, soluble and mitochondrial systems are largely ignored.

2. Regulation at the molecular level

Regulation at the molecular level is to be understood mainly as modulation of enzymatic activity by structural changes in cytochrome P-450. Such structural changes may be effected by low-molecular weight compounds (substrate or cosubstrate) as well as by interacting components of the monooxygenase system (cytochrome P-450 reductase, cytochrome b_5 and lipid).

The low substrate specificity of several cytochrome P-450 forms implies a dominant regulatory function of the substrate which exhibits two characteristics.

(i) Control of enzymatic activity arises from the ability of the substrate to actively induce structural changes in the cytochrome P-450 molecule which may result in changes of the spin equilibrium, the redox potential and the binding affinity towards interacting components, thus modulating the electron transfer reactions. These effects may be differently interrelated in different isozymes. A spin shift activity has been likewise observed for the interaction with cytochrome P-450 reductase (FRENCH et al., 1980), cytochrome b_5 (BENDZKO et al., 1982; TAMBURINI and GIBSON, 1983) and lipid (RUCKPAUL et al., 1982). The active substrate control towards the enzyme is functionally reflected in the activation of molecular oxygen and ultimately in the substrate conversion. In this way the enzyme regulates the electron supply for dioxygen activation and the substrate/oxygen interaction as well.

(ii) Substrate control of the enzymatic activity is further effective if the substrate is subject to biotransformation. The intrinsic chemical properties of the substrate carbon skeleton largely determine the peculiarities of the substrate/oxygen interaction and thus the mode of dioxygen cleavage (homo-/heterolytic) and substrate C-H splitting (without or with intermediate radical formation). Moreover, the reaction pathway, which may proceed by substrate oxidation or/and oxidative side reactions in different ratios is controlled in this way. Finally, the structural specificities of the substrate determine the rate and regio- and stereospecificity of the conversion process.

The intrinsic chemical properties of the substrate which control its effector activity are not the same as those which determine its conversion. Thus a wide divergence is exhibited in, for example perfluorinated hydrocarbons and SKF 525A, the action of which, despite their distinct effector activity, results in uncoupling (STAUDT et al., 1974) or even inhibition of the enzyme.

2.1. The importance of the axial heme iron ligands for the hemoprotein function

The specific function of hemoproteins is mainly determined by the axial heme iron ligands. Contrary to oxygen-binding proteins (myoglobin, hemoglobin) in which the fifth heme coordination position is occupied by a histidine residue, cytochrome P-450 is distinguished by a cysteine residue in this position (Fig. 3).

Previous indirect evidence for the presence of a thiolate group (RÖDER and BAYER, 1969; MURAKAMI and MASON, 1967) has been recently confirmed by RAMAN resonance studies indicating an iron-sulphur bond (CHAMPION et al., 1982). Further evidence for a thiolate ligand has been provided by determination of the X-ray structure of bacterial P-450 CAM crystals (POULOS et al., 1985, 1986). Sequence homologies in the Cys 357 region additionally suggest that in eukaryotic cytochrome P-450 a thiolate residue occupies the fifth heme iron coordination position.

The thiolate ligand leads not only to the specific function of cytochrome P-450 but also to the typical position of the Soret band of the CO-complex at 450 nm (JUNG et al., 1979). Thus the existence of the 450 nm band and the catalytic activity of the enzyme are correlated. In the denaturated state of the enzyme, i.e. cytochrome P-420, the iron-thiolate bond is strongly distorted with a longer sulfur-iron bond (JUNG et al., 1979) or the thiolate is replaced by another ligand (STERN et al., 1973). Infrared studies comparing the CO stretching frequencies of CO-complexes with different hemoproteins suggest that in the denatured state of cytochrome P-450 an imidazole is bound trans to CO instead of thiolate (BÖHM et al., 1979).

The electronic structure of the heme iron bound thiolate is characterized by $3p\pi$-orbitals completely filled with electrons resulting in its strong π-donor properties. In the dioxygen/heme complex of cytochrome P-450 (the dioxygen is bound in trans position to the thiolate) strong π-back donation from the iron to dioxygen is induced by the low 3pπ-orbital ionization potential of the thiolate ligand and its strong π-orbital overlap with the iron 3dπ-orbitals. The strong electronegativity of dioxygen favors this π-back donation which obviously is a prerequisite for reductive dioxygen splitting. The importance of the thiolate ligand for dioxygen activation is supported by experimental and theoretical analyses (HANSON et al., 1977; JUNG, 1980). MÖSSBAUER studies of the dioxygen/cytochrome P-450 complex (SHARROCK et al., 1976) result in

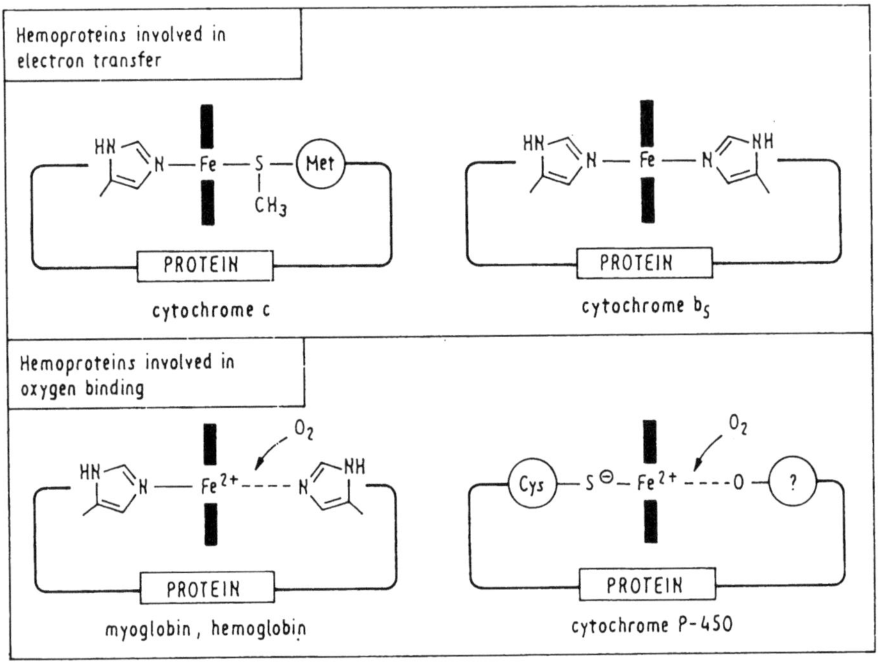

Fig. 3. Axial heme iron ligands of several hemoproteins.

isomer shifts comparable with those in ferric low-spin hemoproteins (LANG and MARSHALL, 1966) indicating a strong charge transfer from the iron to the dioxygen according to the ferric heme superoxide scheme of WEISS (1964) as postulated for oxyhemoglobin. But quantitative differences in the iron-to-dioxygen charge transfer should exist between oxyhemoglobin and cytochrome P-450, which are caused by the different ligands trans to the heme iron-bound dioxygen. Quantum chemical calculations of dioxygen/heme complexes with thiolate and imidazole nitrogen as fifth ligands, indicate a greater loosening of the dioxygen antibonding π-bond in the reduced proton-attached dioxygen/heme complex with the trans thiolate ligand than with the trans imidazole nitrogen ligand. Moreover, in the reduced P-450 dioxygen model the electron supplied is predominantly localized (67%) in the axial π-system as compared to only 32% in the hemoglobin model (JUNG, 1980; REIN et al., 1984).

Figure 4 illustrates the consequences associated with the supply of electrons into the oxygen molecule. The reduction significantly lowers the dioxygen dissociation energy which is linked to an increase in the distance between the two oxygen atoms as prerequisite for the cleavage of the intramolecular oxygen bond. Whether this cleavage occurs homo- or heterolytically is not clear yet (WHITE and COON, 1980; REIN et al., 1984) but may depend on the chemical nature of the substrate.

	oxygen	hydrogen peroxyradical	hydrogen peroxide	hydroxyradical	water
	$O_2 \xrightarrow[H^+]{1e}$	$HO_2^\bullet \xrightarrow[H^+]{1e}$ $\downarrow -H^+$ O_2^- superoxide anion	$H_2O_2 \xrightarrow[H^+, -H_2O]{1e}$ $\downarrow -H^+$ HOO^- hydrogen peroxide anion	$OH^\bullet \xrightarrow[H^+]{1e}$	H_2O
binding distance	1.21 Å	1.28 Å	1.49 Å		
splitting energy *($kJ\,mol^{-1}$)* *ΔH*	494	$276\,[O^-_{2(g)} \rightarrow O^-_{(g)} + O_{(g)}]$ $270\,[HO^\bullet_{2(g)} \rightarrow HO^\bullet_{(g)} + O_{(g)}]$	*heterolytic* $178\,[HO^-_{2(aq)} \rightarrow OH^-_{(aq)} + O_{(g)}]$ $122\,[HO_{2(aq)} + H^+_{(aq)} \rightarrow H_2O_{(l)} + O_{(g)}]$ *homolytic* $208-213\,[H_2O_2 - 2OH^\bullet]$		

$O_{(g)}$, gaseous phase; $O_{(aq)}$, aqueous phase; $H_2O_{(l)}$, liquid phase

Fig. 4. Binding distances and splitting energy on activation of dioxygen.

2.2. The relation between spectral changes and spin state

On binding of a substrate to cytochrome P-450, small but characteristic spectral changes of the heme chromophore in the near UV spectral region (Soret band) and in the visible spectral region are observed (NARASIMHULU et al., 1965; REMMER et al., 1966; SCHENKMAN et al., 1967; HAUGEN and COON, 1976). The substrate-induced spectral changes are titratable and from the difference spectra an apparent binding constant (K_s) can be evaluated. The spectral changes were classified by SCHENKMAN et al. (1967).

The majority of compounds which are metabolized by cytochrome P-450 produce a difference spectrum in the Soret region characterized by a minimum at 417 nm and a maximum at 387 nm; this is denoted as a type I difference spectrum. A number of compounds (alcohols, ketones and certain drugs, such as phenacetin) produce a so-called reverse type I spectrum, with a maximum at about 420 nm and a minimum at 385—390 nm (SCHENKMAN et al., 1972). Another group of compounds shift the Soret band to longer wavelengths. In these cases, in the difference spectrum, substrate specific absorption maxima between 425 and 445 nm and minima between 390 and 420 nm are formed.

These type II spectra are caused by direct interaction of the type II compounds with the heme. For many of these compounds, the majority of which are inhibitors, direct binding to the heme iron is assumed, with displacement of one of the axial ligands from the heme iron (SCHENKMAN et al., 1967; ULLRICH et al., 1975).

Independently, MITANI and HORIE (1969a, b) and WHYSNER et al. (1969, 1970) first suggested a correlation between type I and type II substrate induced spectral and spin state changes. Although the magnetic susceptibility of the substrate-induced spin transition has not yet been measured, the substrate-induced spectral changes of the Soret and the Q bands indicated a spin change in the heme complex. Indeed the spin state related spectral changes of substrate-induced cytochrome P-450 difference spectra could be simulated by methemoglobin with different ligands (REIN et al., 1981) in accordance with a linear correlation seen between the position of the Soret band and the paramagnetic susceptibility of ferric hemoglobin and myoglobin derivatives (SCHELER et al., 1957, HAVEMANN and HABERDITZL, 1958). With increasing paramagnetic susceptibility (μ), the Soret band is shifted to shorter wavelengths and vice versa. Because the Soret band of a mixed spin ferric hemoprotein is composed of two bands with different positions (the high-spin ($S = 5/2$) and the low-spin ($S = 1/2$) band at shorter and longer wavelength, respectively), any change in the spin state alters the observed Soret band(s) (SMITH and WILLIAMS, 1968). The exact position of the Soret band of the respective spin form varies with the hemoprotein. Methemoglobin derivatives with a mixed spin state exhibit only one unresolved Soret band because of the small distance between the relatively broad high-spin (~404 nm) and low-spin (~418 nm) bands. On the other hand the Soret bands of cytochrome P-450 are well separated: high-spin band ~387 nm, low-spin band ~418 nm. Therefore substrate-induced spin shifts are reflected in amplitude changes of both bands which are seen in both the difference and the absolute spectrum and especially in the second derivative spectrum.

As well as optical characterization, the electron paramagnetic resonance method is also suited to spin state determination. The simultaneous appearance of low-spin ($g \simeq 2$) and high-spin ($g \simeq 6$ or 8) signals in the EPR spectrum of a hemoprotein possessing both spin states (REIN and RISTAU, 1965) may be advantageous. However, because EPR measurements of hemoproteins require low temperatures (< 77 K) the application of this method is limited. Because the spin equilibrium is temperature dependent, the actual spin state at physiological temperature is not reflected in the low temperature EPR spectrum. Despite these restrictions a decrease in EPR low-spin signals ($g_x = 1.91$; $g_y = 2.26$; $g_z = 2.44$) of cytochrome P-450 could be observed upon addition of type I substrates (CAMMER et al., 1966; WHYSNER et al., 1969; MITANI and HORIE, 1969b). WATERMAN et al. (1973) found a quantitative correlation between the formation of the type I difference spectrum and the decrease in EPR low-spin signals with cyclohexane when using microsomal liver cytochrome P-450 from phenobarbital pretreated rats. Obviously it is harder to observe

EPR high-spin signals from cytochrome P-450 which are formed concomitantly with the disappearance of the low-spin signals during type I substrate binding. Only at the temperature of liquid helium could the typical high-spin signal of cytochrome P-450 at $g \simeq 8$ be observed (Tsai et al., 1970; Waterman et al., 1973). On the other hand high-spin signals seen in liver microsomes from 3-methylcholanthrene-treated rabbits at temperatures below 20 K by Peisach and Blumberg (1970) were additionally observed at 77 K (Friedrich et al., 1979).

A critical evaluation of the EPR results leads to the conclusion that the substrate-induced spin shift can be determined but in general this method is not suitable for analysing substrate/cytochrome P-450 interactions quantitatively.

Substrate titration experiments of cytochrome P-450 followed in the optical spectrum show that concomitant with a substrate induced (type I) increasing high-spin band, the low-spin band is lowered; conversely with inverse type I substrates the high-spin band decreased concomitant with an increase of the low-spin band. This characteristic behaviour is also observed with isolated isozymes such as P-450 LM2 (Ristau et al., 1979) and indicates that substrate modulates the ratio of the two spin states (high-spin, low-spin). Further substrate-specific effects which influence the structure of the enzyme obviously can be neglected, in agreement with the fact that only small changes occur in the CD-spectrum in the Soret region upon substrate binding of P-450 (Rein et al., 1976b). Substrate specificity is exhibited by the maximum of the induced spin shift ΔA_{max} (387 nm/417 nm) at substrate saturation of the enzyme.

2.3. The spin equilibrium of cytochrome P-450

The physical basis for the existence of spin equilibria in hemoproteins was shown by Griffith and Orgel (1957). These authors showed that octahedral d^5 complexes can exist in only two different magnetic and thermodynamic stable forms, namely either the low-spin state with a total spin of $S = 1/2$ or the high-spin state with a total spin of $S = 5/2$. The temperature dependence of the spin equilibrium indicates a low energy barrier between the two spin conformers which is explicable by ligand field theory. From the Tanabe-Sugano diagram (Tanabe and Sugano, 1954) it can be derived that heme complexes characterized by a spin equilibrium show a relatively small difference in ligand field strengths between the low-spin and the high-spin states. The ligand field strength is near the crossing over point (point π) of the low-spin term $^2T_2$ and the high-spin term $^6A_1$. Point π in the term diagram of d^5-complexes corresponds to the spin-pairing energy for an idealized ferric heme center of O_h symmetry (Rein and Ristau, 1978).

The increase in the high-spin state with higher temperature is observed in hemoproteins exhibiting high-spin/low-spin equilibria such as certain methemoglobin and metmyoglobin derivatives (Scheler et al., 1963; George et al., 1964;

Blanck and Scheler, 1970), peroxidase (Iizuka et al., 1968) and also certain catalase derivatives (Rein et al., 1968). Measuring the dependence of magnetic susceptibility on temperature Beetlestone and George (1964) and Kotani (1969) proved directly the existence of spin equilibria in some methemoglobin and metmyoglobin derivatives.

Based on temperature difference spectra which were obtained independently by Sligar (1976) with soluble bacterial P-450 CAM and by Rein et al. (1976a, 1977) with solubilized microsomal P-450 from phenobarbital-induced rabbit liver, a low-spin/high-spin equilibrium of P-450 could be deduced. Later Cinti et al. (1979) and Sligar et al. (1979) studied the spin equilibrium using microsomal P-450 from uninduced rat liver, and Rein et al. (1982, 1983) used microsomal P-450 from phenobarbital-induced rat liver. A thermodynamic analysis of the spin equilibrium was performed by Sligar (1976) with P-450 CAM, with mammalian P-450 using isozyme P-450 LM2 isolated from phenobarbital induced rabbit liver by Ristau et al. (1978), and with P-450 isolated from phenobarbital-induced rat liver by Tamburini and Gibson (1983) and Gibson and Tamburini (1984). Taken collectively the above authors have provided compelling evidence for a temperature-controlled, spin state equilibrium of cytochrome P-450 for a diverse variety of isozymes.

The temperature-induced difference spectra of cytochrome P-450 exhibit maxima and minima typical for high-spin and low-spin bands in the Soret and visible region as well (Rein et al., 1976a, 1977; Sligar, 1976). Therefore it was concluded that in cytochrome P-450 an equilibrium between two spin conformers exists, i.e. the low-spin state ($S = 1/2$) and the high-spin state ($S = 5/2$). Temperature-induced spectral characteristics revealed that with higher temperature the concentration of the high-spin conformer increases and vice versa. Remarkably, the temperature difference spectra are uniform in their shape independent of the absence or presence of a substrate or even the nature of the substrate used (type I). With type I substrates the amplitudes of the temperature difference spectra are increased reflecting the substrate-induced high-spin shift of the spin equilibrium (Rein et al., 1976a, 1977; Ristau et al., 1979).

The small energy difference between the two spin states is important for the activity of hepatic P-450 because broad substrate specificity, as in P-450 LM2, excludes a large energy profit upon substrate binding. Substrate binding, however, has to provide the energy to force the conformational change of P-450 necessary to start the reaction cycle. Based on the fact that optical spectra of such hemoproteins with intermediate spin values resulting from spin equilibria are composed of two basic spectra (the high-spin and low-spin spectra), the spin equilibrium in cytochrome P-450 has been analysed quantitatively using the optical spectra in the Soret region (Sligar, 1976; Ristau et al., 1979; Cinti et al., 1979; Tamburini and Gibson, 1983). Due to the existence of temperature dependent equilibria in both the absence and presence of substrates (Sligar, 1976; Ristau et al., 1978), two spin equilibria can be formulated.

Temperature dependent spin equilibria:

$$K_1 = \frac{[\text{P-450}^0_{\text{hs}}]}{[\text{P-450}^0_{\text{ls}}]} \qquad K_2 = \frac{[\text{P-450}^0_{\text{hs}} \cdot \text{S}]}{[\text{P-450}^0_{\text{ls}} \cdot \text{S}]} \tag{6, 7}$$

The independent substrate affinities of the two spin conformers (SLIGAR, 1976; RISTAU et al., 1978) result in two substrate-binding equilibria, i.e. the substrate-binding equilibrium of the low-spin conformer and that of the high-spin conformer

Substrate-binding equilibria:

$$K_3 = \frac{[\text{P-450}^0_{\text{ls}}]\,[\text{S}]}{[\text{P-450}^0_{\text{ls}} \cdot \text{S}]} \qquad K_4 = \frac{[\text{P-450}^0_{\text{hs}}]\,[\text{S}]}{[\text{P-450}^0_{\text{hs}} \cdot \text{S}]} \tag{8, 9}$$

Both spin equilibria and substrate-binding equilibria can be described in a thermodynamically closed four-state system (Scheme 1)

Scheme 1

$$\begin{array}{ccc} \text{P-450}^0_{\text{hs}} & \underset{K_4}{\overset{\pm\,\text{S}}{\rightleftharpoons}} & \text{P-450}^0_{\text{hs}} \cdot \text{S} \\ K_1 \updownarrow & & K_2 \updownarrow \\ \text{P-450}^0_{\text{ls}} & \underset{K_3}{\overset{\pm\,\text{S}}{\rightleftharpoons}} & \text{P-450}^0_{\text{ls}} \cdot \text{S} \end{array}$$

The constants K_1—K_4 of the spin and substrate equilibria can be determined. K_1 can be experimentally determined by means of temperature difference spectral titration. This method is rather inadequate for K_2 because of temperature dependent limitations of substrate saturation. Instead , K_2 is obtained by means of the substrate difference spectral titration which follows

$$\Delta E = \frac{c(K_2 - K_1)\,(\varepsilon_{\text{hs}} - \varepsilon_{\text{ls}}) \cdot S}{(1 + K_1)\,(1 + K_2)\,(K_s + S)} \tag{10}$$

where c represents the total enzyme concentration and ε_{hs}, ε_{ls} denote extinction coefficients of the high-spin and low-spin conformers at the respective specific wavelengths, and S is the substrate concentration. The observed mean substrate dissociation constant K_s is also determined by this procedure. At substrate saturation, equation (10) is reduced to

$$\Delta E_{\max} = \frac{c(K_2 - K_1)\,(\varepsilon_{\text{hs}} - \varepsilon_{\text{ls}})}{(1 + K_1)\,(1 + K_2)} \tag{11}$$

Further transformation of equation (11) by relating the observed spectral shift $\Delta E_{\max}$ to enzyme concentration c and extinction coefficient of the total spin transition $\Delta\varepsilon = \varepsilon_{hs} - \varepsilon_{ls}$ results in

$$\frac{\Delta E_{\max}}{c\Delta\varepsilon} = \frac{(K_2 - K_1)}{(1 + K_1)(1 + K_2)} = \Delta\alpha_{\max} \tag{12}$$

where $\Delta\alpha_{\max}$ represents the substrate specific spin equilibrium shift.

The substrate dissociation constants of the low-spin and high-spin conformers (K_3 and K_4) can not be experimentally determined, but they are involved in the observed mean substrate dissociation constant K_s. Thus, the individual constants can be determined using the following equations

$$K_3 = K_s \frac{(1 + K_2)}{(1 + K_1)} \tag{13}$$

and

$$K_4 = \frac{K_1 K_3}{K_2} \tag{14}$$

i.e. the latter equation relates the substrate-induced modification of K_1 to the substrate affinities of the high-spin and low-spin conformers.

The constants $K_1 - K_4$ as determined for cytochromes P-450 from different sources are shown in Table 1.

Table 1. Constants of spin and substrate-binding equilibria of several cytochromes P-450 according to the four-state coupling model

Source of cytochrome P-450	Substrate	K_1	K_2	K_3 (μm)	K_4 (μm)	References
P-450 LM2	benzphetamine	0.08	0.4	370.0	70.0	RISTAU et al. (1979)
P-450 PB-B	benzphetamine	0.07	0.6	250.0	29.0	GIBSON and TAMBURINI (1984)
P-450 CAM	camphor	0.08	15.0	9.0	0.05	SLIGAR (1976)

The difference in the substrate affinity between the two spin conformers is most pronounced in cytochrome P-450 CAM, which is characterized by a substrate-induced high-spin shift of more than 90% (SLIGAR, 1976). In agreement with this large effect, the substrate affinities differ by a factor of about 200. On the other hand smaller substrate-induced high-spin shifts of cytochrome P-450 LM2 agree with smaller asymmetric substrate binding; for example benzphetamine exhibits an approximately five-fold higher affinity to the high-spin conformer of cytochrome P-450 LM2 than to the low-spin conformer (RISTAU et al., 1978). In general, from the four-state model of the enzyme, it follows that the higher affinity of type I substrates to the high-spin conformer is the reason for the spin equilibrium shift towards the high-spin state, and conversely, the observed low-spin shift induced by inverse type I substrates is to be interpreted as a higher affinity of the substrates to the low-spin conformer. The existence of "type O spectra" (PIRRWITZ et al., 1982a) is explicable by the four-state model, too. Despite the high substrate affinity (K_s) determined from the EPR spectrum, an optical substrate difference spectrum is not observed because the respective substrate has the same affinity to both spin conformers.

The study of the temperature dependence of $K_1 - K_4$ allowed the determination of the thermodynamic parameters of the respective partial reactions for cytochrome P-450 LM2 (RISTAU et al., 1978) (Table 2).

Table 2. Equilibria and thermodynamic constants of the thermodynamic model for the interrelationship between substrate-binding (benzphetamine) and spin equilibrium of P-450 LM2

Equilibria	**K** (20 °C)	**ΔH** (kJ · mol⁻¹)	**ΔS** (J · mol⁻¹ · K⁻¹)	**ΔG** (kJ · mol⁻¹)
$K_1 = \frac{[P\text{-}450_{hs}]}{[P\text{-}450_{ls}]}$	0.08	−44.3 ± 0.8	−130 ± 5	−6.3 ± 0.8
$K_2 = \frac{[P\text{-}450_{hs} \cdot S]}{[P\text{-}450_{ls} \cdot S]}$	0.39	−38.0 ± 0.8	−121 ± 5	−2.5 ± 0.8
$K_3 = \frac{[P\text{-}450_{ls}][S]}{[P\text{-}450_{ls} \cdot S]}$	0.37 mM	31.4 ± 2.1	171 ± 9	−18.8 ± 2.5
$K_4 = \frac{[P\text{-}450_{hs}][S]}{[P\text{-}450_{hs} \cdot S]}$	0.07 mM	26.3 ± 2.1	171 ± 9	−23.8 ± 2.5

The low-spin/high-spin transition shows remarkably high entropy values which are a typical for pure spin state equilibria (GOODWIN, 1976) and denoted as configuration entropy in metal complexes (KÖNIG and RITTER, 1974). Caused by the large difference in thermodynamic freedom between the low-spin and high-spin states, this "excess" entropy is especially large in metalloproteins. The observed excess entropy for the spin state transition of cytochrome P-450 LM2 could originate from cooperative breaking of van der Waals heme/protein contacts (Ristau et al., 1979) and/or release of water molecules from the active site. Such an assumption agrees with the interpretation of X-ray data for cytochrome P-450 CAM by Poulos et al. (1986), who suggested that the displacement of the active site water molecules, including the iron-linked water, shifts both the redox potential to more positive values and the spin equilibrium towards the high-spin state. Release of water molecules or at least structural changes of iron-linked water at the spin transition (Rein and Ristau, 1978) should be important for the polarity of the heme environment, which determines the redox potential (Kassner, 1972). The entropy controlled heme/protein interaction, which in agreement with the low energy barrier between the two spin states should be connected with only small local conformational changes in the protein, could represent the molecular basis for the observed spin dependent redox properties of cytochrome P-450.

2.4. The significance of the spin equilibrium in molecular regulation

2.4.1. The redox potential of cytochrome P-450

Cytochromes P-450 are characterized by high negative oxidation-reduction potentials without pronounced isozymic or species differences (Table 3).
This property is related to the presence of both the thiolate ligand and the local nonpolar heme environment. Substrate binding induces a shift of the negative redox potential to more positive values. This was established at an early date for cytochrome P-450 CAM. Concomitantly with a strong (~90%) high-spin shift in camphor binding, the redox potential of this microbial cytochrome P-450 is changed from −303 mV to −173 mV (Sligar, 1976). In contrast the redox potential of cytochrome P-450 LM2 is altered less on substrate binding, with corresponding smaller substrate-induced spin shifts.

Sies and Kandel (1970) observed a small change of 16 mV to more positive values in microsomal cytochrome P-450 with hexobarbital. Guengerich et al. (1975) did not find any significant change in the redox potential of phenobarbital-induced liver cytochrome P-450 upon binding of benzphetamine, which in the absence of substrate exhibits a redox potential of −330 mV. Sligar et al. (1979) measured the redox potential of isolated and partially purified P-450 from uninduced rat liver, both in the presence and absence of type I substrates. A redox potential of −300 mV was found (relative to the standard hydrogen electrode) in the absence of substrates. Contrary to earlier measurements, the addition of hexobarbital or benzphetamine to the partially purified P-450 increased the redox potential distinctly to −237 mV and −225 mV, respectively.

Table 3. Oxidation-reduction potentials of several cytochromes P-450

Source of P-450	Induction/Isozyme	Experimental conditions	E_0' (mV)	References
Pseudomonas putida		− camphor + camphor	−303 −173	SLIGAR, 1976
Rabbit liver	microsomes, phenobarbital induced	without substrate	−340	WATERMAN and MASON, 1972
		− benzphetamine + benzphetamine	−330	GUENGERICH et al., 1975
	P-450 LM2 and microsomes, phenobarbital induced	soluble P-450 LM2 and in vesicles	∼−340	BÄCKSTRÖM et al., 1983
	P-450 LM4	pH 7.4	−360	HUANG et al., 1986
Rat liver	partially purified, uninduced	− substrate + hexobarbital + benzphetamine	−300 −237 −225	SLIGAR et al., 1979
	microsomes, phenobarbital induced	without substrate	−362	
	$P\text{-}450_{PB-B}$	+ DLPC + DLPC, + benzphetamine + DLPC, + cyclohexane from low-spin band from high-spin band + DLPC + DLPC, ethylmorphine	−319 −311 −306 −300 −286 −336 −346	GUENGERICH, 1983*

* selected values

Table 3 — Continued

Source of P-450	Induction/Isozyme	Experimental conditions	E_0' (mV)	References
Bovine adrenals	mitochondrial P-450, partially purified	without external substrate	−400	COOPER et al., 1970
	$P\text{-}450_{scc}$	low spin (predominantly)	−412	LIGHT and ORME-JOHNSON, 1981
		high spin (predominantly)	−297	
		low spin + cholesterol, pH 7.4	−305	
		pH 7.0	−284	LAMBETH and PEMBER, 1983
	$P\text{-}450_{11\beta}$	pH 7.5, + desoxycorticosterone	−286	HUANG et al., 1986

2.4.2. The relation between redox potential and spin state

The treatment of the redox equilibrium of cytochrome P-450 requires consideration not only of the valence states of the iron but also of the substrate-binding and the spin equilibrium. Hence a thermodynamic three-dimensional interaction model for cytochrome P-450 CAM was presented by SLIGAR (1976) which involves the linkage between substrate binding, spin and redox states.

This complete coupling model, however, can be simplified as follows. At least in cytochrome P-450 LM2 and related phenobarbital-inducible low-spin isozymes, substrate-binding affects only the spin equilibrium, and substrate-specific conformational changes of the enzyme can be neglected. This is in accordance with a weak nonspecific enzyme/substrate interaction (MISSELWITZ et al., 1976, 1977), which is reflected in almost unchanged CD-spectra in the Soret region of solubilized cytochrome P-450 from phenobarbital-induced rabbit liver on substrate binding (REIN et al., 1976b). Based on these results the thermodynamic "cube" model of SLIGAR (1976) can be simplified to a planar scheme where P-450 (Fe^{3+}) and P-450 (Fe^{2+}) denote either substrate-free or substrate-saturated ferric and ferrous cytochrome P-450, respectively, K_a^0 and K_a^r are the spin equilibrium constants for both valence states, K_b^{hs} and K_b^{ls} the redox equilibrium constants for high-spin and low-spin cytochrome P-450, respectively (Scheme 2)

Scheme 2

$$\begin{array}{ccc}
\text{P-450 }(Fe^{3+})_{hs} & \underset{\pm e^-}{\overset{K_b^{hs}}{\rightleftharpoons}} & \text{P-450 }(Fe^{2+})_{hs} \\
\Big\Updownarrow K_a^{0} & & \Big\Updownarrow K_a^{r} \\
\text{P-450 }(Fe^{3+})_{ls} & \underset{\pm e^-}{\overset{K_b^{ls}}{\rightleftharpoons}} & \text{P-450 }(Fe^{2+})_{ls}
\end{array}$$

Using this approach an effective redox equilibrium constant K_b^{eff} can be derived

$$K_b^{eff} = \frac{[\text{P-450}(Fe^{2+})_{hs}] + [\text{P-450}(Fe^{2+})_{ls}]}{[\text{P-450}(Fe^{3+})_{hs}] + [\text{P-450}(Fe^{3+})_{ls}]} \tag{15}$$

A further simplification can be introduced due to the fact that the ferrous cytochrome P-450 obviously exists only in the high-spin state ($S = 2$) as demonstrated by several experimental approaches (Mössbauer spectroscopy, Sharrock et al., 1973; magnetic susceptibility, Champion et al., 1975; magnetic circular dichroism spectra, Dawson et al., 1978).

The omission of the P-450 $(Fe^{2+})_{ls}$ term in equation (15) leads to

$$K_b^{eff} = \frac{[\text{P-450}(Fe^{2+})_{hs}]}{[\text{P-450}(Fe^{3+})_{hs}] + [\text{P-450}(Fe^{3+})_{ls}]} \tag{16}$$

$$= K_b^{hs} \cdot \frac{K_a^{0}}{1 + K_a^{0}} \tag{17}$$

$$= K_b^{hs}\, \alpha \tag{18}$$

where α represents the high-spin fraction of the ferric cytochrome P-450, $\alpha = [\text{P-450 }(Fe^{3+})_{hs}]/[\text{P-450 }(Fe^{3+})_{hs}] + [\text{P-450 }(Fe^{3+})_{ls}]$. The redox potential of cytochrome P-450 can now be calculated by use of the Nernst equation after introduction of the derived relationships

$$E_m = \frac{RT}{F} \ln K_b^{eff} = \frac{RT}{F} \ln K_b^{hs} + \frac{RT}{F} \ln \alpha \tag{19}$$

A linear dependence of the redox potential E_m on the logarithm of the high-spin fraction α as predicted by the above equation was shown using isolated cytochrome P-450 LM2 in the presence of different tertiary amines (Fig. 5) (Rein et al., 1989). The substrate free cytochrome P-450 LM2 exhibits a redox potential of -359 mV which is changed to more positive values by substrates resulting in e.g. $-$ 317 mV in presence of benzphetamine. The linear relationship $E_m/-\log \alpha$ clearly demonstrates the regulation of a cytochrome P-450 redox equilibrium via the spin state of the heme iron. The redox potential of the pure high-spin state of cytochrome P-450 LM2 was determined by extrapolation with -296 mV[1]. This value is more negative than the extrapolated redox potential of the high-spin state of the microbial cytochrome P-450 CAM of -175 mV (SLIGAR et al., 1979). This difference obviously reflects species-specific redox potentials. Therefore a common redox potential of the high-spin state for different cytochromes P-450 as has been proposed previously (REIN et al., 1979; SLIGAR et al., 1979) is not justified. The existence of species and isozyme differences of the redox potential is supported by further data listed in Table 3.

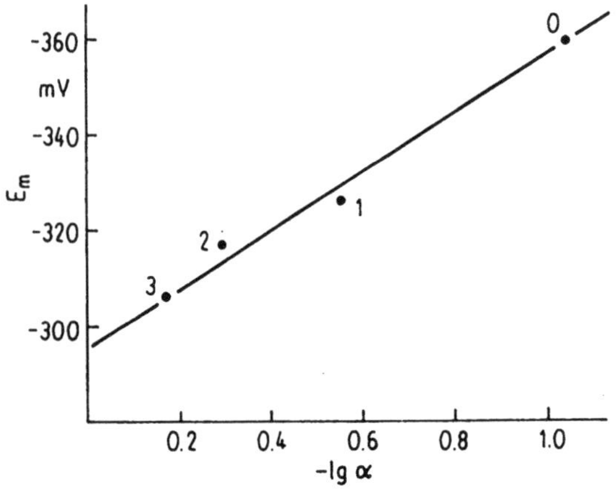

Fig. 5. Correlation between midpoint redox potential E_m and high spin fraction α of cytochrome P-450 LM2.
0 = without substrate,
1 = $(\text{Ø} \cdot CH_2 \cdot CH_2)_2 \cdot NCH_3$,
2 = $\text{Ø} \cdot CH_2 \cdot CHCH_3 \cdot NCH_3 \cdot CH_2 \cdot \text{Ø}$,
3 = $(\text{Øo} - Cl \cdot CH_2)_2 \cdot NCH_3$.
The spin contents were determined according to RISTAU et al., 1978.

2.4.3. The spin/redox couple as regulator of the reduction reaction

According to the simplified redox coupling model:

$$\begin{array}{ccc} \text{P-450 } (Fe^{3+})_{hs} & \underset{\pm e^-}{\overset{K_b^{hs}}{\rightleftharpoons}} & \text{P-450 } (Fe^{2+})_{hs} \\ K_a^0 \updownarrow & & \\ \text{P-450 } (Fe^{3+})_{ls} & & \end{array} \qquad (20)$$

the reduction of ferric cytochrome P-450 would exhibit a spin state specificity in that the high-spin fraction P-450 $(Fe^{3+})_{hs}$ is reduced directly, the low-spin fraction P-450 $(Fe^{3+})_{ls}$, on the other hand, would be sequentially reduced via realignment of the spin equilibrium. The regulation of the reduction velocity

[1] As derived from equation (19) the redox potential for the low-spin state $K_a^0 = 0$ results in $E_m = -\infty$.

via the substrate-dependent high-spin state presupposes that the preceding substrate binding occurs more rapidly. The rate of this partial reaction depends on the chemical nature of the respective substrate. The rate constants for type I compounds have been reported to range from 10^4 to 10^5 M^{-1} s^{-1} (Blanck and Smettan, 1978). As reduction occurs on a time scale of seconds, substrate binding is not rate-limiting for the subsequent reduction of the substrate complex. This reaction scheme raises the question regarding the relaxation kinetics of the spin equilibrium. A slow relaxation should be rate-limiting with respect to the reduction of the low-spin fraction, resulting in biphasic reduction kinetics, as suggested by Backes et al. (1982).

By means of a laser temperature jump perturbation technique, the spin transition in cytochrome P-450 LM2 was found to proceed in the nanosecond range (Ziegler et al., 1982). As this is some six orders of magnitude faster than the reduction velocity, the latter is not limited by the relaxation of the spin equilibrium. The overall P-450 reduction rate v_r can be specified in terms of distinct rate constants k_{hs} and k_{ls} for the high-spin and the low-spin states, respectively, as follows (Blanck et al., 1983)

$$v_r = \text{P-450}\,(Fe^{3+})_{hs} \cdot k_{hs} + \text{P-450}\,(Fe^{3+})_{ls} \cdot k_{ls} \tag{21}$$

Introduction of α for the high-spin fraction and $(1 - \alpha)$ for the low-spin fraction yields

$$v_r = \text{P-450}\,(Fe^{3+})\,\big(\alpha \cdot k_{hs} + (1 - \alpha) \cdot k_{ls}\big) \tag{22}$$

Due to the fast spin equilibrium, the relative high-spin content α remains constant throughout the reduction process. Thus the reaction must exhibit an exponential time function

$$v_r = \text{P-450}\,(Fe^{3+}) \cdot k_{app}; \quad k_{app} = \alpha \cdot k_{hs} + (1 - \alpha) \cdot k_{ls} \tag{23}$$

and the overall rate constant of the monophasic reaction k_{app} must exhibit a linear correlation with the high-spin fraction.

In earlier investigations, a substrate-induced acceleration of the cytochrome P-450 reduction was reported by several authors (reviewed by Bjorkhem, 1977; Gander and Mannering, 1980). These results reflect the increase of k_{app} with a substrate-induced increase in α. Imai et al. (1977), using highly purified cytochrome P-450 from phenobarbital-induced rabbit liver and benzphetamine at different sub-saturation concentrations, observed a gradual increase of k_{app} with the substrate concentration, corresponding to the gradual increase of α (Rein et al., 1979). For a series of benzphetamine derivatives under saturation conditions and microsomal cytochrome P-450 from phenobarbital-induced rats, a linear k_{app}/α correlation was seen, demonstrating the existence of an unreactive low-spin state ($k_{ls} \simeq 0$), in accordance with the above model (Blanck et al., 1983).

Investigations on reconstituted cytochrome P-450 LM2 systems (thus avoiding complications of multiple isozymes) confirmed the k_{app}/α correlation but revealed a rather distinct $k_{ls} \neq 0$ (SCHWARZE et al., 1985). This result may indicate an alternative reduction process via a relatively unfavourable ferrous low-spin state (cf. Scheme 2). FISHER and SLIGAR (1985), using cytochrome P-450 CAM and several camphor derivatives as substrates, likewise observed a correlation between reduction and high-spin fraction α but with a significantly wider range of k_{app} values. Obviously species and substrate specificities modulate the spin state dependence of the cytochrome P-450 reduction.

Moreover, under substrate saturation, i.e. in the pure high-spin state, the mitochondrial cytochromes P-450 11β and P-450 scc exhibit high negative redox values of −286 mV and −284 mV, respectively. The high-spin cytochrome P-450 LM4 and the related cytochrome P-450 from rat liver (HUANG et al., 1986; GUENGERICH, 1983) (see Table 3) are also characterized by similar high negative redox values. On the basis of these data, it has to be concluded that the observed redox/spin state correlations (SLIGAR et al., 1979) cannot be generalized to all cytochrome P-450 forms. Obviously specific substrate/cytochrome P-450 interactions may induce conformational changes of the hemoprotein which are more important in regulating the reduction rate than substrate-induced spin shifts.

2.5. The control function of the substrate

As already mentioned, the substrate can regulate enzymatic activity in several ways. Due to the multi-step reaction cycle substrate control can occur at several partial reactions. Furthermore, different functions of the substrate have to be differentiated: as effector it is mainly affecting the enzyme; as substrate in the proper sense it is affected by the enzyme. Both of these interactions are mutual, implying that there are also effects of the passive component upon the active partner. In order to avoid unreliable speculation, the following results mainly on the effector function of the substrate will be discussed.

Any approach to a detailed understanding of the molecular mechanism of this regulation process requires a thorough description of the interacting components: the substrate and the active center of the enzyme. Having discussed the functional importance of the spin equilibrium which is substrate-specifically shifted to the enzymatically active form we now turn to the question of what are the decisive physicochemical and sterochemical properties of the substrate and of the binding site.

2.5.1. Active site characteristics

Efficiency of substrate conversion depends on the structural and physicochemical properties of the active site of the enzyme, which determine the strength and specificity of its interaction with the substrate. Our knowledge of the spatial structure of the active site is restricted because no membrane-bound cytochrome P-450 has yet been crystallized. Therefore indirect evidence such as active site titrations (with substrates and inhibitors), negative modelling and appropriate physical methods have been used for structural studies on the active site.

Because all type I substrates of cytochrome P-450 exhibit lipophilic character, the binding site for these substrates is assumed to be a hydrophobic domain of the polypeptide chain which should be located near the heme. Which amino acid residues constitute the active site is still not completely clear. BrCN-derived heme peptides consisting of 41—51 amino acid residues of different P-450 enzymes (P-450 CAM, P-450 LM2, P-450 LM4, P-450 scc, P-450 11β) show that all of them contain at least one residue each of cysteine, histidine and tyrosine which could be heme iron ligands (Dus, 1980). Experiments involving chemical modification of tyrosines suggest that Tyr 380 is located in the active site of P-450 LM2 (Jänig et al., 1985). Neighbouring amino acid residues of this tyrosine should also be involved in the substrate-binding site. Based on the amino acid sequence of P-450 LM2 (Tarr et al., 1983) models of the peptide backbone in the region of Tyr-380 have been constructed (Schwarze et al., 1987). Different benzphetamine derivatives could be fitted to the models, showing that Tyr-380 and Phe-377 may be involved in substrate binding.

Further evidence for the lipophilic character of the binding site results from the thermodynamic data for substrate binding. Such data were obtained for partially purified P-450 from phenobarbital-induced rat and rabbit liver (Misselwitz et al., 1976, 1979) and for P-450 LM2 with benzphetamine as substrate (Ristau et al., 1978). In all cases the binding of type I substrates was characterized by high entropy values which yield 170 J mol^{-1} K^{-1} for the binding of benzphetamine to P-450 LM2. The unfavourable positive binding enthalpy of 30 kJ mol^{-1} is more than compensated by the large entropy value. Thus the binding of type I substrates to P-450 (at least to the unspecific phenobarbital inducible P-450 forms of the liver) is an entropy-driven process, characteristic of predominantly hydrophobic interactions.

From X-ray studies it is known that in P-450 CAM not only is the heme deeply imbedded in the protein molecule, but also the substrate binding site is located in the interior of the macromolecule (Poulos et al., 1985, 1986). Although spatial structure of microsomal P-450 forms are lacking the heme environment of P-450 LM2 should have a more open structure than P-450 CAM. This assumption is supported by a fast exchange rate of solvent protons between the paramagnetic ferric heme iron and the bulk solvent in P-450 LM2 (Rein et al., 1976c) which is two times faster than that in P-450 CAM (Griffin and Peterson, 1975). Moreover, proton magnetic relaxation measurements of P-450 LM2 using stereochemical NMR probes indicate a very accessible heme (Rein et al., 1976c). The geometry of the active site of P-450 LM2 has been examined using spin-labelled substrate analogues (Pirrwitz et al., 1982b). Thus with the spin-labelled type I substrate 1,1,2,3-tetraethyl-2-oxy-5-amino-isoindoline as a stereochemical probe the diameter of the heme pocket of P-450 LM2 was shown to be less than 10 Å. In the enzyme/substrate-analogue complex the distance of the paramagnetic nitroxide group both from the heme iron and from the protein surface is greather than 7 Å. Because the spin label itself is included in the binding of the substrate analogue to the enzyme (PIRRWITZ et al., 1977, 1982b), it can be suggested that these distances reflect the dimensions of the substrate-binding site of cytochrome P-450 LM2.

As well as the static properties of the active site, its flexibility may also influence binding properties, especially with regard to substrate specificity. SCHWARZ et al. (1984) have studied the dynamics of a spin-labelled substrate analogue. By saturation transfer EPR, they determined an effective rotational correlation time τ_R of 40 ns for P-450 LM2 bound n-propylisocyanide. This value is at least three orders of magnitude lower than the correlation time of the macromolecule (180 μs in liposomes, 480 μs in microsomes). The correlation time determined was independent of the degree of purification but varied with temperature, indicating a dependence on phase transitions of the phospholipids. The mobility of the substrate analogue in the bound state thus reflects a relatively high conformational flexibility of the substrate-binding region of cytochrome P-450 LM2 which enables the enzyme to react with a large number of substrates differing in stereochemical structure.

WHITE et al. (1984) have drawn a similar conclusion that there is considerable movement of the substrate molecule in the active center of cytochrome P-450 LM2. They compared the product patterns of three different substrates (5-exo-hydroxycamphor, 5-hydroxyadamantanone and 1-adamantanol), which are hydroxylated by both cytochrome P-450 LM2 and cytochrome P-450 CAM. The isosteric products of hydroxylation by cytochrome P-450 CAM suggested attack at a topologically congruent position due to a rigid enzyme/substrate complex. This conclusion is in agreement with recent data on the three-dimensional structure of cytochrome P-450 CAM in substrate-bound (POULOS et al., 1985) and substrate-free states (POULOS et al., 1986). Comparison of these structures revealed that no detectable conformational change occurred during camphor binding, but that the flexibility in certain regions of the molecule decreased in the camphor-bound state. On the other hand, the non-isosteric products formed by cytochrome P-450 LM2 were similar to the product distribution during solution-phase hydroxylations, indicating a rather high flexibility in the active site of cytochrome P-450 LM2.

2.5.2. Structure/activity relationships

The ability of a substrate to act as an effector of the enzyme is based on its capacity to convert the enzyme into its biologically active conformation (intrinsic activity of the substrate; ARIENS, 1972). Rather unspecific hydrophobic interactions are mainly involved in substrate binding to cytochrome P-450. Electronic and stereochemical properties of a substrate, on the other hand, lead to a high degree of substrate specificity and consequently should mainly determine substrate recognition and its intrinsic activity.

The presence in microsomes of different cytochrome P-450 isozymes which act on substrates of different structure seriously impaired or even precluded finding correlations between stereochemical and electronic properties of substrates and their binding properties to microsomal P-450 and of their capacity to shift the spin equilibrium and so control the catalytic activity.

This problem has been partially overcome by using cytochrome P-450 LM2 and a homologous series of benzphetamine derivatives (PETZOLD et al., 1985). ^{13}C-NMR measurements of the N-methyl carbon and the quaternary phenyl carbon atom of the derivatized benzphetamines were used to monitor changes in the chemical shift in the aliphatic and aromatic moieties of those compounds. In the NMR spectrum a high field chemical shift of the quaternary aromatic carbon atom has been shown by WEHRLI and WIRTHLIN (1976) to indicate a lower electron density. Based on this finding PETZOLD et al. (1985) assigned an increased binding affinity of the derivatized benzphetamines and their lowered solubility in aqueous media to a lowered electron density at the quaternary aromatic carbon atom and also at the carbon atom to which the N-methyl group is bound. Further, the capacity of the substrate to induce an enzymatically active conformation was shown to depend strongly on structural parameters of the substrate and in the case of cytochrome P-450 LM2, on a geometry in which the substrate is in a bulky, tightly packed form which by ROSSI (1983) has called a 'twisted butterfly conformation' and attributed to metyrapone, DDT and phenobarbital. The corresponding regulation of enzymic activity in the same series of substrates has been examined by SCHWARZE et al. (1985).

To clarify the role of structural features of the substrate in its interaction with the enzyme, WHITE and MCCARTHY (1986) used a set of ten *para*-substituted toluene derivatives (H, F, Br, Cl, I, nitro, methyl, cyano, isopropyl and t-butyl). The enzymic parameters determined were the apparent dissociation constant (K_D) for the enzyme/substrate complex, the interaction energy (ΔG^{int}) between the substrate-binding and the spin equilibrium, and the catalytic rate constant (k_{cat}) for hydroxylation. These were correlated with several physicochemical constants of the individual substrate molecules such as the Hansch constant π for hydrophobicity, the Hammett value σ, the dielectric constant ε, and the molar volume V. A linear combination of the molar volume and the hydrophobic Hansch constant allowed prediction of the apparent dissociation constant K_D, whereas the spin state interaction energies were best predicted by a linear combination of the Hansch constant and the reciprocal of the dielectric constant. A deuterium isotope effect of 2.6 (d_8-toluene/d_0-toluene) showed that hydrogen abstraction was in part rate-limiting with this series of substrates.

In a recent study on the phenobarbital-inducible isozyme from rat liver (main fraction), quantitative structure/activity relationships of a homologous series of ten alkylbenzenes with cytochrome P-450 were determined by using molecular orbital calculations (LEWIS et al., 1986). From the good correlation found with the substrate-bound spin equilibrium constant K_2, the authors concluded that the electron-accepting potential of the hydrocarbons is an important structural feature. A hydrocarbon which is a good electron donor and a poor electron acceptor favours the high-spin form of a phenobarbital inducible cytochrome P-450 isozyme.

Summing up the results of experiments aimed at establishing quantitative structure/activity relationships for rationalization of drug design strategies, it is clear that the data are far from perfect. All analyses have adopted the approach of trying to elucidate structure/activity relationships by correlating functional data from substrate/enzyme interactions with accessible parameters describing characteristics of the substrate molecule. This approach suffers from disregarding the multistep character of the reaction cycle in which different properties of the substrate determine its suitability as a substrate. In other words the properties of any substrate which enable it to induce an enzymatically active conformation in the enzyme (including parameters such as binding constant, spin shift, reduction rate, hydrophobicity) may be different from those parameters which determine the susceptibility of the substrate to conversion by the enzyme (as e. g. electron-accepting or-donating properties at the atom or group to be hydroxylated). Even in the recent papers of LEWIS et al. (1986) and WHITE and McCARTHY (1986) this necessary differentiation has not been made.

One needs to be careful in extrapolating the above results to other isozymes and one should be further aware that the accumulated data are still inadequate to provide full understanding of the detailed principles of regulation by the substrate. Nevertheless, the findings on cytochrome P-450 LM2 revealed several lines of progress: (i) Certain indications about the geometry and the physicochemical character of the active site have been obtained despite the unknown three-dimensional structure. (ii) The decisive role of the substrate in regulation of P-450 activity has been demonstrated at several steps of the reaction cycle; namely control of the reduction reaction by shifting the spin equilibrium as effector; control of the conversion rate via distinct electronic and stereochemical properties, selection of alternative pathways by switching on the monooxygenase or the oxidative pathway which will be discussed in the next section.

2.6. The regulation of alternate pathways of oxygen activation

Cytochromes P-450 are unique in exhibiting a functional versatility due to their broad substrate specificity and oxygen-activating capability. This versatility on the one hand is characterized by different monooxygenase reactions which have been classified into six categories and are discussed in detail by GUENGERICH in this volume. On the other hand the functional versatility is reflected in the capability of cytochromes P-450 to catalyze not only monooxygenase (24) but also oxidase (25a—c) and peroxidase reactions (26)

$$RH + O_2 + NADPH + H^+ \rightarrow ROH + H_2O + NADP^+ \quad (24)$$

$$2O_2 + NADPH \rightarrow 2O_2^- + NADP^+ + H^+ \quad (25a)$$

$$O_2 + NADPH + H^+ \rightarrow HOOH + NADP^+ \quad (25b)$$

$$O_2 + 2\,NADPH + 2\,H^+ \rightarrow 2\,H_2O + 2NADP^+ \quad (25c)$$

$$RH + XOOH \rightarrow ROH + XOH \quad (26)$$

The oxidase reactions can be differentiated according to the number of electrons transferred to oxygen, resulting in the formation of either superoxide anion (1 electron, equation 25a), hydrogen peroxide (2 electrons, equation 25b) or water (4 electrons, equation 25c). The possibility that cytochrome P-450 can function as a peroxidase (equation 26) was initially proposed by HRYCAY and O'BRIEN (1972) and has been supported by the detection of a free radical species formed on interaction of cumene hydroperoxide with cytochrome P-450, comparable to that observed for compound I of peroxidase or the higher valence state of iron for hemoglobin and myoglobin (RAHIMTULA et al., 1974). This so-called shunt mechanism of cytochrome P-450 is of no physiological importance. Nothing is known about its regulation except that cytochrome P-420 exhibits no peroxidase activity.

Taking into account potential biological and particularly toxicological implications of reactive oxygen metabolites, the oxidase activity of cytochrome P-450 deserves special consideration with regard to regulation mechanisms. In considering what factors regulate the oxidase and the monooxygenase pathways three main questions must be answered (for more details see Chapter 4):

(1) By which mechanism is the formation of the activated oxygen species determined?

(2) What role does the substrate play in regulating both pathways?

(3) How much does isozyme specificity influence substrate control?

In 1957 GILLETTE et al. observed the formation of H_2O_2 by rabbit liver microsomes during demethylation of monomethyl-4-aminoantipyrine in the presence of NADPH. This finding led to efforts to explain the discrepancy between the NADPH oxidized on the one hand and the sum of oxygen consumed and product formed on the other, and so to clarify the mechanism for generation of hydrogen peroxide during NADPH oxidation. An understanding of this mechanism required the identification of which activated oxygen species produced the hydrogen peroxide. The detection of a ferrous oxygenated cytochrome P-450 complex (ISHIMURA et al., 1971; TYSON et al., 1972; ESTABROOK et al., 1971) stimulated further research on this point. Experimental results from several groups yielded two differing conclusions. The main difference between these lay in whether the superoxide anion was assumed to act as one-electron reduced source for the generation of the hydrogen peroxide (ESTABROOK and WERRINGLOER, 1977; KUTHAN and ULLRICH, 1982) or the second electron reduction step occurred with conversion of the ferrous dioxygen/cytochrome P-450 complex to a peroxide anion complex of ferric cytochrome P-450 (HEINEMEYER et al., 1980).

Based on the measured stoichiometry of O_2^- to H_2O_2 (close to 2:1 in the absence of any substrate) and the finding that substrates can modify the rates of decomposition of the ferrous dioxygen complex of cytochrome P-450 and thereby the formation of H_2O_2 via O_2^-, KUTHAN and ULLRICH (1982) drew the conclusion that protonation and subsequent release of a two-electron reduced oxygen species does not represent a significant pathway for the formation of H_2O_2. Due to dissociation and dismutation, the superoxide anion radical disproportionates according to equation (14) into hydrogen peroxide and oxygen

$$2O_2^- + 2H^+ \rightarrow H_2O_2 + O_2 \qquad (27)$$

Several authors have proved either the influence of the substrate on the generation of activated oxygen species or the importance of isozymes in this process by investigating reconstituted systems (NORDBLOM and COON, 1977; MORGAN et al., 1982; OPRIAN et al., 1983; GORSKY et al., 1984). These studies revealed that in the absence of substrates that can be hydroxylated different isozymes generate different amounts of H_2O_2. The relation between the oxidase and the monooxygenase pathways in an individual isozyme is further regulated by the respective substrate. What structural or physicochemical peculiarities of the substrate or the isozyme are decisive in determining which pathway the reaction follows is unknown as yet. A recent paper by PARKINSON et al. (1986) on an induced uncoupling by alkylation of Cys-292 by means of 2-bromo-4'-nitroacetophenone in cytochrome P-450c suggests a conceivable mechanism by which a substrate may regulate different pathways.

In addition to the formation of H_2O_2 there exists a four-electron reduction pathway leading to water. In the presence of pseudo substrates (perfluoroalkanes) which have been shown not to be hydroxylated (STAUDT et al., 1974) the formation of H_2O_2 was not enhanced. Instead, on the basis of a ratio of NADPH oxidation to oxygen consumption of 2:1, water was assumed to be the final reduction product of dioxygen via the oxidatic pathway. From studies on microsomes without exogenous substrates and a ratio NADPH oxidized/H_2O_2 generated greater than 2, ZHUKOV and ARCHAKOV (1982) interpreted their results likewise as four-electron reduction of O_2 to H_2O. Irrespective of isozyme specific H_2O_2 generation, GORSKY et al. (1984) observed an increased NADPH/O_2 ratio with a stoichiometry of about 2 in accordance with the proposed water formation. OPRIAN et al. (1983) examined the dependence of water formation on the ratio *P-450*/oxygen. At a four-fold molar excess of cytochrome *P-450* LM4 over oxygen, formation of water occurs, whereas at a two-fold molar excess of *P-450* LM4 over oxygen, H_2O_2 is formed, indicating an increased NADPH consumption per oxygen under oxygen deficiency. To what extent that pathway accounts for the consumption of cellular energy remains an open question. Proof that water is the product requires confirmation by experiments with $^{18}O_2$.

The toxicological relevance of such oxygen activating pathways has been shown recently by GONDER et al. (1985). These authors studied the susceptibility of responsive and non-responsive mice towards oxygen radicals and H_2O_2 in the presence of high oxygen pressure. Such strains which are susceptible to hepatic microsomal enzyme induction by aromatic hydrocarbons were more sensitive to toxic effects of 100% oxygen exposure than were genetically unresponsive mice, despite the fact that high oxygen concentrations also induce antioxidant enzymes such as catalase and superoxide dismutase in several species. Non-responsive mice survived significantly longer with less lung damage. The authors concluded that such genetic differences in susceptibility to oxidative stress may have long-term implications in therapeutics and patient care in humans.

The capacity of cytochrome P-450 to convert not only a great variety of compounds via detoxifying and toxifying pathways but also to produce toxic activated oxygen species via alternate pathways, demands reevaluation of that functional versatility.

3. Regulation at the membrane level

The polysubstrate monooxygenase system of the mammalian liver cell is located in the endoplasmic reticulum. The successful reconstitution of an enzymically active monooxygenase system (LU et al., 1969a, b; LU and LEVIN, 1974) showed that phospholipids are required for the catalytic activity (STROBEL et al., 1970). The structural peculiarities of the endoplasmic reticulum therefore will exert a functional control upon the enzyme system (INGELMAN-SUNDBERG, 1986).

3.1. Lipid composition and protein constituents of the endoplasmic reticulum

The endoplasmic reticulum is composed of a variety of lipid species. About 55% PC, 20–25% PE, 5–10% PS and PI, respectively, and 4–7% sphingomyelin are the main constituents (DEPIERRE and DALLNER, 1975). The fatty acid content comprises mainly the 16:0, 16:1, 18:0, 18:1, 18:2, 20:4, 22:6 compounds (LEE and SNYDER, 1973), where the first number refers to the number of carbon atoms and the second to the number of double bonds in the fatty acid.

The structural organization of the membrane lipids provides a highly ordered matrix for the incorporation of proteins. In general, the solubility of lipids in water is low. Therefore they tend to associate into supramolecular aggregates, the critical micelle formation concentration (CMC) lying in the micromolar range. The association and its specificities depend on the chain lengths of the fatty acids.

About 60—70% by weight of the endoplasmic reticulum consists of protein and 30—40% of phospholipid. The protein moiety is composed of about 10% cytochrome P-450 which increases to about 20% after phenobarbital induction. The cytochrome P-450/lipid molar ratio amounts to about 1:200 to 1:120 (DePierre and Dallner, 1975; DePierre and Ernster, 1975, 1977). The stoichiometry of reductase and cytochrome P-450 was determined with a molar ratio of 1:15 independent of induction (Shephard et al., 1983).

Both proteins are attached to the membrane of the endoplasmic reticulum by hydrophobic anchors which comprise 10% of the flavoprotein molecule (Black et al., 1979) and one or two transmembranal segments of the cytochrome P-450 molecule (Sakaguchi et al., 1981; DeLemos-Chiarandini et al., 1987; Nelson and Strobel, 1988) contrary to previous suggestions assuming at least 8 transmembranal segments (Bar-Nun et al., 1980). With respect to the lateral distribution of both proteins within the membrane bilayer, a macroscopic asymmetry has been excluded (Schulze and Staudinger, 1975). The formation of domains of cooperating proteins, however, was proved (Schulze et al., 1972). By means of ferritin-labelled antibodies, a cluster-like association of cytochrome P-450 (Matsuura et al., 1978) and of reductase (Morimoto et al., 1976) within microsomes has been detected. Generally a heterogeneous distribution of the membrane proteins across the cytoplasmic surface of the membrane must be assumed (Morimoto et al., 1976; DePierre and Ernster, 1977; Matsuura et al., 1978).

3.2. Membrane fluidity and mobility of membrane components

The interaction of the components of the monooxygenase system depends strongly on the fluidity of the membrane matrix, which is characterized by a phase transition from the gel state into the fluid liquid-crystalline state of the constituting lipids. This process is characterized by the phase transition temperature T_c, which increases with increasing chain length of the fatty acid and with its degree of saturation (Chapman, 1975; Lichtenberg et al., 1983). The functional significance of unsaturated fatty acids results from their ability to decrease T_c and thus increase the membrane fluidity. The constitutive membrane lipids of the monooxygenase system are mainly in the fluid state at the mammalian physiological temperature, and calorimetric investigations have proved the absence of any phase transition in that range (Mabrey et al., 1977).

The mobility of the membrane lipids is rather high. The lateral diffusion coefficients D_L are $10^{-7} - 10^{-9}$ cm^2 s^{-1} (Kuo and Wade, 1979). Data for reconstituted vesicles irrespective of their cytochrome P-450 or reductase loading are within this range (Schwarz et al., 1984). The transversal exchange in phospholipid bilayers, on the other hand, proceeds on a time scale of minutes or even days (ROTHMAN and DAWIDOWICZ, 1975). The latter changes can therefore be eliminated as being rate-determining in the catalytic process of the enzyme system, even though the incorporation of cytochrome P-450 enhances the transversal lipid transfer (BÖSTERLING and TRUDELL, 1982a; BARSUKOV et al., 1982).

The lateral diffusion of the membrane proteins is likewise rapid. In DMPC-reconstituted vesicles the diffusion coefficient of cytochrome P-450 at 25 °C amounts to $D_L = 10^{-8}\ cm^2\ s^{-1}$, and similar results were obtained with egg yolk PC (WU and YANG, 1984). These data are consistent with the respective parameters of other membrane proteins (CHERRY, 1979). A local lateral diffusion coefficient D_L^{loc} of $10^{-9}\ cm^2\ s^{-1}$ has been determined by means of rotation investigations for cytochrome P-450 and a cytochrome P-450/reductase complex in reconstituted lipid vesicles (GUT et al., 1982; KAWATO et al., 1982). These data have been used to derive a collision rate of the electron exchange proteins within vesicular systems of around $10^3\ s^{-1}$ (KAWATO et al., 1982; WU and YANG, 1984). Since the electron transfer from the cytochrome P-450 reductase towards cytochrome P-450 proceeds within seconds (DUPPEL and ULLRICH, 1976; PETERSON et al., 1976; TANIGUCHI et al., 1979), a diffusion limitation of that reaction seems rather unlikely, though the entropy factor may decrease the effective collision rate by orders of magnitude, as is usual for macromolecules. A limitation is obvious from studies on cytochrome P-450 reduction in DMPC-reconstituted systems (TANIGUCHI et al., 1980; SMETTAN et al., 1984). Below the phase-transition temperature (23 °C), the rate constant of the rapid cytochrome P-450 reduction is significantly decreased, in correspondence with a decrease in the lateral diffusion coefficient of more than one order of magnitude (WU and YANG, 1984). Microsomal incorporation or removal of cholesterol by means of vesicle exchange techniques (ARCHAKOV et al., 1983), on the other hand, showed that the observed changes in viscosity of the membrane did not alter the characteristics of the NADPH-dependent cytochrome P-450 reduction, whereas the NADH-dependent reduction of cytochrome b_5 was significantly affected.

The rotational diffusion of the microsomal proteins indicates the formation of cytochrome P-450 oligomers and of complex formation with reductase as well, which would facilitate the electron transfer process (PETERSON et al., 1976). First investigations by transient dichroism methods (RICHTER et al., 1979; MCINTOSH et al., 1980), or with delayed fluorescence polarization techniques (GREINERT et al., 1979), indicated a significant rotational mobility of cytochrome P-450 in both microsomes and in reconstituted proteoliposomes. Studies on the absorption anisotropy after flash photolysis of the CO complex of cytochrome P-450 yielded a mean rotational relaxation time for rat liver microsomes of 120 μs at 20 °C, and for PC/PE/PS-(or PA-)reconstituted vesicles of 95 μs (KAWATO et al., 1982). With decreasing lipid/protein ratio, cytochrome P-450 was gradually immobilized. Coreconstitution of equimolar amounts of cytochrome P-450 and reductase resulted in a significant decrease in the relaxation time towards 40 μs. This value has been taken to indicate the formation of a monomolecular 1:1 complex between these proteins (GUT et al., 1982). Similar measurements using eosin isothiocyanate covalently bound to rabbit liver cytochrome P-450 LM2 incorporated into PC/PE/PA-reconstituted vesicles yielded a rotational relaxation time of 110 μs at 25 °C (GREINERT et al., 1979). Using eosin maleimide and rat liver cytochrome P-450, similar values of 50—100 μs were determined (KAWATO et al., 1982). Thus there is reasonable agreement with respect to different species and different methods which indicates the formation of both cytochrome P-450 aggregates and cytochrome P-450/reductase complexes.

Similar characteristics of the protein association in membranes have been determined by applying saturation transfer EPR. Labelling of sulfhydryl groups of rabbit liver microsomal cytochrome P-450 with a spin-labelled maleimide revealed an effective rotational correlation time of 480 μs at 20 °C (SCHWARZ et al., 1982b). The corresponding value for the rotation of cytochrome P-450 LM2 in reconstituted vesicles was 180 μs (SCHWARZ et al., 1982a). The correspondence of the data is evident, while differences may be due to varying experimental conditions.

The association of the membrane proteins depends not only on the lipid/protein ratio but also on the conformational state and the degree of immersion of the proteins in the membrane. Thus substrate binding increases the rotation time of cytochrome P-450, while reduction leads to a decrease (GREINERT et al., 1982a, b).

In general, the relatively high values of the cytochrome P-450 rotation time indicate a significantly lower protein mobility in the lipid matrix than in aqueous solution, which reflects not only the increased viscosity of the medium but also a protein association. For cytochrome P-450 LM2 in buffer solution a correlation time of 220 ns has been determined (SCHWARZ et al., 1982b). As compared to a theoretical value of the monomer in water solution of 21 ns the tenfold increase is consistent with a hexameric aggregate of cytochrome P-450 LM2, as supported by different methods (COON et al., 1976). Considering the increased viscosity of the lipid membrane, a correlation time for the monomer of 21 μs was calculated, consistent with corresponding data for other membrane proteins (Cherry, 1979). On the basis of the experimental data of 50—100 μs for cytochrome P-450 in microsomes and reconstituted proteoliposomes, an association of 6—12 molecules into clusters is suggested to represent the structural basis of cytochrome P-450 function in vesicles.

3.3. Structural peculiarities of protein/lipid interactions

The structural features of the membrane-bound cytochrome P-450 system constitute a variety of intermolecular interaction facilities, which allow a multitude of regulation mechanisms on a molecular level. Whereas the protein/protein interaction mainly concerns reactions between monomeric species but modified by protein/protein association, the protein/lipid interaction is mainly based on supramolecular lipid structures which are formed above the CMC. That refers to biological cytochrome P-450 systems in the endoplasmic reticulum or its biochemical equivalent, the microsome, and equally to reconstituted vesicle systems. DLPC, on the other hand, which is most frequently used in reconstitution experiments, was shown to be active already below the CMC (MIWA and LU, 1981). Moreover, this lipid forms vesicles above the CMC as shown by electron microscopy (HAUSER and BARRATT, 1973), but no vesicle formation was observed for the protein-containing reconstituted system (AUTOR et al., 1973). Reconstitution experiments with this lipid, therefore, should contribute to discriminate molecular and supramolecular lipid effects.

First insights into the mechanism of cytochrome P-450/lipid interactions were obtained by various spectroscopic techniques. CHIANG and COON (1979) observed that in the presence of phospholipids the ellipticity of the peptide chromophore of cytochrome P-450 in the far UV region increases, indicating an increase in the helical content of the protein. Derivative spectroscopic analyses revealed that both phospholipids and detergents provoke a tyrosine signal on interaction with cytochrome P-450 (RUCKPAUL et al., 1980). In contrast, other amphiphilic or hydrophobic compounds such as fatty acids and alkanes, did not elicit that signal. The spectral changes can be explained by an increase of the nonpolar character of the immediate environment of the tyrosine residues. This may be caused by either a conformational change of the protein or by a change in its surrounding milieu, thus influencing tyrosine residues at the surface of the macromolecule. The latter seems more likely because of the relatively low specificity of the effect, as well as the absence of a signal in the presence of the nonmicellar partners of interaction. The investigations by CHIANG and COON (1979), however, and measurements by ARCHAKOV (1982) point to at least some additional conformational changes of cytochrome P-450 in the interaction with lipids. CHIANG and COON (1979) further observed that in the near UV region (Soret band) in the presence of DLPC a cytochrome P-450 difference spectrum is formed, which indicates a shift towards the high spin state of the cytochrome P-450 spin equilibrium. A similar result could be seen for partially purified solubilized rat cytochrome P-450 which on addition of a microsomal lipid extract exhibited a spin shift from 17% to 53% high-spin content at 20 °C; the addition of oleic acid resulted in a shift towards 40—45% high-spin (GIBSON et al., 1980). The incorporation of cytochrome P-450 LM2 into preformed vesicles similarly provoked a high-spin shift which was favoured by negatively charged lipids (RUCKPAUL et al., 1982). Triton N-101 and Tween 20 proved unable to effect that conversion. Likewise proteoliposomes, obtained by cholate gel filtration (SCHWARZE et al., 1985) or octylglucoside dialysis (PETZOLD et al., 1985) exhibited the lipid-induced spin shift.

The interaction of cytochrome P-450 with substrates or reductase also results in a high-spin shift of the spin equilibrium. By means of this common effect, the mutual interaction of substrates, reductase, and lipid can be analysed and even quantitated (FRENCH et al., 1980). The binding of DLPC to cytochrome

P-450 LM2 exhibits biphasic binding characteristics, with an apparent dissociation constant for the high-affinity site of 3—6 μM, and of the low-affinity site of 50—70 μM. Benzphetamine or reductase addition favoured the lipid binding. The lipid, on the other hand, increased the benzphetamine and reductase affinities towards cytochrome P-450. The apparent K_D = 180 μM for benzphetamine in the presence of DLPC was decreased on addition of reductase to 110 μM, and likewise the respective reductase parameter K_D = 115 μM was decreased to 42 μM. Thus the results indicate a mutually favoured binding to cytochrome P-450.

These findings provide evidence that conformational and reactivity variations are induced in cytochrome P-450 on interaction with lipids. On the other hand, rather less is known about the interaction of lipids with the reductase. The likely reason is that only the hydrophobic portion of the flavoprotein (6.1 kDa) would be directly involved, whereas communication to the hydrophilic head (71.2 kDa) (BLACK et al., 1979; BLACK and COON, 1982) and thus to the main part of the molecule, would be less affected. The corresponding interaction could therefore lie below the limit of detection. Nevertheless, by means of CD measurements in the 200—300 nm range, an increase in the helical content of the protein from 28% without lipid up to 41% on interaction with lipid has recently been observed; DLPC, DLPE, DLPS proved less effective, but DMPC and DPPC were highly active (MAGDALOU et al., 1985).

3.4. Functional aspects of the component interactions in the membrane

The results discussed so far refer mainly to structural alterations of the protein moiety in interaction with lipids. The functional implications and the mechanisms of these protein/protein- and protein/phospholipid-interactions will be dealt with now. First of all, the successful reconstitution of the enzyme system proved the requirement for phospholipid to produce the functional activity, especially under physiological stoichiometry of the components (LU and COON, 1968; STROBEL et al., 1970; LU and WEST, 1972). The extraction of lipids by organic solvents, on the other hand, decreased the catalytic activity of cytochrome P-450 (VORE et al., 1974). The investigations showed a correlation between catalytic activity and the NADPH-dependent cytochrome P-450 reduction in that the rate of electron transfer decreased on delipidation, due to an alteration in the contribution of the fast and slow partial reaction to the biphasic process. Further correlations between cytochrome P-450 reduction and substrate conversion or NADPH oxidation (IMAI et al., 1977; IYANAGI et al., 1981) suggest an important regulatory role in catalysis for the first electron transfer process. The second electron transfer reaction or even a step beyond further controls the catalytic activity of the enzyme system. It is generally accepted that late steps in the catalytic cycle represent the rate limitation (MATSUBARA et al., 1976; IMAI et al., 1977; ISHIMURA, 1978; TANIGUCHI et al., 1979; WHITE and COON, 1980). The cytochrome P-450 substrate binding reaction as well as the O_2 binding process, on the other hand, have been shown to be not rate-limiting (Blanck et al., 1976; Imai et al., 1977).

The regulation of the catalytic function by both electron transfer steps obviously depends on pH. A rate limitation by the first step at low pH changes to the second transfer with increasing pH (Werringloer and Kawano, 1980b; Ruf and Eichinger, 1982). Studies on protein/lipid interaction in electron transfer thus have already to start with studying the first electron transfer and its regulation per se. This step may further reflect common structural features with respect to both transfer processes.

3.4.1. Characteristics of the cytochrome P-450 reduction

The general pattern of cytochrome P-450 reduction in the first electron transfer process is of a biphasic reaction, composed of two independent first-order partial reactions. This was observed in early investigations on microsomes, in solubilized systems, and in reconstituted systems (reviewed by Blanck et al., 1984a). The lipid component can be effectively substituted by detergents (Lu et al., 1974a; Ingelman-Sundberg, 1977a); these systems also exhibit biphasic kinetics in the cytochrome P-450 reduction process (Taniguchi et al., 1979; Ruckpaul et al., 1982). Although as many as four partial reactions could be observed at rat microsomes (Ruf, 1980), most studies of cytochrome P-450 reduction have used a biphasic approximation.

In microsomes about 50—70% of the total cytochrome P-450 is reduced in the rapid partial reaction under anaerobic conditions. The slow partial reaction differs in its specific rate constant by a factor 10—20 (Gillette et al., 1973; Matsubara et al., 1976; Peterson et al., 1976; Blanck et al., 1979a). The fast reaction obviously represents the physiologically relevant process. The relative amounts of the two partial reactions differ depending on the stoichiometry of the constituting components, on the substrate, on the aggregation state of the system, and on the presence or absence of oxygen.

The biphasic reaction pattern of the NADPH-dependent cytochrome P-450 reduction may be described by 4 interaction models (Fig. 6):

1. The electron transfer between reductase and cytochrome P-450 proceeds via statistical collision of the randomly distributed proteins. The increase in rate constant with increasing relative reductase content and decreasing relative lipid amount is taken to indicate rate limitation by the lateral mobility of the membrane. The biphasic characteristics of cytochrome P-450 reduction, however, is not explained by this random distribution model (Yang, 1975, 1977; Duppel and Ullrich, 1976; Yang et al., 1977, 1978).
2. A more likely interpretation of the cytochrome P-450 specificity is provided by the cluster model (Franklin and Estabrook, 1971; Stier and Sackmann, 1973; Peterson et al., 1976; Stier, 1976). According to this concept cytochrome P-450 is assumed to homologously associate for the main part and moreover to be capable of a further heterologous association with reductase. The cytochrome P-450/reductase clusters formed would thus

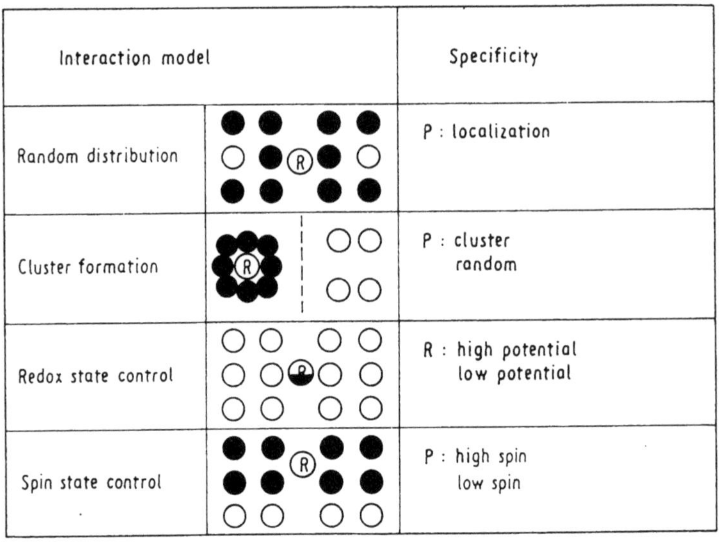

Fig. 6. Model approaches of the reductase/P-450 interaction in the first electron transfer process (P-450 reduction). (Abbreviations cf. p. 37).

represent a translation-free electron transfer system with a more favourable cytochrome P-450 reduction than with a randomly distributed portion of cytochrome P-450, which, being structurally impaired, would exhibit the slow partial reaction. Alternatively, reductase-free cytochrome P-450 associates (rotamers) are proposed to interact with randomly distributed reductase and the biphasic reaction characteristics is suggested to be an expression of negative kinetic cooperativity in the cytochrome P-450 oligomers (Greinert et al., 1982a, b; Stier et al., 1982). An additional reductase association at the microsomal stoichiometry is open as yet. At stoichiometric excess of reductase, on the other hand, a dissociation of the cytochrome P-450 associates and the formation of 1:1 binary complexes between both proteins could be evidenced (Gut et al., 1982).

3. A differentiation of the cytochrome P-450 reduction by reaction with different redox states of the reductase is further proposed (Oprian et al., 1979). This redox state model is based on the dependence of the phase distribution of the reduction process on the NADPH concentration. This model, however, must be questioned because of the maintained biphasicity of the reaction in the presence of excess NADPH and reductase.
4. The sequential spin state model considers a specific reactivity of the high-spin and the low-spin component of the cytochrome P-450 population. The high-spin fraction is assumed to exhibit the rapid partial reaction, the low-spin fraction, on the other hand, is considered to be reduced via a rate-limiting relaxation into the high-spin state conformation (Backes et al.,

1980, 1982; TAMBURINI et al., 1984). The measurement of the reduction kinetics in the high-spin band at 650 nm directly proved the favoured electron transfer towards that component (BACKES et al., 1985). However, the relaxation of the spin equilibrium has been measured to proceed in the nanosecond range (ZIEGLER et al., 1982, 1983), thus excluding rate limitation by this step. In consequence, an additional consecutive conformational change, being spectrally inactive, is proposed to represent the rate-limiting step in the sequential spin state model (TAMBURINI et al., 1984).

For further treatment of the cytochrome P-450 reduction at the membrane level a threefold control by

- substrates via the spin equilibrium,
- reductase interaction,
- phospholipid interaction,

has been differentiated. Substrate control represents a rather molecular phenomenon with respect to cytochrome P-450. This regulation therefore has been discussed in section 2 of this chapter. Main topics of the following treatment are thus the intermolecular component interactions of the cytochrome P-450 system.

3.4.2. Regulation of the electron transfer by reductase/cytochrome P-450 interaction

The reductase/cytochrome P-450 interaction in general is determined by the stoichiometry of both proteins, which in microsomes is about one order of magnitude in favour of cytochrome P-450 ($P/R^1 \geq 10$). Thus the reductase must cycle throughout the cytochrome P-450 population because of an as yet unproved intermolecular electron transfer between cytochrome P-450 molecules. Consequently the reductase would represent the "enzyme" for the "substrate" cytochrome P-450 in the reaction. This implies a Michaelis-Menten type mechanism. By means of detergent disintegration of the enzyme system and variation of the R/P stoichiometry, a monophasic first order process has been observed (BLANCK et al., 1979a, b; ROHDE et al., 1983)

$$R_{red} + P_{ox} \underset{k_{-1}}{\overset{k_1}{\rightleftharpoons}} R_{red}P_{ox} \xrightarrow{k} R_{ox}P_{red} \underset{k_{-2}}{\overset{k_2}{\rightleftharpoons}} R_{ox} + P_{red} \qquad (28)$$

in which the formation of the donor/acceptor complex $R_{red}P_{ox}$ is followed by the rate-limiting electron exchange (k) and the dissociation of the producomplex $R_{ox}P_{red}$. The rate of electron transfer may be regulated by intramolecular transfer steps in either protein, or by the intermolecular transfer.

[1] Abbreviations used in section 3.4: P: cytochrome P-450; R: NADPH-dependent cytochrome P-450 reductase; RP: complex between R and P (molar ratio: 1:1); P_{ox}, P_{red}, R_{ox}, R_{red}: P and R in the reduced or oxidized state.

The rate of the reduction of cytochrome P-450 is thus determined by

$$v = k\,[R_{red}P_{ox}] \tag{29}$$

i.e. the stoichiometric profile reflects the formation of the bimolecular electron exchange complex between reductase and cytochrome P-450 depending on the dissociation constant K_{RP}.

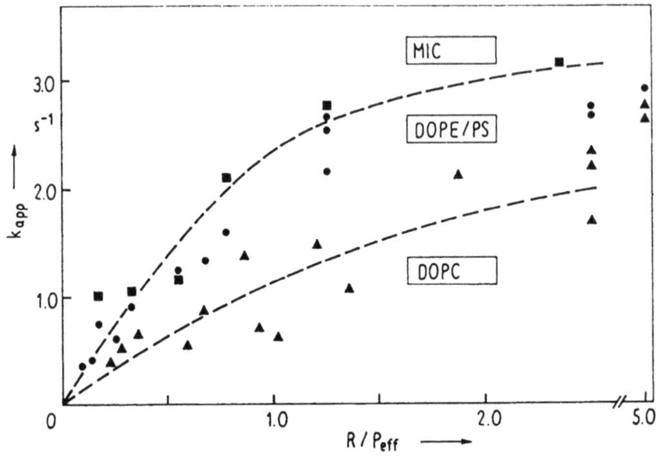

Fig. 7. Anaerobic NADPH-dependent P-450 LM2 reduction in reconstituted vesicles. Dependence of the rate constant k_{app} of the rapid partial reaction on the protein stoichiometry R/P_{eff}. (For details cf. Blanck et al., 1984b).

With reconstituted cytochrome P-450 LM2 systems, evidence for the same mechanism has been obtained. The rapid partial reaction of the biphasic first-order cytochrome P-450 reduction process which would predominate in the catalytic process, exhibited a corresponding stoichiometric profile with clear saturation characteristics (Blanck et al., 1984b). Kinetic titration of the RP-complex formation within liposomes prepared by use of a microsomal lipid mixture (MIC) revealed a dissociation constant $K_{RP} = 0.048$ μM in the presence of benzphetamine (Fig. 7).

3.4.3. Phospholipid control of electron transfer

With regard to the phospholipid specificity of cytochrome P-450 reduction, similar characteristics were observed with different liposomal preparations. The reaction kinetics depend on the specific lipid, with a dissociation constant $K_{RP} = 0.47$ μM for DOPC and $K_{RP} = 0.051$ μM for a 3:1 (w:w) mixture of DOPE/PS. This indicated that negatively charged lipids favour the formation of the electron exchange complex (Fig. 7) (Blanck et al., 1984b).

A complex formation between the two proteins in the presence of DLPC can be directly observed by difference spectroscopy with dissociation constants $K_{RP} = 0.12$ µM in the absence of substrate and $K_{RP} = 0.04$ µM in the presence of benzphetamine (FRENCH et al., 1980).

The rate constant k, on the other hand, turned out to be independent of the lipid species (BLANCK et al., 1984b). This result agrees with the observation of unchanged redox potentials of cytochrome P-450 in the presence of different phospholipids (GUENGERICH et al., 1975; BÄCKSTRÖM et al., 1983; GUENGERICH, 1983).

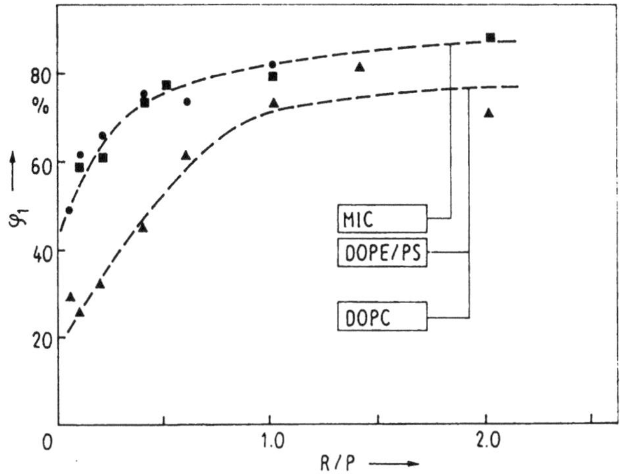

Fig. 8. Anaerobic NADPH-dependent P-450 LM2 reduction in reconstituted vesicles Extent of the rapid partial reaction φ_1 (%) on the protein stoichiometry R/P. (For details cf. BLANCK et al., 1984b).

The regulation of cytochrome P-450 reduction by phospholipids is further reflected in the specificity of the fractional distribution of both partial reactions. With increasing acidity of the headgroups of the lipids, as indicated by the increased electrophoretic mobility of the vesicles (INGELMAN-SUNDBERG et al., 1983), the physiologically relevant rapid cytochrome P-450 reduction process is favoured (INGELMAN-SUNDBERG et al., 1983; BLANCK et al., 1984b) (Fig. 8).

Thus the role of phospholipid may be further understood as a control of the functional association of cytochrome P-450 molecules to form cluster-like structures as a structural basis for efficient electron transfer.

The charge dependence of the lipid control in the cytochrome P-450 reduction process suggests charge-dependent electrostatic interactions between the electron exchanging proteins. Such interactions have been observed in the cytochrome b_5 system (LOVERDE and STRITTMATTER, 1968; DAILEY and STRITTMATTER, 1979), between cytochrome b_5 and cytochrome P-450 (JANS-

SON et al., 1985; TAMBURINI and SCHENKMAN, 1986), and in the cytochrome P-450 system, as well (BERNHARDT et al., 1983, 1984; MAKOWER et al., 1984). Chemical modification of the N-terminal amino group and of a functionally linked lysine of cytochrome P-450 LM2 resulted in restricted electron transfer and catalytic activity. High ionic strength in general perturbs electron exchange (BERNHARDT et al., 1984). A negatively charged vesicle surface may therefore interact with complementary groups on the cytochrome P-450 molecule or, on the other hand, on the reductase, as deduced for the membrane-neighbouring part of the amino acid sequence which connects the hydrophilic head and the membrane-bound anchor of the molecule (BLACK and COON, 1982). These electrostatic interactions would improve recognition and ordering or improve alignment of the interacting proteins.

The endogenous interaction of reductase and cytochrome P-450 at physiological stochiometry obviously requires a further mode of protein association, with at least a homologous aggregation of cytochrome P-450 molecules to build up functional clusters. The distintegration of cytochrome P-450 systems by means of different detergents (BLANCK et al., 1979a; INGELMAN-SUNDBERG, 1977b; DEAN and GRAY, 1982; ROHDE et al., 1983; WAGNER et al., 1984; DAVYDOV et al., 1985) significantly decreases the electron transfer rate and catalytic activity. Homologous protein association by a predominantly electrostatic interaction seems less plausible. The control function of the lipid seen at the supramolecular level points to further regulation mechanisms. The phospholipid seems to provide the structural features for protein association into functional clusters which leads to the formation of a catalytically active reductase/cytochrome P-450 1:1 complex. That control must provide

- the correct local concentration of particular phospholipids to provide conditions for partitioning of cytochrome P-450,
- the correct association of cytochrome P-450, based on lipid specific association constants, but with high mobility of the cytochrome P-450 molecules to allow rapid successive reductase/cytochrome P-450 interaction in an active 1:1 complex.

Protein-induced phase separations of membrane lipids have been the subject of several investigations (GREINERT et al., 1982a; STIER et al., 1982; BAYERL et al., 1984, 1985, 1986). By means of ^{31}P-NMR and photoselection techniques it was shown that a non-lamellar lipid phase — absent in protein-free liposomes — is formed in liver microsomes and reconstituted vesicles. This phase is suggested to be formed by the lamellophobic PE (GREINERT et al., 1982a; STIER et al., 1982) which, like PA and PS, forms the hexagonal form II (LUZZATI et al., 1968) if there is an unsaturated fatty acid located in position 2 of the glycerol. The wedgeshaped PE structure is proposed to adapt cytochrome P-450 specifically to the bilayer membrane by a half-micelle structure of this lipid. Correlation of cytochrome P-450 and the PE content in microsomes has been reported for the increased level of protein and lipid in male rats (BELINA et al., 1974), in age dependence (KAMATASKI et al., 1981; STIER et al., 1982) and on drug induction (STIER et al., 1982). The high transverse molecular mobility (VAN DEN BESSELAAR et al., 1978) is suggested to depend on the perturbation of the endoplasmic reticulum bilayer membrane (STIER et al., 1982).

The microsomal reductase participates in the cytochrome P-450/PE interaction. The isotropic ^{31}P-NMR signal disappears after trypsin treatment, and the lateral diffusion coefficient decreases in parallel from 10^{-7} cm^2 s^{-1} to 10^{-8} cm^2 s^{-1} at 37 °C (BAYERL et al., 1984). The cleavage of the hydrophilic part of the protein molecule leaves a hydrophobic anchor part which evidently cannot generate the interaction signal. In further investigations by use of hexane phosphonic acid diethyl ester as a molecular ^{31}P-NMR probe the specific interaction of cytochrome P-450 and reductase with PE was confirmed (BAYERL et al., 1985). Recently, by means of nonionic detergents, a preferential removal of PC over PE from microsomes without significant effects on the functional properties of the enzyme system at the PC solubilization level has lent further support to a specific protein/PE interaction (BAYERL et al., 1986).

The acidic phospholipid PA likewise appears to show a specific cytochrome P-450 interaction. By means of the cholate dialysis technique vesicles of PC/PA were reconstituted with cytochrome P-450 LM2. In the absence of the protein a phase separation of the lipid into domains of gel phase which contained enriched PA, and a surrounding fluid matrix of mainly PC was observed. The PA phase transition at T_c disappeared on cytochrome P-450 incorporation, indicating a preferential PA/cytochrome P-450 interaction (BÖSTERLING et al., 1981).

A specific PS/cytochrome P-450 interaction is not yet proven. The catalytic activity of reconstituted vesicles containing PS is significantly increased in comparison to egg yolk PC or DOPC systems (INGELMAN-SUNDBERG and JOHANSSON, 1980b; INGELMAN-SUNDBERG et al., 1981). The increase in catalytic activity of reconstituted vesicle systems with the acidity of the lipid headgroup would indicate the charge to be more significant than the lamellophobic molecule shape (KISSEL et al., 1979; INGELMAN-SUNDBERG and JOHANSSON, 1980b; INGELMAN-SUNDBERG et al., 1981).

3.4.4. Electron transfer and substrate conversion

The cytochrome P-450 reduction reaction and the catalytic activity show similar dependences on spin equilibrium, R/P-stoichiometry, and lipid specificity. Quantitative relations between cytochrome P-450 reduction and substrate conversion V_{max} have been derived recently (SCHWARZE et al., 1985), and have been confirmed by several detailed investigations.

The substrate conversion is preceded by the introduction of the second electron, which is widely accepted to represent the rate-limiting step in the overall reaction. The reaction scheme under anaerobic conditions may thus be approximated by

$$R + P \underset{}{\overset{K_{RP}}{\rightleftharpoons}} RP \underset{k_{-1}}{\overset{k_1}{\rightleftharpoons}} RP^- \xrightarrow{k_2} RP^{--} \longrightarrow R + P \qquad (30)$$

That equation considers reductase R and cytochrome P to build up the 1:1 interaction complexes RP, RP^-, RP^{--} which contain cytochrome P-450 up to a twofold reduced state. The approximation further proposes rapid kinetics with respect to the preequilibration step (K_{RP}) and the decay of the two-electron reduced complex RP^{--} to release products, H_2O, and oxidized protein (substrate, products, H_2O omitted for simplicity). It is further proposed that in excess of NADPH the electron transfer towards free reductase is not rate limiting. According to the scheme the reductase is cycling between RP and RP^- thus leading to the steady state equation

$$k_1 [RP] = (k_{-1} + k_2) [RP^-] \qquad (31)$$

for the catalytic process. The substrate conversion rate V_{max} is then given by

$$V_{max} = k_2 [RP^-] = k_2 \frac{k_1}{k_{-1} + k_2} [RP] \qquad (32)$$

This equation indicates a linear dependence of V_{max} on the interaction complex RP, as shown for the apparent rate constant of cytochrome P-450 reduction, k_{app} (Blanck et al., 1984b). Therefore titrating RP in dependence on the ratio R/P by means of the substrate conversion rate V_{max} must likewise reveal the dissociation constant K_{RP} in the investigated system, as shown by Dignam and Strobel (1977), Vermilion and Coon (1978), Miwa et al. (1979), Oprian et al. (1979), French et al. (1980), Miwa and Lu (1981, 1984) and Müller-Enoch et al. (1984).

The importance of a proper structural association and alignment of the electron exchange proteins (reductase and cytochrome P-450) in the catalytic process has been recently supported by the isolation of a cytochrome P-450 monooxygenase in *Bacillus megaterium* which in a single polypeptide chain (119 kDa) contains 1 mol each of FAD and FMN per mol of heme. This self-sufficient monooxygenase exhibits the highest catalytic activity for a cytochrome P-450 system yet reported (4600 nmol of fatty acid hydroxylated per nmol of cytochrome P-450 and min) (Narhi and Fulco, 1986).

The regulation of V_{max} according to equation (32) depends on the rate-limiting constant k_2 of the second electron transfer and on the steady state concentration of the one electron-reduced intermediate 1:1 complex of reductase and cytochrome P-450 (RP^-). The quantity of that complex depends on the spin state of the cytochrome P-450 population through the rate constant of the first electron transfer k_1, on the rate constants k_{-1} and k_2, and on the amount of the oxidized 1:1 complex of reductase and cytochrome P-450 (RP). The latter quantity is determined by the respective concentrations of reductase and cytochrome P-450 and by the lipid controlled dissociation constant of the complex. Thus equation (32) comprises the rate-limiting control of the second electron transfer as well as the spin state, the reductase/cytochrome P-450, and the phospholipid control of the first transfer reaction. For microsomes as well as for cytochrome P-450 LM2 reconstituted systems with a series of benzphetamine analogues, k_1 increases linearly with the high-spin amount α of cytochrome P-450 (BLANCK et al., 1983; SCHWARZE et al., 1985). The transformation of this spin state regulation towards V_{max} is elucidated by the modified equation (33)

$$V_{max} = k_2 [RP^-] = k_2 \frac{k_1 \cdot \text{const.}}{k_1 + k_{-1} + k_2} \tag{33}$$

k_1 was proposed to be the only rate constant which depends on the spin state of a stoichiometrically determined system. Because $k_1 > k_2$ the variable regulation term of V_{max} amounts to $k_1/(k_1 + k_{-1})$, and the dependence of V_{max} on k_1 therefore must exhibit hyperbolic characteristics, a linear dependence at $k_1 < k_{-1}$, and saturation behaviour at $k_1 > k_{-1}$. The respective pattern of the benzphetamine series indicated an approximate $k_1 \leq k_{-1}$ system, that was supported by an independent proof of $k_1 \sim k_{-1}$ (RUF and EICHINGER, 1982). Since the variable regulation term reflects the amount of the intermediate ternary complex of cytochrome P-450, substrate, and oxygen, that quantity must increase with k_1. Experimental evidence for that relation has been presented (IYANAGI et al., 1981).

3.5. Cytochrome b_5 dependent regulation

The regulation of electron transfer and thus of the catalytic activity of cytochrome P-450 systems as treated so far involves the interaction of the three main components of the liver microsomal monooxygenase system. The cytochrome P-450 system in the endoplasmic reticulum, however, exhibits an additional interaction with the NADH-dependent cytochrome b_5 electron transfer chain, which may be involved to generate a synergistic enhancement of the enzymatic activity (CONNEY et al., 1957; NILSSON and JOHNSON, 1963; COHEN and ESTABROOK, 1971a, b, c; HILDEBRANDT and ESTABROOK, 1971; CORREIA and MANNERING, 1973a, b; ARCHAKOV and DEVICHENSKY, 1975; ARCHAKOV et al., 1975a, b). Cytochrome b_5 contributes the second electron to the oxyferrous complex of some cytochrome P-450s, whereas the first electron is predominantly provided by the cytochrome P-450 reductase (WHITE and COON, 1980), as indicated by the facts that:

- the NADPH-dependent reduction of microsomal cytochrome P-450 is not accelerated by addition of NADH (HILDEBRANDT and ESTABROOK, 1971),

- the addition of cytochrome b_5 antibodies does not disturb that reaction (Noshiro at al., 1980),
- the NADH-dependent reduction of cytochrome P-450 is relatively slow (~10%) compared to the NADPH transfer chain reaction (Archakov et al., 1975a; Miyake et al., 1975).

The donor function of cytochrome b_5 is not obligatory (Lu et al., 1974b, c; Imai and Sato, 1977). The protein exhibits a substrate specificity: cytochrome b_5 antibodies inhibit the monooxygenation of 7-ethoxycoumarin, benzo[a]pyrene and benzphetamine to different extents but do not inhibit the conversion of aniline (Noshiro et al., 1979). Furthermore, isozyme specificities were observed in reconstituted systems (Jansson and Schenkman, 1973, 1977; Hrycay and Estabrook, 1974), and an obligatory requirement for a distinct cytochrome P-450 isozyme could even be shown (Sugiyama et al., 1979, 1980; Miki et al., 1980).

The electron transfer between cytochrome b_5 and cytochrome P-450 has been the subject of several more recent investigations (Werringloer and Kawano, 1980a; Bonfils et al., 1981; Noshiro et al., 1981; Pompon and Coon, 1984), which enabled the integration of the cytochrome b_5-dependent reactions into a general reaction scheme for the catalytic function of the cytochrome P-450 system in terms of regulation specificities. The kinetic studies so far lend further support to the rate-limiting role of the second electron transfer to the terminal protein, which may interfere with synergistic effects of the cytochrome b_5 transfer chain.

Of the proteins of the cytochrome b_5 chain, the FAD-containing cytochrome b_5 reductase (Strittmatter and Velick, 1956a, b) and cytochrome b_5 appear to be randomly distributed in the endoplasmic reticulum and to be functionally controlled by translational diffusion in the two dimensional membrane plane (Rogers and Strittmatter, 1974). The perturbation by cholesterol of cytochrome b_5 reduction supports that result (Archakov et al., 1983).

The random lateral distribution of the proteins within the membrane lipid allows the potential complexation of cytochrome b_5 with cytochrome P-450. In the presence of DLPC a 1:1 complex is formed, as indicated by an increase in the high-spin content of cytochrome P-450 LM2 (Bonfils et al., 1981). The apparent dissociation constant K_D is 2.3 μM, and decreases in the presence of benzphetamine to 0.4 μM. Complementary studies proved the stabilization of the substrate complex of cytochrome P-450 with benzphetamine by cytochrome b_5, the respective dissociation constants being 220 μM and 50 μM. The investigations demonstrated an obligatory role of lipid for maximal interaction of the components and different binding sites on the cytochrome P-450 molecule for cytochrome b_5 and for benzphetamine. The mutual facilitation of the compounds towards complex formation parallels the corresponding observation for cytochrome P-450 (French et al., 1980). The results for the cytochrome P-450/cytochrome b_5 couple support conformational changes inducing interactions (Hlavica, 1984): CD measurements in the far UV region revealed concomitant conformational changes in one or both proteins, and the dissociation constant determined was $K_D = 3.2\ \mu M$. Further support for 1:1 complex formation has been presented (INGELMAN-SUNDBERG et al., 1980; CHIANG, 1981; BENDZKO et al., 1982; BÖSTERLING and TRUDELL, 1982b; VATSIS et al., 1982; TAMBURINI and GIBSON, 1983; INGELMAN-SUNDBERG and JOHANSSON, 1980a, 1984; TANIGUCHI et al., 1984). Phospholipid is also required for an appropriate interaction to take place (INGELMAN-SUNDBERG and JOHANSSON, 1980a).

Whereas binary complex formation between cytochrome b_5 and cytochrome P-450 is a prerequisite for an effective second electron transfer, the catalytic action, on the other hand, requires the additional involvement of the cytochrome P-450 reductase. The ternary cytochrome P-450 reductase/cytochrome P-450/cytochrome b_5 complex has been shown to be the catalytically active species, with the component in deficit being rate limiting (INGELMAN-SUNDBERG et al., 1980; INGELMAN-SUNDBERG and JOHANSSON, 1984).

The interactions of the proteins in and between the two electron transfer chains are obviously electrostatic in nature. They control the coupling of cytochrome b_5 reductase/cytochrome b_5 (LOVERDE and STRITTMATTER, 1968; DAILY and STRITTMATTER, 1979), of cytochrome P-450 reductase/cytochrome b_5 (DAILY and STRITTMATTER, 1980), of cytochrome P-450/cytochrome b_5 (TAMBURINI et al., 1985), and of cytochrome P-450 reductase/cytochrome P-450 (cf. above).

By means of specifically reconstituted bilayer systems from purified components of rabbit liver, a systematic study of the first electron transfer towards cytochrome P-450 was performed. The investigations revealed that the electron transport via cytochrome P-450 reductase represents the effective reaction: the rate constants of the NADPH- and the NADH-supported processes were $1.1\ s^{-1}$ and $0.5\ s^{-1}$, respectively (TANIGUCHI et al., 1984). The rapid NADH-supported cytochrome b_5 reductase/cytochrome b_5 transfer, the rate constant of which was determined as $30\ s^{-1}$, was ineffective because of the slow cytochrome b_5/cytochrome P-450 transfer at $0.046\ s^{-1}$. Thus the observed cytochrome b_5 synergism must be attributed to the second electron transfer process (Fig. 9).

The second electron is transferred towards the one-electron reduced cytochrome P-450 species in the 1:1 complex with reduced cytochrome b_5, the dissociation constant of which is $K_D = 7.5\ \mu M$, falling to $2.2\ \mu M$ in the presence of benzphetamine (BONFILS et al., 1981). The photoreduced cytochrome P-450 after O_2 binding shows first order rate constants for the electron transfer from cytochrome b_5 (as measured by reoxidation of either cytochrome b_5 or cytochrome P-450) of $2.5\ s^{-1}$ and $4-7\ s^{-1}$ in the presence of the substrate. These data in the order of the rate of first electron transfer between cytochrome P-450 reductase and cytochrome P-450 at reductase saturation (BLANCK et al., 1984b) indicate the competence of the second electron transfer in cytochrome b_5-dependent synergism of cytochrome P-450 driven reactions.

The reoxidation of the cytochrome P-450/oxygen complex can proceed alternatively by donation of an electron to the oxidized cytochrome b_5 (POM-

PON and COON, 1984). The multiphasic reaction exhibits rate constants consistent with the electron acceptance process. Thus in general the decay of the reduced cytochrome P-450/oxygen complex is increased by interaction with cytochrome b_5. A specificity has been observed with respect to distinct cytochrome P-450 isozymes: cytochrome P-450 LM4 preferentially donates an electron to cytochrome b_5, whereas cytochrome P-450 LM2 prefers to accept an electron from cytochrome b_5. Reversible electron exchange between cytochrome P-450 LM2 and cytochrome b_5 in a covalent complex is proposed to occur in the catalytic cycle: the first electron input generates ferrous cytochrome P-450, followed by immediate transfer of this electron to the oxidized cytochrome b_5; after a further reduction of the ferric cytochrome P-450 and binding of molecular oxygen, the ferrous cytochrome b_5 returns the electron to the oxyferrous cytochrome P-450 (TAMBURINI and SCHENKMAN, 1987).

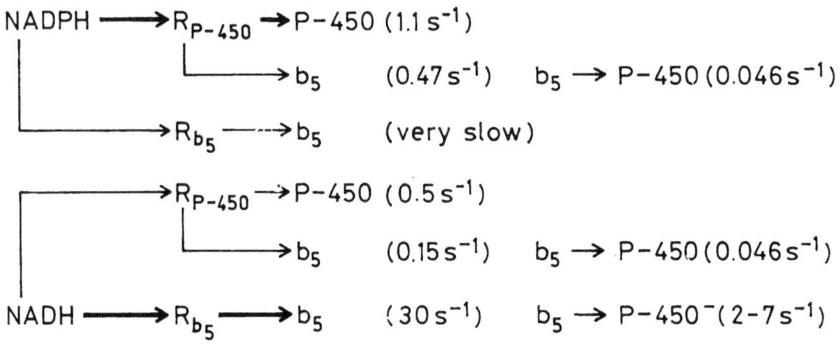

Fig. 9. Electron transfer reactions (according to BONFILS et al., 1981; TANIGUCHI et al., 1984).

Besides the electron carrier role of cytochrome b_5 in cytochrome P-450 catalyzed reactions the former protein may further exert an effector function. The decay of the cytochrome P-450 ternary complex has been shown to be favoured in the presence of cytochrome b_5 but with no evidence of a redox coupling of the two proteins (GUENGERICH et al., 1976). Moreover, in a reconstituted bypass system, the cumene hydroperoxide-supported conversion 4-chloroaniline is modified by cytochrome b_5 addition: the apparent K_D values of the substrate and of the oxygen donor complex as well, are decreased, and the rate of complex formation is increased; the temperature function of the ternary complex formation by reaction of substrate-bound cytochrome P-450 with the oxygen donor, furthermore, is significantly altered (HLAVICA, 1984). The obligatory cytochrome b_5 activation of one distinct P-450 isozyme points further to the effector action of that protein by means of the induction of conformational changes in the terminal hemoprotein oxygenase (SUGIYAMA et al., 1979, 1980; MIKI et al., 1980).

The control of the oxidase pathway of the monooxygenase system additionally may depend on cytochrome b_5. Since the oxidative generation of H_2O_2 proceeds mainly by dissociation of the 1-electron reduced cytochrome P-450/oxygen complex and subsequent dismutation (HILDEBRANDT and ROOTS, 1975; HILDEBRANDT et al., 1975; KUTHAN et al., 1978; KUTHAN and ULLRICH, 1982), the amount of the latter species in the steady state would determine that pathway. Decreased H_2O_2 generation at low pH and at low ionic strength, in line with a decreased steady state level of oxycytochrome P-450 could be observed, due to an increased reductive action of cytochrome b_5 (INGELMAN-SUNDBERG, 1980a; WERRINGLOER and KAWANO, 1980a; NOSHIRO et al., 1981).

The partition between productive and abortive pathways in dependence on cytochrome b_5 has been dealt with recently by POMPON (1987). Cytochrome b_5 could be shown to increase effectively product formation and to decrease H_2O_2 and H_2O generation. A steady state kinetic model with two branch points for water, hydrogen peroxide and product formation is proposed which allows to predict cytochrome b_5 effects.

3.6. Isozyme interactions and protein compartmentation

The functional interaction of the cytochrome P-450 monooxygenase system with components of the cytochrome b_5 system is evidently based on protein/protein interactions. A further interaction may be generated in microsomes by isozyme/isozyme interactions or even by interactions with the multitude of other membrane proteins. Recent investigations on the microsomal metabolism of warfarin demonstrated specific mutual inhibitions of rat microsomal isozymes induced by different agents, which could not be compensated for by additional reductase (KAMINSKY and GUENGERICH, 1985). Moreover, an isozyme specificity was exhibited with respect to the reductase interaction. The cluster hypothesis of reductase/cytochrome P-450 interaction mentioned earlier has been further put into question by recent results on the regiospecific hydroxylation of steroids (ESTABROOK et al., 1985). By means of differential inhibition it was shown that specific reactions may proceed independently of each other, thus exhibiting the failure of a competition for reductase in molecular deficit. The findings are discussed as being due to a structural hetero-distribution of the interacting proteins. The heterogeneously composed lipid matrix can obviously provide the structural basis for a specific localization or compartmentation of the membrane proteins.

4. Regulation at the cellular level

Cellular regulation of cytochrome P-450 activity is characterized by a high degree of complexity and is mainly concerned with induction. Induction in mammals is subject to various influences such as biosynthesis and turnover of cytochrome P-450 as well as metabolic and diffusional processes. Tissue, sex and age dependences have been observed. Inducers of cytochrome P-450 may also initiate induction of post-oxidative enzymes. Moreover, regulation at the cellular level includes inducibility and thus genetic polymorphism can result in individual differences with regard to biotransformation of xenobiotics.

A thorough discussion of all these subjects would exceed the extent of this article. Hormonal control of cytochrome P-450 is discussed by Morgan and Gustafsson later in this volume, and induction will be dealt with extensively in the following volume of this series. This section, therefore will be limited to some selected general aspects of induction of cytochrome P-450, as the most important enzyme in the control of biotransformation of xenobiotics at the cellular level.

Due to the capability of cytochrome P-450 to transform xenobiotics to inactive products or to convert them — dependent on their chemical nature — to reactive intermediates, induction of cytochrome P-450 has become of increasing importance with regard to biological consequences of biotransformation. The rate and pathways of xenobiotic biotransformation decisively depends on the concentration and the pattern of cytochrome P-450 isozymes in different tissues. 61 different cytochrome P-450 enzymes (55 endoplasmic, 3 mitochondrial and 3 soluble in origin) have been recognized recently (cf. Estabrook and Vereczkey, 1985). Many of them occur as isozymes firmly evidenced as distinct gene products by protein and DNA sequencing which recently has been reviewed by Black and Coon (1986).

Therefore cellular regulation of cytochrome P-450 activity essentially depends on regulation of induction. Available information strongly implies that induction is controlled transcriptionally whereas its modulation proceeds at the level of expression. Tissues are characterized by different activities of expression which even exist among differently localized hepatocytes. Significant effects on the expression of certain cytochrome P-450 genes have been observed with endotoxins, interferons and steroid hormones (Wolf, 1986).

Additionally, activity of inducible and constitutive isozymes can be regulated by exogenous substances and steroid hormones acting as positive or negative effectors. Thus flavones stimulate the biotransformation of zoxazolamine in vivo (Lasker et al., 1982). Flavone and 7,8-benzoflavone stimulate the catalytic activity of cytochrome P-450 LM3c at the biotransformation of benzo(a)-pyrene. For nomenclature of cytochrome P-450 isozymes the reader is referred to Guengerich (chapter 3, this volume). It has been shown that certain metabolites of progesterone may influence the 6β-hydroxylase activity of cytochrome P-450 LM3b both negatively and positively proving the competitive inhibition of the 6β-hydroxylation as most effective kind of influence. Like cytochrome P-450 LM3b, the LM3c form shows homotropic cooperativity in the biotransformation of steroids thus indicating an allosteric regulation (Johnson et al., 1983). The induction process may be connected with promotor releasing enhancer functions (Jones et al., 1986).

4.1. Historical background of induction

The induction of drug metabolizing enzymes was discovered in the early 50's by Miller et al. (1952) performing investigations on chemical carcinogenesis. These studies were designed to investigate the role of biotransformation of dimethylaminoazobenzene-induced carcinogenesis, as initially described by György et al. (1941). Richardson and Cunningham (1951) reported that liver tumors develop much slower in rats which were treated simultaneously with 3'-methyl-4-dimethylaminoazobenzene and 20-methylcholanthrene than in those rats which were treated with the azo compound only. Since an increased oxidative demethylation from methylated amino azo dyes was found in liver homogenates from rats and mice after pretreatment with aromatic hydrocarbons (Brown et al., 1954) the lower carcinogenic effect of two simultaneously administered carcinogens could be explained by an increased activity of the biotransformation system. The mechanism by which the polycyclic aromatic hydrocarbons (PAH) inhibit the formation of hepatic tumors is based on their capability to increase the activity of metabolizing enzymes which degrade the azo dye into non-carcinogenic derivatives. PAHs (e.g. 3-methylcholanthrene) have a carcinogenic action on the skin and subcutaneous tissues, but there is little or no effect on the liver.

This was the first experimental evidence for the inducibility of a NADPH- and O_2-dependent enzyme in liver microsomes. Extending these findings, a correlation between the inductive effect of certain PAHs on oxidative demethylation and its inhibitory effect on the hepatocarcinogenicity of 3'-methyl-4-dimethylaminoazobenzene has been demonstrated by Conney et al. (1956).

Many similar observations following the above experimental approach are all based on the same cellular control mechanism, i. e. the capability of the cell to respond to certain substances (inducers) with the biosynthesis of specific enzymes. Due to this adaptive process either the length and intensity of drug action or the extent of toxic effects originating from biotransformation of xenobiotics are altered.

4.2. Classification of inducers and mechanisms of induction

Up to now 3 classes of cytochrome P-450 inducers are known which induce different families of isozymes with various substrate specificities (Whitlook, 1986).

(1) the phenobarbital type (including most of inducing drugs)

(2) the methylcholanthrene type (PAHs with different efficiency, TCDD)

(3) the steroid type (most potent inducer: pregnenolone-16α-carbonitrile (PCN)) (Gonzales et al., 1985).

Due to recent findings, however, this classification has to be extended. JUVONEN et al. (1985) found a pyrazole inducible form in mice different from other so far known monooxygenases. As investigations on this subject are still in progress no final conclusions can be drawn as yet (GUENGERICH, this volume).

The molecular mechanisms of induction are far from being understood in full detail. Available data suggest a transcriptionally controlled mechanism irrespective of the inducer type. Due to the toxic consequences, the induction of the PAH converting isozymes has been particularly extensively studied and has led to an evidenced sequence of partial reactions during the induction process. As prototype of this kind of inducers TCDD has been used. After passage through the cell membrane TCDD forms a complex with a cytosolic receptor in the liver cell. Subsequently the TCDD receptor complex is translocated from the cytosol to the nucleus where it binds site-specifically near the cytochrome P-450c gene. The binding stimulates the transcription of specific genes including cytochrome P-450c. The transport of the polyA-mRNA from the nucleus into the cytosol is suggested to represent the rate-limiting step of biosynthesis (IVERSEN et al., 1986). The subsequent translation finally leads to selective biosynthesis of the specific isozyme into which mitochondrial derived heme is inserted to form the holoenzyme. The increased amount of the cytochrome P-450 isozyme results in an enhanced biotransformation of respective substrates with carcinogenic or mutagenic effects (NEBERT et al., 1975; POLAND et al., 1976; NEBERT, 1979).

The structural gene for human cytochrome P_1-450 which is closely associated with aryl hydrocarbon hydroxylase (AHH) activity has been assigned to chromosome 15 q22—q24 using mitogen stimulated lymphocytes (AMSBAUGH et al., 1986). This finding is in agreement with another recent report using somatic cell hybrids where the human TCDD-inducible cytochrome P_1-450 gene has been assigned to chromosome 15 (HILDEBRANDT et al., 1985). The regulatory gene(s) which previously have been attributed to human chromosome 2 required for the expression of AHH activity is not necessarily located at the same chromosome as the structural gene (BROWN et al., 1976).

In contrast to the detailed induction mechanism initiated by TCDD far less information on the mechanism(s) of phenobarbital and PCN induction is available. Recently PIKE et al. (1985) proved a transcriptionally controlled PB induction by showing the same extent of induced mRNA in the nuclei as in the cytoplasm. These findings exclude changes in the rates of processing, transport or degradation of mRNA as reason for the enhanced formation of the isozyme. SCHUETZ et al. (1984a, b) have shown that PCN and other glucocorticoides (dexamethasone, α-methylprednisolone) increased the concentration of cytochrome P-450 PCN by stimulating the synthesis of P-450 PCN directed mRNA.

The induction process not only enhances the overall concentration of cytochrome P-450 but considerably alters the pattern of isozymes. Phenobarbital for example causes a 2.6-fold increase in the overall concentration of cytochrome P-450 in the liver of male rats, whereas the percentage of the specifically inducible isozyme related to total cytochrome P-450 is increased from 5% to as much as 86% (PHILLIPS et al., 1983). This means that at the same time the content of all other isozymes decrease from 85% to 14%. It is still unclear to date whether the distribution pattern of the isozymes which are not induced by phenobarbital is maintained or whether these isozymes are altered individually thus changing their mutual ratio and pattern. Such an assumption is tempting because the induction by β-naphthoflavone (BNF) does not only lead to a 1.4-fold increase in the total content of cytochrome P-450 in the microsomal membrane but concomitantly causes a repression of the translatable mRNA responsible for the phenobarbital specific isozyme (PHILLIPS et al., 1983). HARADA and OMURA (1983) found somewhat lower values but in the same range.

These finding are remarkable in so far as they provide the biochemical basis to understand the functional complexity of the monooxygenase system with respect to pharmacological and toxicological consequences by its multiplicity and thus may explain the linkage between the activity of the cytochrome P-450 system and chemical carcinogenesis.

4.3. Relations between induction and chemical carcinogenesis

Studies by NEBERT and coworkers have provided compelling evidence for the correlation between the inducibility of cytochrome P-450 (NEBERT and JENSEN, 1979) and the occurrence of chemical carcinogenesis. They used mouse strains, which contrary to normal mice (responsive), are insensitive to induction of cytochrome P_1-450 by PAH (nonresponsive). Relating the AHH activity of the liver to 3-methylcholanthrene induced fibrosarcomas revealed a significant lower frequency of tumors in nonresponsive mice as compared to responsive animals. Similar results were found with lung tumors after intra-tracheal administration of 3-MC. Extending these results NEBERT's group (NEBERT et al., 1975; POLAND et al., 1976) have shown that the cytosolic TCDD binding receptor is lacking in nonresponsive mice (THORGEIRSSON and NEBERT, 1977), thus rationalizing the observed lower tumor frequency in nonresponsive mice.

Despite of possible additional influences (e.g. DNA repair mechanisms, immunological status, sensitivity to viral infections), the above results indicate that the occurrence of chemically induced tumors is closely associated with the inducible AHH activity.

In addition to the direct xenobiotic-initiated, transcriptionally controlled regulation of induction, another indirect linkage between chemical carcinogenesis and induction has been described including augmentation of transcription and promotor release. Tumor promoting effects of phenobarbital were already observed by PERAINO et al. (1971). Feeding of rats with phenobarbital after feeding with 2-acetylaminofluorene (AAF) resulted in a significant increase in the incidence of hepatomas whereas the simultaneous administration of both compounds reduced the hepatocarcinogenic effects of AAF.

BUCHMAN et al. (1986) have studied the influence of 3-MC type induction (treatment with 3,3', 4,4'-tetrachlorobiphenyl) and PB type induction (treatment with 2,2', 4,4', 5,5'-hexachlorobiphenyl) on diethylnitrosamine (DEN) initiated carcinogenesis in rats. Treatment with both types of inducers subsequent to DEN administration extensively enhanced the number of preneoplastic islets in liver as compared to DEN treatment only. Obviously both types of PCBs are potent promoters of DEN-induced preneoplastic foci in rat liver thus confirming earlier findings at the tumor promoting effects of PBBs (JENSEN et al., 1983).

JONES et al. (1986) have shown that the process of induction may be connected with an enhancer function, releasing a promotor. Studying a dioxin responsive genomic element which is flanking the 5'-end of the cytochrome P-450 gene in mouse hepatoma cells the authors found this element to regulate the mouse mammary tumor virus promotor. The TCDD/receptor complex is required to initiate the biosynthesis of this element. The responsiveness of the element towards TCDD is maintained at transfection into cells from either a heterologous mouse tissue or a heterologous species (human). The dioxin-responsive genetic element together with TCDD receptors constitute a TCDD-regulated enhancer system. The authors conclude that the system augments gene expression by increasing the frequency of transcription initiation.

With respect to this category of indirect influences of induction, a related observation was made by FARIS and CAMPBELL (1981, 1983) on the long-term effects of induction. These authors found that neonatal exposure to phenobarbital may irreversibly alter the pattern of cytochrome P-450 catalyzed biotransformations in adult male rats. The mechanism by which this apparent imprinting proceeds remains to be elucidated.

4.4. Polymorphism of induction

A correlation between chemical carcinogenesis and the cytochrome P-450 dependent activation of certain xenobiotics has been revealed in investigations on laboratory animals. It was therefore of interest to establish whether these results could be extrapolated to humans, and what role the genetically determined inducibility plays as a risk factor with regard to biotransformation of xenobiotics.

KELLERMANN et al. (1973a, b) have provided evidence for the existence of a correlation between AHH activity and tumor incidence in man. Analysing human lymphocytes they have shown a trimodal distribution of AHH inducibility to exist for the caucasian population of the USA. Patients with bronchial carcinoma predominantly belong to those groups with mean or high AHH inducibility. Numerous investigators have attempted to reproduce the results, with varying degrees of success (cf. PFEIL et al., 1983). The main reasons for the poor reproducibility include seasonal fluctuations (PAIGEN et al., 1981) drug-related and alimentary influences, and effects from secondary environmental stimuli (e.g. smoking). PELKONEN et al. (1980) have compiled these results from various authors, and concluded that in 70% of the cases, there is a correlation between AHH activity and the occurrence of lung cancer.

Unequivocal evidence for the genetic determination of AHH activity in man has been obtained. A study involving 16 pairs of twins has yielded a considerable hereditary component for AHH inducibility in human mitogen stimulated lymphocyte cultures (ATLAS et al., 1976; OKUDA et al., 1977). AMSBAUGH et al. (1986) found that the benzanthracene-induced AHH activity in lymphocytes corresponds with the amount of cytochrome P_1-450 mRNA. This finding indicates a causal structure/function relationship and more or less excludes any contribution from other non-cytochrome P_1-450 enzymes to the individual differences.

This tends to suggest genetic polymorphism as being responsible for the different biotransformation activities. Indeed, anomalous patterns of cytochrome P-450 isozymes can produce defects in biotransformation. Due to its enzymatically different attack, three types of genetic polymorphism in cytochrome P-450 have been differentiated: C-, S- and N-hydroxylations. To the most frequently occurring first group debrisoquine, spartein and bufuralol could be attributed. Recently a connection between bronchocarcinoma and debrisoquine metabolism has been suggested by AYESH et al. (1984, 1985). Persons with bronchocarcinoma appear all as extensive metabolizers and perhaps predominantly homozygotes for extensive metabolism (cf. KALOW, 1987). PLUMMER et al. (1986), however, have shown that debrisoquine hydroxylase does not metabolize or activate any of the known precarcinogens, so that the connection being seen cannot be a major cause but rather should be regarded as a risk factor contributing to bronchocarcinoma. But despite of this the knowledge of each more risk factor will improve our predictions of the causes of cancer.

5. Concluding remarks and outlook

Regulation of the monooxygenase system at the **molecular level** proceeds predominantly through interaction of cytochrome P-450 with substrates. The target for the substrate controlled regulation is provided by an equilibrium of two conformers having different functional properties. Substrates shift the spin equilibrium towards the high-spin conformation. This reaction is connected with a redox potential shift to a more positive value resulting in a greater electromotive force for a subsequent electron transfer from the reductase to the monooxygenase. A correlation between spin and redox state and the reduction rate, as well, could be evidenced and indicates the regulatory significance of the substrate dependent spin equilibrium. This regulatory function could be shown not to be limited to the first electron transfer but extended through the whole reaction cycle resulting in an additional correlation with the substrate conversion by use of a homologous series of substrates. Based on thermodynamic data the spin-redox coupling could be rationalized in a corresponding model. In the polysubstrate cytochrome P-450 LM2, changes of both the spin and the redox equilibrium are mainly entropy driven processes. Conformational changes have been evidenced to be rather small.

The regulation of the P-450 system at the **membrane level** depends on the specificities of the membrane proteins and of the membrane constituting lipids, as well, which offer the structural matrix for the catalytic reaction. The membrane bilayer provides the "volume" for the incorporated proteins, determines their translational and rotational mobility, and may exert diffusional limitations towards protein/protein interactions. The functional properties of the proteins may be further modified by immersion depth and orientation, as well. A heterogenously composed lipid matrix provides the structural basis for protein compartmentation, hetero-distribution and association, thus generating functional specificities of protein/protein interaction. The electron supply towards P-450 is obviously rate limiting in enzymatic activity. The physiologically relevant NADPH-dependent electron transfer proceeds within functionally active associates of P-450 and reductase (cluster). The electron transfer is controlled by 1:1 complex formation between the exchange proteins. The head group of the lipid exerts a control function in that negatively charged species additionally favour cluster and complex formation. The NADH-dependent cytochrome b_5 system is capable of interaction with the monooxygenase system. The main function of the b_5 system is a synergistic action with respect to the substrate conversion, in that it delivers the second electron in the oxygen activation process, dependent on the P-450 species and the substrate. Besides the redox coupling of both systems, b_5 exerts an effector function towards P-450. Furthermore the b_5 system controls the reaction pathways of the P-450 system.

Induction of cytochrome P-450 as **cellular regulation mechanism** has proved to be physiologically most important due to its accessibility for external influences. Different classes of inducers are capable of changing the isozymic pattern with regard to concentration and distribution. By the induction intensity or length of drug effects can be modified and the conversion of xenobiotics to toxic or carcinogenic compounds can be enhanced or accelerated as well. The extent of these effects is genetically determined and has been shown to depend on polymorphism. A correlation between the individual cytochrome P-450 inducibility and the carcinogenic risk has been evidenced in both laboratory animals and in humans. The determination of the inducibility by means of in vitro assays (e.g. AHH activity assay in human lymphocytes; biotransformation activity of debrisoquine) proves the possibility of prophylactic diagnosis and other biomonitoring applications. Besides utilization for medicinal purposes an application of cytochrome P-450 systems for biocatalytic processes (e.g. conversion to useful compounds, biomonitoring, decontamination) seems possible on the basis of present knowledge.

Investigations of monooxygenase systems over about three decades have provided a huge amount of experimental data. Some of them have been selected in this article and may open a window for a comprehensive view. Other data offer a fascinating challenge to stimulate further efforts to open more windows. Unsolved problems identified in this article will provide leads for further research which the reader is invited to take part in.

The regulation at the molecular level by a spin/redox coupling mechanism has been evidenced for P-450 LM2. But what do we know about the regulation of high-spin P-450s? Is the catalytic activity of a P-450 species rather determined by a substrate specific conformational state or the respective redox potential and what coupling relations exist between both properties? Alternate pathways of monooxygenases have been described. By which properties of the substrate are they controlled and what determines the isozyme specificity of the different pathways is unclear at present. What properties of the substrate are decisive in the susceptibility towards hydroxylation and what conclusions can be drawn for drug design strategies?

The organization of the P-450 system within the membrane is described by different models, but no details exist concerning the composition of proposed clusters. Are they homogeneously composed or mixed up from different isozymes and is the reductase included? Another as yet undescribed area is the isozyme specificity of the P-450/reductase interaction.

The molecular basis of the induction mechanism has just begun to be understood for TCDD but what mechanisms do exist for other types of inducers has to be clarified. It is not clear if inducibility, genetic polymorphism and promotor susceptibility are interrelated and at what level are they regulated. Which structural and chemical peculiarities determine whether a compound will act as substrate or as an inducer is unknown, as well as the question of regulation mechanisms of P-450 dependent transformations of endogenous substrates.

In particular, we are only just in the beginning to understand the multiple regulation mechanisms originating from the structural multiplicity and the functional versatility of the system under study and their physiological implications. Many efforts are still necessary to tackle this challenge.

6. References

AMSBAUGH, S. C., J.-H. DING, D. C. SWAN, N. C. POPESCU, and Y. T. CHEN, (1986), Cancer Res. **46**, 2423–2427.

ARCHAKOV, A. I. and V. M. DEVICHENSKY, (1975), Arch. Biochem. Biophys. **166**, 313–317.

ARCHAKOV, A. I., V. M. DEVICHENSKY, and A. V. KARYAKIN, (1975a), Arch. Biochem. Biophys. **166**, 295–307.

ARCHAKOV, A. I., V. M. DEVICHENSKY, I. I. KARUZINA, and A. V. KARYAKIN, (1975b), Arch. Biochem. Biophys. **166**, 308–312.

ARCHAKOV, A. I., (1982), in: Cytochrome P-450, Biochemistry, Biophysics and Environmental Implications, (E. HIETANEN, M. LAITINEN, O. HÄNNINEN, eds.), Elsevier/North Holland, Amsterdam, 487–495.

Archakov, A. I., E. A. Borodin, G. E. Dobretsov, E. I. Karasevich, and A. V. Karyakin, (1983), Eur. J. Biochem. **134**, 89–95.
Ariens, J., (1972), Drug Design, Vol. 1, Academic Press, New York.
Atlas, S. A., E. S. Vesell, and D. W. Nebert, (1976), Cancer Res. **36**, 4619–4630.
Autor, A. P., R. M. Kaschnitz, J. K. Heidema, and M. J. Coon, (1973), Mol. Pharmacol. **9**, 93–104.
Ayesh, R., J. R. Idle, J. C. Ritchie, M. C. Crothers, and M. R. Hetzel, (1984), Nature 312, 169–170.
Ayesh, R. and J. R. Idle, (1985), in: Microsomes and drug oxidation (A. R. Boobis, J. Caldwell, F. DeMateis, and C. R. Elcombe, eds.), Taylor and Francis, London, 340–346.
Backes, W. L., S. G. Sligar, and J. B. Schenkman, (1980), Biochem. Biophys. Res. Commun. **97**, 860–867.
Backes, W. L., S. G. Sligar, and J. B. Schenkman, (1982), Biochemistry **21**, 1324 to 1330.
Backes, W. L., P. P. Tamburini, I. Jansson, G. G. Gibson, S. G. Sligar, and J. B. Schenkman, (1985), Biochemistry **24**, 5130–5136.
Bäckström, D., M. Ingelman-Sundberg, and A. Ehrenberg, (1983), Acta Chem. Scand. Ser. B **37**, 891–894.
Bar-Nun, S., G. Kreibich, M. Adesnik, L. Alterman, M. Negishi, and D. D. Sabatini, (1980), Proc. Natl. Acad. Sci. USA **77**, 965–969.
Barsukov, L. I., V. I. Kulikov, G. I. Bachmanova, A. I. Archakov, and L. D. Bergelson, (1982), FEBS Lett. **144**, 337–340.
Bayerl, T., G. Klose, K. Ruckpaul, K. Gast, and A. Möps, (1984), Biochim. Biophys. Acta **769**, 399–403.
Bayerl, T., G. Klose, K. Ruckpaul, and W. Schwarze, (1985), Biochim. Biophys. Acta **812**, 437–446.
Bayerl, T., G. Klose, J. Blanck, and K. Ruckpaul, (1986), Biochim. Biophys. Acta **858**, 285–293.
Beetlestone, J. G. and P. George, (1964), Biochemistry **3**, 707–714.
Belina, H., S. D. Cooper, R. Farkas, and G. Feuer, (1974), Biochem. Pharmacol. **24**, 301–303.
Bendzko, P., S. A. Usanov, W. Pfeil, and K. Ruckpaul, (1982), Acta Biol. Med. Germ. **41**, K1–K8.
Bernhardt, R., N. T. Ngoc Dao, H. Stiel, W. Schwarze, J. Friedrich, G.-R. Jänig, and K. Ruckpaul, (1983), Biochim. Biophys. Acta **745**, 140–148.
Bernhardt, R., A. Makower, G.-R. Jänig, and K. Ruckpaul, (1984), Biochim. Biophys. Acta **785**, 186–190.
Bjorkhem, I., (1977), Pharm. Ther. Part I, 327–348.
Black, S. D., J. S. French, C. H. Williams, and M. J. Coon, (1979), Biochem. Biophys. Res. Commun. **91**, 1528–1535.
Black, S. D. and M. J. Coon, (1982), J. Biol. Chem. **257**, 5929–5938.
Black, S. D. and M. J. Coon, (1986), in: Cytochrome P-450, (P. R. Ortiz de Montellano, ed.), Plenum, New York, 161–216.
Blake, R. C. and M. J. Coon, (1981), J. Biol. Chem. **256**, 12127–12133.
Blanck, J. and W. Scheler, (1970), Acta Biol. Med. Germ. **25**, 29–39.
Blanck, J., G. Smettan, G.-R. Jänig, and K. Ruckpaul, (1976), Acta Biol. Med. Germ. **35**, 1455–1463.
Blanck, J. and G. Smettan, (1978), Pharmazie **33**, 321–324.
Blanck, J., J. Behlke, G.-R. Jänig, D. Pfeil, and K. Ruckpaul, (1979a), Acta Biol. Med. Germ. **38**, 11–21.
Blanck, J., K. Rohde, and K. Ruckpaul, (1979b), Acta Biol. Med. Germ. **38**, 23 to 32.
Blanck, J., H. Rein, M. Sommer, O. Ristau, G. Smettan, and K. Ruckpaul, (1983), Biochem. Pharmacol. **32**, 1683–1688.

Blanck, J., G. Smettan, and S. Greschner, (1984a), in: Cytochrome P-450, (K. Ruckpaul, H. Rein, eds.), Akademie-Verlag, Berlin, 111–162.
Blanck, J., G. Smettan, O. Ristau, M. Ingelman-Sundberg, and K. Ruckpaul, (1984b), Eur. J. Biochem. **144**, 509–523.
Böhm, S., H. Rein, G. Butschak, G. Scheunig, H. Billwitz, and K. Ruckpaul, (1979), Acta Biol. Med. Germ. **38**, 249–255.
Bösterling, B., J. R. Trudell, and H. J. Galla, (1981), Biochim. Biophys. Acta **643**, 547–556.
Bösterling, B. and J. R. Trudell, (1982a), Biochim. Biophys. Acta **689**, 155–160.
Bösterling, B. and J. R. Trudell, (1982b), J. Biol. Chem. **257**, 4783–4787.
Bonfils, C., C. Balny, and P. Maurel, (1981), J. Biol. Chem. **256**, 9457–9465.
Brown, R. R., J. A. Miller, and E. C. Miller, (1954), J. Biol. Chem. **209**, 211–222.
Brown, S., F. J. Wiebel, H. V. Gelboin, and J. D. Minna, (1976), Proc. Natl. Acad. Sci. USA **73**, 4628–4632.
Buchman, A., W. Kunz, C. R. Wolf, F. Oesch, and L. W. Robertson, (1986), Cancer Letters **32**, 243–253.
Cammer, W., J. B. Schenkman, and R. W. Estabrook, (1966), Biochem. Biophys. Res. Commun. **23**, 264–268.
Champion, P. M., E. Munck, P. Debrunner, T. Moss, J. Lipscomb, and I. C. Gunsalus, (1975), Biochim. Biophys. Acta **376**, 579–582.
Champion, P. M., B. R. Stallard, G. C. Wagner, and I. C. Gunsalus, (1982), J. Amer. Chem. Soc. **104**, 5469–5473.
Chapman, D., (1975), Quart. Rev. Biophys. **8**, 185–235.
Cherry, R. J., (1979), Biochim. Biophys. Acta **559**, 289–327.
Chiang, J. Y. L. and M. J. Coon, (1979), Arch. Biochem. Biophys. **195**, 178–187.
Chiang, J. Y. L., (1981), Arch. Biochem. Biophys. **211**, 662–673.
Cinti, D. L., S. G. Sligar, G. G. Gibson, and J. B. Schenkman, (1979), Biochemistry **18**, 36–42.
Cohen, B. S. and R. W. Estabrook, (1971a), Arch. Biochem. Biophys. **143**, 37–45.
Cohen, B. S. and R. W. Estabrook, (1971b), Arch. Biochem. Biophys. **143**, 46–53.
Cohen, B. S. and R. W. Estabrook, (1971c), Arch. Biochem. Biophys. **143**, 54–65.
Conney, A. H., E. C. Miller, and J. A. Miller, (1956), Cancer Res. **16**, 450–459.
Conney, A. H., R. R. Brown, J. A. Miller, and E. C. Miller, (1957), Cancer Res. **17**, 628–633.
Coon, M. J., D. A. Haugen, F. P. Guengerich, J. L. Vermilion, and W. L. Dean, (1976), in: The Structural Basis of Membrane Function, (Y. Hatefi and L. Djavadi-Ohaniance, eds.), Academic Press, New York, London, 409–427.
Coon, M. J. and K. P. Vatsis, (1978), in: Polycyclic Hydrocarbons and Cancer: Environment, Chemistry and Metabolism, (H. V. Gelboin and P. O. P. Ts'O, eds.), Academic Press, New York, Vol. 1, 335–360.
Cooper, D. Y., H. Schleyer, and O. Rosenthal, (1970), Ann. N. Y. Acad. Sci. **174**, 205–211.
Correia, M. A. and G. J. Mannering, (1973a), Mol. Pharmacol **9**, 455–469.
Correia, M. A. and G. J. Mannering, (1973b), Mol. Pharmacol. **9**, 470–485.
Dailey, H. A. and P. Strittmatter, (1979), J. Biol. Chem. **254**, 5388–5396.
Dailey, H. A. and P. Strittmatter, (1980), J. Biol. Chem. **255**, 5184–5189.
Davydov, D. R., A. V. Karyakin, B. Binas, B. I. Kurganov, and A. I. Archakov, (1985), Eur. J. Biochem. **150**, 155–159.
Dawson, J., J. Trudell, R. Linder, G. Barth, E. Bunnenberg, and C. Djerassi, (1978), Biochemistry **17**, 33–42.
Dean, W. L. and R. D. Gray (1982), J. Biol. Chem. **257**, 14679–14685.
DeLemos-Chiarandini, C., A. B. Frey, D. D. Sabatini, and G. Kreibich, (1987), J. Cell. Biol. **104**, 209–219.
DePierre, J. W. and G. Dallner, (1975), Biochim. Biophys. Acta **415**, 411–472.
DePierre, J. W. and L. Ernster, (1975), FEBS Lett. **55**, 18–21.

DePierre, J. W. and L. Ernster, (1977), Ann. Rev. Biochem. **46**, 201—262.
Dignam, J. D. and H. W. Strobel, (1977), Biochemistry **16**, 1116—1123.
Duppel, W. and V. Ullrich, (1976), Biochim. Biophys. Acta **426**, 399—407.
Dus, K. M., (1980), in: Biochemistry, Biophysics and Regulation of Cytochrome P-450, (J. Å. Gustafsson et al., eds.), Elsevier/North Holland Biomedical Press, Amsterdam, 129—132.
Estabrook, R. W., A. G. Hildebrandt, J. Baron, K. J. Netter, and K. Leibman, (1971), Biochem. Biophys. Res. Commun. **42**, 132—139.
Estabrook, R. W. and J. Werringloer, (1977), in: Microsomes and Drug Oxidations, (V. Ullrich, I. Roots, A. G. Hildebrandt, R. W. Estabrook, and A. H. Conney, eds.), Pergamon Press, Oxford, 748—757.
Estabrook, R. W. and L. Vereczkey, (1985), Biokemia (Hungaria) **9**, 145—155.
Estabrook, R. W., C. Martin-Wixtrom, and J. Sheets, (1985), Abstracts of the 5th International Conference on Biochemistry, Biophysics and Induction of Cytochrome P-450, Budapest, 27.
Faris, R. A. and T. C. Campbell, (1981), Science **211**, 719—721.
Faris, R. A. and T. C. Campbell, (1983), Cancer Res. **43**, 2576—2583.
Fisher, M. T. and S. G. Sligar, (1985), J. Amer. Chem. Soc. **107**, 5018—5019.
Franklin, M. R. and R. W. Estabrook, (1971), Arch. Biochem. Biophys. **143**, 318 to 329.
French, J. S., F. P. Guengerich, and M. J. Coon, (1980), J. Biol. Chem. **255**, 4112 to 4119.
Friedrich, J., G. Butschak, G. Scheunig, O. Ristau, H. Rein, K. Ruckpaul, and G. Smettan, (1979), Acta Biol. Med. Germ. **38**, 207—216.
Gander, J. E. and G. J. Mannering, (1980), Pharmac. Therap. **10**, 191—221.
George, P., J. Beetlestone, and J. S. Griffith, (1964), Rev. Mod. Physics **36**, 441—451.
Gibson, G. G., D. L. Cinti, S. G. Sligar, and J. B. Schenkman, (1980), J. Biol. Chem. **255**, 1867—1873.
Gibson, G. G. and P. P. Tamburini, (1984), Xenobiotica **14**, 27—47.
Gillette, J. R., B. B. Brodie, and B. N. LaDu, (1957), J. Pharm. Exp. Therap. **119**, 532—540.
Gillette, J., H. Sasame, and B. Stripp, (1973), Drug Metab. Dispos. **1**, 164—175.
Gonder, J. C., R. A. Proctor, and J. A. Will, (1985), Proc. Natl. Acad. Sci. USA **82**, 6315—6319.
Gonzalez, F. J., D. W. Nebert, J. P. Hardwick, and C. B. Kasper, (1985), J. Biol. Chem. **260**, 7435—7441.
Goodwin, H. A., (1976), Coord. Chem. Rev. **18**, 293—333.
Gorsky, L. D., D. R. Koop, and M. J. Coon, (1984), J. Biol. Chem. **259**, 6812—6817.
Greinert, R., H. Staerk, A. Stier, and A. Weller, (1979), J. Biochem. Biophys. Methods **1**, 77—83.
Greinert, R., S. A. E. Finch, and A. Stier, (1982a), Xenobiotica **12**, 717—726.
Greinert, R., S. A. E. Finch, and A. Stier, (1982b), Biosci. Rep. **2**, 991—994.
Griffin, B. W. and J. A. Peterson, (1975), J. Biol. Chem. **250**, 6445—6451.
Griffith, J. S. and L. E. Orgel, (1957), Quart. Rev. Chem. Soc. (London) **11**, 381—399.
Guengerich, F. P., D. P. Ballou, and M. J. Coon, (1975), J. Biol. Chem. **250**, 7405 to 7414.
Guengerich, F. P., D. P. Ballou, and M. J. Coon, (1976), Biochem. Biophys. Res. Commun. **70**, 951—956.
Guengerich, F. P., (1977), J. Biol. Chem. **252**, 3970—3979.
Guengerich, F. P., (1983), Biochemistry **22**, 2811—2820.
Gut, J., C. Richter, R. J. Cherry, K. H. Winterhalter, and S. Kawato, (1982), J. Biol. Chem. **257**, 7030—7036.
György, P., E. C. Poling, and H. Goldblatt, (1941), Proc. Soc. Exp. Biol. Med. **47**, 41—44.

HANSON, J. K., S. G. SLIGAR, and I. C. GUNSALUS, (1977), Croatica Chem. Acta **49**, 237–250.
HARADA, N. and T. OMURA, (1983), J. Biochem. **93**, 1361–1373.
HAUGEN, D. A. and M. J. COON, (1976), J. Biol. Chem. **251**, 7929–7939.
HAUSER, H. and M. D. BARRATT, (1973), Biochem. Biophys. Res. Commun. **53**, 399–405.
HAVEMANN, R. and W. HABERDITZL, (1958), Z. Physik. Chem. **209**, 135–161.
HAYAISHI, O., (1974), Molecular Mechanisms of Oxygen Activation, Academic Press, New York, London.
HEINEMEYER, G., S. NIGAM, and A. G. HILDEBRANDT, (1980), Naunyn Schmiedeberg's Arch. Pharmacol. **314**, 201–210.
HILDEBRANDT, A. G. and R. W. ESTABROOK, (1971), Arch. Biochem. Biophys. **143**, 66–79.
HILDEBRANDT, A. G. and I. ROOTS, (1975), Arch. Biochem. Biophys. **171**, 385–397.
HILDEBRANDT, A. G., M. TJOE, and I. ROOTS, (1975), Biochem. Soc. Transact. **3**, 807 to 811.
HILDEBRAND, C. E., F. J. GONZALEZ, O. W. MCBRIDE, and D. W. NEBERT, (1985), Nucleic Acids Res. **13**, 2009–2016.
HLAVICA, P., (1984), Arch. Biochem. Biophys. **228**, 600–608.
HRYCAY, E. G. and P. J. O'BRIEN, (1972), Arch. Biochem. Biophys. **153**, 480–494.
HRYCAY, E. G. and R. W. ESTABROOK, (1974), Biochem. Biophys. Res. Commun. **60**, 771–778.
HUANG, Y. Y., T. HARA, S. G. SLIGAR, M. J. COON, and T. KIMURA , (1986), Biochemistry **25**, 1390–1394.
ICHIKAWA, Y. and T. YAMANO, (1972), J. Biochem. **71**, 1053–1063.
IIZUKA, T., M. KOTANI, and T. YONETANI, (1968), Biochim. Biophys. Acta **167**, 257 to 267.
IMAI, Y., R. SATO, and T. IYANAGI, (1977), J. Biochem. (Tokyo) **82**, 1237–1246.
IMAI, Y. and R. SATO, (1977), Biochem. Biophys. Res. Commun. **75**, 420–426.
INGELMAN-SUNDBERG, M. (1977a), Biochim. Biophys. Acta **488**, 225–234.
INGELMAN-SUNDBERG, M., (1977b), in: Microsomes and Drug Oxidations, (V. ULLRICH, I. ROOTS, A. G. HILDEBRANDT, R. W. ESTABROOK, eds.), Pergamon Press, New York, 67–75.
INGELMAN-SUNDBERG, M. and I. JOHANSSON, (1980a), Biochem. Biophys. Res. Commun. **97**, 582–589.
INGELMAN-SUNDBERG, M. and I. JOHANSSON, (1980b), Biochemistry **19**, 4004–4011.
INGELMAN-SUNDBERG, M. I. JOHANSSON, A. BRUNSTRÖM, G. EKSTRÖM, T. HAAPARANTA, and J. RYDSTRÖM, (1980), in: Biochemistry, Biophysics and Regulation of Cytochrome P-450, (J. A. GUSTAFSSON, A. CARLSTEDT-DUKE, M. MODE, J. RAFTER, eds.), Elsevier/North Holland Biomedical Press, Amsterdam, 299–306.
INGELMAN-SUNDBERG, M., T. HAAPARANTA, and J. RYDSTRÖM, (1981), Biochemistry **20**, 4100–4106.
INGELMAN-SUNDBERG, M., J. BLANCK, G. SMETTAN, and K. RUCKPAUL, (1983), Eur. J. Biochem. **134**, 157–162.
INGELMAN-SUNDBERG, M. and I. JOHANSSON, (1984), Acta Chem. Scand. B **38**, 845–851.
INGELMAN-SUNDBERG, M., (1986), in: Cytochrome P-450, Structure, Mechanism and Biochemistry, (P. R. ORTIZ DE MONTELLANO, ed.), Plenum, New York, 119–160.
ISHIMURA, Y., V. ULLRICH, and J. A. PETERSON, (1971), Biochem. Biophys. Res. Commun. **42**, 140–146.
ISHIMURA, Y., (1978), in: Cytochrome P-450, (R. SATO and T. OMURA, eds.), Kodansha Ltd., Tokyo, Academic Press, New York, 209–227.
IVERSEN, P. L., R. N. HINES, and E. BRESNICK, (1986), Bio-Essays **4**, 15–19.
IYANAGI, T., T. SUZAKI, and S. KOBAYASHI, (1981), J. Biol. Chem. **256**, 12933–12939.
JÄNIG, G. R., R. KRAFT, A. MAKOWER, H. RABE, and K. RUCKPAUL, (1985), in: Cytochrome P-450, Biochemistry, Biophysics and Induction, (L. VERECZKEY and K. MAGYAR, eds.), Akademiai Kiado, Budapest, 53–56.

Jansson, I. and J. B. Schenkman, (1973), Mol. Pharmacol. **9**, 840—845.
Jansson, I. and J. B. Schenkman, (1977), Arch. Biochem. Biophys. **178**, 89—107.
Jansson, I., P. P. Tamburini, L. V. Favreau, and J. B. Schenkman, (1985), Drug Metab. Dispos. **13**, 453—458.
Jensen, R. K., S. D. Sleight, S. D. Aust, J. I. Goodman, and J. E. Trosko, (1983), Toxicol. Appl. Pharmacol. **71**, 163—176.
Johnson, E. F., G. E. Schwab, and H. H. Dieter, (1983), J. Biol. Chem. **258**, 2785—2788.
Jones, P. B. C., D. R. Galeazzi, J. M. Fisher, and J. P. Whitlock, (1985), Science **227**, 1499—1502.
Jones, P. B. C., L. K. Durrin, D. R. Galeazzi, and J. R. Whitlock, (1986), Proc. Natl. Acad. Sci. USA **83**, 2802—2806.
Jung, C. and O. Ristau, (1978), Pharmazie **33**, 329—331.
Jung, C., J. Friedrich, and O. Ristau, (1979), Acta Biol. Med. Germ. **38**, 363—377.
Jung, C., (1980), Doctoral Thesis, Academy of Sciences of the GDR, Berlin.
Juvonen, R., P. Kaipainen, and M. Lang, (1985), in: Cytochrome P-450, Biochemistry, Biophysics and Induction, (L. Vereczkey and K. Magyar, eds.), Akademiai Kiado, Budapest, 331—334.
Kalow, W., (1987), Eur. J. Clin. Pharmacol **31**, 633—641.
Kamataki, T., K. Maeda, Y. Yamazoe, T. Nagai, and R. Kato, (1981), Biochem. Biophys. Res. Commun. **116**, 1—7.
Kaminsky, L. S. and F. P. Guengerich, (1985), Eur. J. Biochem. **149**, 479—489.
Kassner, R. J., (1972), J. Amer. Chem. Soc. **95**, 2674—2677.
Kawato, S., J. Gut, R. J. Cherry, K. Winterhalter, and C. Richter, (1982), J. Biol. Chem. **257**, 7023—7029.
Kellermann, G., M. Luyten-Kellermann, and C. R. Shaw, (1973a), Amer. J. Human Genet. **25**, 327—331.
Kellermann, G., C. R. Shaw, and M. Luyten-Kellermann, (1973b), N. Engl. J. Med. **289**, 934—937.
Kissel, M. A., S. A. Usanov, D. I. Metelitza, and A. A. Akhrem, (1979), Biol. Khim. Moscow **5**, 1558—1563.
König, E. and G. Ritter, (1974), in: Mössbauer Effect Methodology, (I. J. Gruverman, C. W. Seidel, D. K. Dieterly, eds.), Plenum Press, New York, Vol. **9**, 3—13.
Kotani, M., (1969), Ann. New York, Acad. Sci. **158**, 20—29.
Kuo, A. L. and C. G. Wade, (1979), Chem. Phys. Lipids **25**, 135—139.
Kuthan, H., H. Tsuji, H. Graf, V. Ullrich, J. Werringloer, and R. W. Estabrook, (1978), FEBS Lett. **91**, 343—345.
Kuthan, H. and V. Ullrich, (1982), Eur. J. Biochem. **126**, 583—588.
Lang, G. and W. Marshall, (1966), Proc. Phys. Soc. (London) **87**, 3—34.
Lambeth, J. D. and S. O. Pember, (1983), J. Biol. Chem. **258**, 5596—5602.
Lasker, J. M., M. T. Huang, and A. H. Conney, (1982), Science **216**, 1419—1421.
Lee, T. C. and F. Snyder, (1973), Biochim. Biophys. Acta **291**, 71—82.
Lewis, D. F. V., P. P. Tamburini, and G. G. Gibson, (1986), Chemico-biol. Interact. **58**, 289—300.
Lichtenberg, D., R. J. Robson, and E. A. Dennis, (1983), Biochim. Biophys. Acta **737**, 285—304.
Light, D. R. and N. R. Orme-Johnson, (1981), J. Biol. Chem. **256**, 343—350.
Loverde, A. and P. Strittmatter, (1968), J. Biol. Chem. **243**, 5779—5787.
Lu, A. Y. H. and M. J. Coon, (1968), J. Biol. Chem. **243**, 1331—1332.
Lu, A. Y. H., K. W. Junk, and M. J. Coon, (1969a), J. Biol. Chem. **244**, 3714—3721.
Lu, A. Y. H., H. W. Strobel, and M. J. Coon, (1969b), Biochem. Biophys. Res. Commun. **36**, 545—551.
Lu, A. Y. H. and S. B. West, (1972), Mol. Pharmacol. **8**, 490—500.
Lu, A. Y. H. and W. Levin, (1974), Biochim. Biophys. Acta **344**, 205—240.
Lu, A. Y. H., W. Levin, and R. Kuntzman, (1974a), Biochem. Biophys. Res. Commun. **60**, 266—272.

Lu, A. Y. H., W. Levin, H. Selander, and D. M. Jerina, (1974b), Biochem. Biophys. Res. Commun. **61**, 1348–1355.
Lu, A. Y. H., S. B. West, M. Vore, D. Ryan, and W. Levin, (1974c), J. Biol. Chem. **249**, 6701–6709.
Luzzati, V., T. Gulik-Krzywicki, and A. Tardieu, (1968), Nature (London) **218**, 1031–1043.
Mabrey, S., G. Powis, J. B. Schenkman, and T. R. Tritton, (1977), J. Biol. Chem. **252**, 2929–2933.
Magdalou, J., C. Thirion, M. Balland, and G. Siest, (1985), Int. J. Biochem. **17**, 1103–1107.
Makower, A., R. Bernhardt, H. Rabe, G. R. Jänig, and K. Ruckpaul, (1984), Biomed. Biochim. Acta **43**, 1333–1341.
Malmström, B. G., (1982), Ann. Rev. Biochem. **51**, 21–59.
Mason, H. S., W. L. Fowlks, and E. Peterson, (1955), J. Amer. Chem. Soc. **77**, 2914–2915.
Matsubara, T., J. Baron, L. L. Peterson, and J. A. Peterson, (1976), Arch. Biochem. Biophys. **172**, 463–469.
Matsuura, S., Y. Fujii-Kuriyama, and Y. Tashiro, (1978), J. Cell. Biol. **78**, 503–519.
McIntosh, P. R., S. Kawato, R. B. Freedman, and R. J. Cherry, (1980), FEBS Lett. **122**, 54–58.
Miki, N., T. Sugiyama, and T. Yamano, (1980), J. Biochem. **88**, 307–316.
Miller, E. C., J. A. Miller, and R. R. Brown, (1952), Cancer Res. **12**, 282–283.
Misselwitz, R., G. R. Jänig, H. Rein, E. Buder, D. Zirwer, and K. Ruckpaul, (1976), Acta Biol. Med. Germ. **35**, K19–K25.
Misselwitz, R., G. R. Jänig, H. Rein, E. Buder, D. Zirwer, and K. Ruckpaul, (1977), Acta Biol. Med. Germ. **36**, K35–K41.
Mitani, F. and S. Horie, (1969a), J. Biochem. **65**, 269–280.
Mitani, F. and S. Horie, (1969b), J. Biochem. **66**, 139–149.
Miwa, G. T., S. B. West, M. TuanHuang, and A. Y. H. Lu, (1979), J. Biol. Chem. **254**, 5695–5700.
Miwa, G. T. and A. Y. H. Lu, (1981), Arch. Biochem. Biophys. **211**, 454–458.
Miwa, G. T. and A. Y. H. Lu, (1984), Arch. Biochem. Biophys. **234**, 161–166.
Miyake, Y., Y. Nakamura, N. Takayama, and K. Horike, (1975), J. Biochem. **78**, 773–783.
Morgan, E. T., D. R. Koop, and M. J. Coon, (1982), J. Biol. Chem. **257**, 13951–13957.
Morimoto, T., S. Matsuura, S. Sasaki, Y. Tashiro, and T. Omura, (1976), J. Cell. Biol. **68**, 189–201.
Müller-Enoch, D., P. Churchill, S. Fleischer, and F. P. Guengerich, (1984), J. Biol. Chem. **259**, 8174–8182.
Murakami, K. and H. S. Mason, (1967), J. Biol. Chem. **242**, 1102–1110.
Narasimhulu, S., D. Y. Cooper, and O. Rosenthal, (1965), Life Sci. **4**, 2101–2107.
Narhi, L. O. and A. J. Fulco, (1986), J. Biol. Chem. **261**, 7160–7169.
Nebert, D. W., J. R. Robinson, A. Niwa, K. Kumaki, and A. P. Poland, (1975), J. Cell. Physiol. **85**, 393–414.
Nebert, D. W. (1979), Mol. Cell. Biochem. **27**, 27–46.
Nebert, D. W. and N. M. Jensen, (1979), Critical Rev. Biochem. **6**, 401–437.
Nelson, D. R. and H. W. Strobel, (1988), J. Biol. Chem. **236**, 6038–6050.
Nilsson, A. and B. C. Johnson, (1963), Arch. Biochem. Biophys. **101**, 494–498.
Nordblom, G. D. and M. J. Coon, (1977), Arch. Biochem. Biophys. **180**, 343–347.
Noshiro, M., N. Harada, and T. Omura, (1979), Biochem. Biophys. Res. Commun. **91**, 207–213.
Noshiro, M., H. H. Ruf, and V. Ullrich, (1980), in: Biochemistry, Biophysics and Regulation of Cytochrome P-450, (J. A. Gustafsson, J. Carlstedt-Duke, A. Mode, J. Rafter, eds.), Elsevier/North Holland Biomedical Press, Amsterdam, 351–354.
Noshiro, M., V. Ullrich, and T. Omura, (1981), Eur. J. Biochem. **116**, 521–526.

Okuda, T., E. S. Vesell, E. Plotkin, R. Tarone, R. C. Bast, and H. V. Gelboin, (1977), Cancer Res. **37**, 3904–3911.
Oprian, D. D., K. P. Vatsis, and M. J. Coon, (1979), J. Biol. Chem. **254**, 8895–8902.
Oprian, D. D., L. D. Gorsky, and M. J. Coon, (1983), J. Biol. Chem. **258**, 8684–8691.
Otsuka, J., (1970), Biochim. Biophys. Acta **214**, 233–235.
Paigen, B., E. Ward, A. Reilly, L. Houten, H. L. Gurtoo, J. Minowada, K. Steenland, M. B. Havens, and P. Sartori, (1981), Cancer Res. **41**, 2757–2761.
Parkinson, A., P. E. Thomas, D. E. Ryan, L. D. Gorsky, J. E. Shively, J. M. Sayer, D. M. Jerina, and W. Levin, (1986), J. Biol. Chem. **261**, 11487–11495.
Peisach, J. and W. E. Blumberg, (1970), Proc. Natl. Acad. Sci. USA **67**, 172–179.
Pelkonen, O., N. T. Kärki, and E. A. Sotaniemi, (1980), in: Human Cancer, its Characterization and Treatment, (W. Davis, K. R. Harrap, G. Statopoulos, eds.), Excerpta Medica, Amsterdam, 48–57.
Peraino, C., R. J. M. Fry, and E. Staffeldt, (1971), Cancer Res. **31**, 1506–1512.
Peterson, J. A., R. E. Ebel, D. H. O'Keeffe, T. Matsubara, and R. W. Estabrook, (1976), J. Biol. Chem. **251**, 4010–4016.
Petzold, D. R., H. Rein, D. Schwarz, M. Sommer, and K. Ruckpaul, (1985), Biochim. Biophys. Acta **829**, 253–261.
Pfeil, D., J. Friedrich, J. Lampe, and K. Ruckpaul, (1983), Dtsch. Gesundheitswesen **38**, 281–285.
Phillips, I. R., E. A. Shephard, R. M. Bayney, S. F. Pike, B. R. Rabin, R. Heath, and N. Carter, (1983), Biochem. J. **212**, 55–64.
Pike, S. F., E. A. Shephard, B. R. Rabin, and I. R. Phillips, (1985), Biochem. Pharmacol. **34**, 2489–2494.
Pirrwitz, J., G. Lassmann, H. Rein, O. Ristau, G. R. Jänig, and K. Ruckpaul, (1977), FEBS Lett. **83**, 15–18.
Pirrwitz, J., D. Schwarz, E. V. Elenkova, B. Atanasov, H. Rein, and K. Ruckpaul, (1982a), in: Microsomes, Drug Oxidations and Drug Toxicity (R. Sato and R. Kato, eds.), Japan Sci. Soc. Press, Tokyo; Wiley-Interscience, New York, 111.
Pirrwitz, J., D. Schwarz, H. Rein, O. Ristau, G. R. Jänig, and K. Ruckpaul, (1982b), Biochim. Biophys. Acta **708**, 42–48.
Plummer, S., A. R. Boobis, and D. S. Davies, (1986), Arch. Toxicol. **58**, 165–170.
Poland, A. P., E. Glover, and A. S. Kende, (1976), J. Biol. Chem. **251**, 4936–4946.
Pompon, D. and M. J. Coon, (1984), J. Biol. Chem. **259**, 15377–15385.
Pompon, D., (1987), Biochemistry **26**, 6429–6435.
Poulos, T. L., B. C. Finzel, I. C. Gunsalus, G. C. Wagner, and J. Kraut, (1985), J. Biol. Chem. **260**, 16122–16130.
Poulos, T. L., B. C. Finzel, and A. J. Howard, (1986), Biochemistry **25**, 5314–5322.
Rahimtula, A. D., P. J. O'Brien, E. G. Hrycay, J. A. Peterson, and R. W. Estabrook, (1974), Biochem. Biophys. Res. Commun. **60**, 695–702.
Rein, H. and O. Ristau, (1965), Biochim. Biophys. Acta **94**, 516–524.
Rein, H., O. Ristau, F. Hackenberger, and F. Jung, (1968), Biochim. Biophys. Acta **167**, 538–546.
Rein, H., O. Ristau, J. Friedrich, G. R. Jänig, and K. Ruckpaul, (1976a), Proceedings of the Conference "Cytochrome P-450 — Structural Aspects", Primosten-Yugoslavia 1976; Croat. Chem. Acta **49**, 251–261 (1977).
Rein, H., G. R. Jänig, W. Winkler, and K. Ruckpaul, (1976b), Acta Biol. Med. Germ. **35**, K41–K50.
Rein, H., S. Maricic, G. R. Jänig, S. Vuk-Pavlovic, B. Benko, O. Ristau, and K. Ruckpaul, (1976c), Biochim. Biophys. Acta **446**, 325–330.
Rein, H., O. Ristau, J. Friedrich, G. R. Jänig, and K. Ruckpaul, (1977), FEBS Lett. **75**, 19–22.
Rein, H. and O. Ristau, (1978), Pharmazie **33**, 325–328.
Rein, H., O. Ristau, R. Misselwitz, E. Buder, and K. Ruckpaul, (1979), Acta Biol. Med. Germ. **38**, 187–200.

Rein, H., O. Ristau, and K. Ruckpaul, (1981), in: Industrial and Environmental Xenobiotics, (I. Gut, M. Cikrt, G. L. Plaa, eds.), Springer Verlag Berlin, Heidelberg, New York, 147–159.

Rein, H., J. Blanck, M. Sommer, O. Ristau, K. Ruckpaul, and W. Scheler, (1982), in: Cytochrome P-450 Biochemistry, Biophysics and Environmental Implications, (E. Hietanen, M. Laitinen, O. Hänninen, eds.), Elsevier/North Holland Biomedical Press, Amsterdam, 585–588.

Rein, H., M. Sommer, J. Blanck, J. Friedrich, and K. Ruckpaul, (1983), Zbl. Pharm. **122**, 1289–1292.

Rein, H., C. Jung, O. Ristau, and J. Friedrich, (1984), in: Cytochrome P-450 Structural and Functional Relationships, (K. Ruckpaul and H. Rein, eds.), Akademie-Verlag Berlin, 163–249.

Rein, H., O. Ristau, J. Blanck, and K. Ruckpaul, (1989), in: Cytochrome P-450, Biochemistry and Biophysics, (I. Schuster, ed.), Taylor and Francis, London, New York, Philadelphia.

Remmer, H., J. B. Schenkman, R. W. Estabrook, H. Sasame, J. R. Gillette, S. Narasimhulu, D., Y. Cooper, and O. Rosenthal, (1966), Mol. Pharmacol. **2**, 187 to 190.

Richardson, H. L. and L. Cunningham, (1951), Cancer Res. **11**, 274.

Richter, C., K. H. Winterhalter, and R. J. Cherry, (1979), FEBS Lett. **102**, 151–154.

Ristau, O., H. Rein, G. R. Jänig, and K. Ruckpaul, (1978), Biochim. Biophys. Acta **536**, 226–234.

Ristau, O., H. Rein, S. Greschner, G. R. Jänig, and K. Ruckpaul, (1979), Acta Biol. Med. Germ. **38**, 177–185.

Röder, A. and E. Bayer, (1969), Eur. J. Biochem. **11**, 89–92.

Rogers, J. H. and P. Strittmatter, (1974), J. Biol. Chem. **249**, 895–900.

Rohde, K., J. Blanck, and K. Ruckpaul, (1983), Biomed. Biochim. Acta **42**, 651 to 662.

Rossi, M., (1983), J. Med. Chem. **24**, 1246–1252.

Rothman, I. E. and E. A. Dawidowicz, (1975), Biochemistry **14**, 2809–2816.

Ruckpaul, K., H. Reih, D. P. Ballou, and M. J. Coon, (1980), Biochim. Biophys. Acta **626**, 41–56.

Ruckpaul, K., H. Rein, J. Blanck, O. Ristau, and M. J. Coon, (1982), Acta Biol. Med. Germ. **41**, 193–203.

Ruf, H. H., (1980), in: Biochemistry, Biophysics and Regulation of Cytochrome P-450, (J. A. Gustafsson, J. Carlstedt-Duke, A. Mode, J. Rafter, eds.), Elsevier/North Holland Biomedical Press, Amsterdam, 355–358.

Ruf, H. H. and V. Eichinger, (1982), in: Cytochrome P-450, Biochemistry, Biophysics and Environmental Implications, (E. Hietanen, M. Laitinen, O. Hänninen, eds.), Elsevier/North Holland Biomedical Press, Amsterdam, New York, Oxford, 597–600.

Sakaguchi, M., K. Mihara, and R. Sato, (1981), Proc. Natl. Acad. Sci. USA. **81**, 3361 to 3364.

Scheler, W., G. Schoffa, and F. Jung, (1957), Biochem. Z. **329**, 232–246.

Scheler, W., J. Blanck, and W. Graf, (1963), Naturwiss. **50**, 500–501.

Schenkman, J. B., H. Remmer, and R. W. Estabrook, (1967), Mol. Pharmacol. **3**, 113–123.

Schenkman, J. B., D. L. Cinti, S. Orrenius, P. Moldeus, and R. Kaschnitz, (1972), Biochemistry **11**, 4243–4251.

Schuetz, E. G., S. A. Wrighton, J. L. Barwick, and P. S. Guzelian, (1984a), J. Biol. Chem. **259**, 1999–2006.

Schuetz, E. G. and P. S. Guzelian, (1984b), J. Biol. Chem. **259**, 2007–2012.

Schulze, H. J., V. M. Pönnighaus, and H. Staudinger, (1972), Hoppe-Seyler's Z. Physiol. Chem. **353**, 1195–1204.

SCHULZE, H. J. and H. STAUDINGER, (1975), Naturwissenschaften **62**, 331–340.
SCHWARZ, D., J. PIRRWITZ, M. J. COON, and K. RUCKPAUL, (1982a), Acta Biol. Med. Germ. **41**, 425–430.
SCHWARZ, D., J. PIRRWITZ, and K. RUCKPAUL, (1982b), Arch. Biochem. Biophys. **216**, 322–328.
SCHWARZ, D., K. GAST, H. W. MEYER, U. LACHMANN, M. J. COON, and K. RUCKPAUL, (1984), Biochem. Biophys. Res. Commun. **121**, 118–125.
SCHWARZ, D., J. PIRRWITZ, H. REIN, and K. RUCKPAUL, (1984), Biomed. Biochim. Acta **43**, 295–307.
SCHWARZE, W., J. BLANCK, O. RISTAU, G. R. JÄNIG, K. POMMERENING, H. REIN, and K. RUCKPAUL, (1985), Chemico-biol. Interactions **54**, 127–141.
SCHWARZE, W., (1987), unpublished results.
SHARROCK, M., E. MUNCK, P. G. DEBRUNNER, V. MARSHALL, J. D. LIPSCOMB, and I. C. GUNSALUS, (1973), Biochemistry **12**, 258–265.
SHARROCK, M., P. G. DEBRUNNER, C. SCHULZ, J. D. LIPSCOMB, V. MARSHALL, and I. C. GUNSALUS, (1976), Biochim. Biophys. Acta **420**, 8–26.
SHEPHARD, E. A., I. R. PHILLIPS, R. M. BAYNEY, S. F. PIKE, and B. R. RABIN, (1983), Biochem. J. **211**, 333–340.
SIES, H. and M. KANDEL, (1970), FEBS Lett. **9**, 205–210.
SLIGAR, S. G., (1976), Biochemistry **15**, 5399–5406.
SLIGAR, S. G., D. L. CINTI, G. G. GIBSON, and J. B. SCHENKMAN, (1979), Biochem. Biophys. Res. Commun. **90**, 925–932.
SMETTAN, G., P. A. KISSELEV, M. A. KISSEL, A. A. AKHREM, and K. RUCKPAUL, (1984), Biomed. Biochim. Acta **43**, 1073–1082.
SMITH, D. W. and R. J. P. WILLIAMS, (1968), Biochem. J. **110**, 297–301.
STAUDT, H., F. LICHTENBERGER, and V. ULLRICH, (1974), Eur. J. Biochem. **46**, 99–106.
STERN, J. O., J. PEISACH, W. E. BLUMBERG, A. Y. H. LU, and W. LEVIN, (1973), Arch. Biochem. Biophys. **156**, 404–413.
STIER, A. and E. SACKMANN, (1973), Biochim. Biophys. Acta **311**, 400–408.
STIER, A., (1976), Biochem. Pharmacol. **25**, 109–113.
STIER, A., S. A. E. FINCH, R. GREINERT, M. HÖHNE, and R. MÜLLER, (1982), in: Liver and Aging, (K. KITANI, ed.), Elsevier/North Holland Biomedical Press, Amsterdam, 3–14.
STRITTMATTER, P. and S. F. VELICK, (1956a), J. Biol. Chem. **221**, 253–264.
STRITTMATTER, P. and S. F. VELICK, (1956b), J. Biol. Chem. **221**, 277–286.
STROBEL, H. W., A. Y. H. LU, J. HEIDEMA, and M. J. COON, (1970), J. Biol. Chem. **245**, 4851–4854.
SUGIYAMA, T., N. MIKI, and T. YAMANO, (1979), Biochem. Biophys. Res. Commun. **90**, 715–720.
SUGIYAMA, T., N. MIKI, and T. YAMANO, (1980), J. Biochem. **87**, 1457–1467.
TAMBURINI, P. P. and G. G. GIBSON, (1983), J. Biol. Chem. **258**, 13444–13452.
TAMBURINI, P. P., G. G. GIBSON, W. L. BACKES, S. G. SLIGAR, and J. B. SCHENKMAN, (1984), Biochemistry **23**, 4526–4533.
TAMBURINI, P. P., R. E. WHITE, and J. B. SCHENKMAN, (1985), J. Biol. Chem. **260**, 4007–4015.
TAMBURINI, P. P. and J. B. SCHENKMAN, (1986), Arch. Biochem. Biophys. **245**, 512–522.
TAMBURINI, P. P. and J. B. SCHENKMAN, (1987), Proc. Natl. Acad. Sci. USA **84**, 11 to 15.
TANABE, Y. and S. SUGANO, (1954), J. Phys. Soc. (Japan) **9**, 753–776.
TANIGUCHI, H., Y. IMAI, T. IYANAGI, and R. SATO, (1979), Biochim. Biophys. Acta **550**, 341–356.
TANIGUCHI, H., Y. IMAI, and R. SATO, (1980), in: Microsomes, Drug Oxidations and Chemical Carcinogenesis, (M. J. COON, A. H. CONNEY, R. W. ESTABROOK, H. V. GELBOIN, J. R. GILLETTE, P. J. O'BRIEN, eds.), Academic Press, New York, 537–540.
TANIGUCHI, H., Y. IMAI, and R. SATO, (1984), Arch. Biochem. Biophys. **232**, 585–596.

TARR, G. E., S. D. BLACK, V. S. FUJITA, and M. J. COON, (1983), Proc. Natl. Acad. Sci. USA **80**, 6552–6556.
THORGEIRSSON, S. S. and D. W. NEBERT, (1977), Adv. Canc. Res. **149**, 149–193.
TSAI, R. C. Y., I. C. GUNSALUS, J. PEISACH, W. BLUMBERG, W. ORME-JOHNSON, and H. BEINERT, (1970), Proc. Natl. Acad. Sci. USA **66**, 1157–1163.
TYSON, C. A., J. D. LIPSCOMB, and I. C. GUNSALUS, (1972), J. Biol. Chem. **247**, 5777 to 5784.
ULLRICH, V. and W. DUPPEL, (1975), in: The Enzymes (P. D. BOYER, ed.), Academic Press, New York, Vol. **11**, 253–297.
ULLRICH, V., W. NASTAINCZYK, and H. H. RUF, (1975), Biochem. Soc. Trans. **3**, 803–807.
VAN DEN BESSELAAR, A. M. H. P., N. DEKRUIJFF, H. VAN DEN BOSCH, and L. L. M. VANDEENEN, (1978), Biochim. Biophys. Acta **510**, 242–255.
VATSIS, K. P., A. D. THEOHARIDAS, D. KUPFER, and M. J. COON, (1982), J. Biol. Chem. **257**, 11221–11229.
VERMILION, J. L. and M. J. COON, (1978), J. Biol. Chem. **253**, 2694–2704.
VORE, M., J. G. HAMILTON, and A. Y. H. LU, (1974), Biochem. Biophys. Res. Commun. **56**, 1038–1044.
WAGNER, S. L., W. L. DEAN, and R. D. GRAY, (1984), J. Biol. Chem. **259**, 2390–2395.
WATERMAN, M. R. and H. MASON, (1972), Arch. Biochem. Biophys. **150**, 57–63.
WATERMAN, M. R., V. ULLRICH, and R. W. ESTABROOK, (1973), Arch. Biochem. Biophys. **155**, 355–360.
WEHRLI, F. W. and T. WIRTHLIN, (1976), Interpretation of Carbon — 13 NMR Spectra, Heyden and Sons, London, 22–47.
WEISS, J. J., (1964), Nature (London) **202**, 83–84.
WERRINGLOER, J. and S. KAWANO, (1980a), in: Microsomes, Drug Oxidation and Chemical Carcinogenesis, (M. J. COON, A. H. CONNEY, R. W. ESTABROOK, H. V. GELBOIN, J. R. GILLETTE, P. J. O'BRIEN, eds.), Academic Press, New York, 469–476.
WERRINGLOER, J. and S. KAWANO, (1980b), in: Biochemistry, Biophysics and Regulation of Cytochrome P-450, (J. A. GUSTAFSSON, J. CARLSTEDT-DUKE, A. MODE, J. RAFTER, eds.), Elsevier/North Holland, Amsterdam, 359–362.
WHITE, R. E. and M. J. COON, (1980), Ann. Rev. Biochem. **49**, 315–356.
WHITE, R. E., S. G. SLIGAR, and M. J. COON, (1980), **255**, 11108–11111.
WHITE, R. E., M. MC CARTHY, K. D. EGEBERG, and S. G. SLIGAR, (1984), Arch. Biochem. Biophys. **228**, 493–502.
WHITE, R. E. and M. MCCARTHY, (1986), Arch. Biochem. Biophys. **246**, 19–32.
WHITLOCK, J. P., (1986), Ann. Rev. Pharmacol. Toxicol. **26**, 333–369.
WHYSNER, J. A., J. RAMSEYER, G. M. KAZMI, and B. W. HARDING, (1969), Biochem. Biophys. Res. Commun. **36**, 795–801.
WHYSNER, J. A., J. RAMSEYER, and B. W. HARDING, (1970), J. Biol. Chem. **245**, 5441–5449.
WOLF, C. R., (1986), Trends in Genetics **2**, 209–214.
WU, E. S. and C. S. YANG, (1984), Biochemistry **23**, 28–33.
YANG, C. S., (1975), FEBS Lett. **54**, 61–64.
YANG, C. S., (1977), J. Biol. Chem. **252**, 293–298.
YANG, C. S., F. S. STRICKHART, and L. P. KICHA, (1977), Biochim. Biophys. Acta **465**, 362–370.
YANG, C. S., F. S. STRICKHART, and L. P. KICHA, (1978), Biochim. Biophys. Acta **509**, 326–337.
ZHUKOV, A. A. and A. I. ARCHAKOV, (1982), Biochem. Biophys. Res. Commun. **109**, 813–818.
ZIEGLER, M., J. BLANCK, and K. RUCKPAUL, (1982), FEBS Lett. **150**, 219–222.
ZIEGLER, M., J. BLANCK, S. GRESCHNER, K. LENZ, A. LAU, and K. RUCKPAUL, (1983), Biomed. Biochim. Acta **42**, 641–649.

Chapter 2
Catalytically Active Metalloporphyrin Models for Cytochrome P-450.

D. Mansuy and P. Battioni

1. Introduction

1.1. The use of metalloporphyrins to understand cytochrome P-450 dependent reactions. Static and dynamic models

The cytochromes P-450 should be considered as important elements in the adaptation of living organisms to their ever-changing environment. Because of this role, cytochrome P-450-dependent monooxygenases must be able to catalyze the oxidation of almost any organic compound, even if this compound is very inert from a chemical point of view. Accordingly, a distinctive characteristic of these monooxygenases, when compared to other enzymes involved in oxidative reactions, is their ability to oxidize even very inert molecules such as alkanes. In fact, it seems that whenever a living organism has been faced with the difficult problem of the oxidation of an inert C—H bond for the oxidative metabolism of an endogenous or exogenous compound, it has answered in almost all cases by the biosynthesis of a cytochrome P-450. Several recent reviews have described the properties of these cytochromes (WHITE et al., 1980; GUENGERICH et al., 1984a; REIN et al., 1984; MIWA et al., 1986; ORTIZ DE MONTELLANO, 1986a). A great deal is known about the structure and function of cytochromes P-450 thanks to a huge number of spectroscopic studies that have been devoted to these hemoproteins and also thanks to a recent X-ray structure of the microbial cytochrome P-450 CAM that catalyzes the 5-exo hydroxylation of camphor (POULOS et al., 1985). Actually, the nature of the iron environment in the various intermediates involved in the catalytic cycle of cytochrome P-450 will be discussed in the following section. However, because of the high molecular weight of cytochromes P-450 (around 50 kD), it is still difficult to study the detailed mechanisms of substrate oxidation by these cytochromes and the nature of the iron-metabolite complexes which are formed during the oxidation of particular classes of substrates. A possible approach to try to solve these problems is to use biomimetic chemical systems based on iron-porphyrins. In fact, during these last ten years, a great deal of work has been done in that direction and one can classify the studies that have been performed on iron-porphyrin biomimetic systems into three groups depending upon the objective of these studies.

A first objective was to prepare and to completely characterize iron-porphyrin complexes able to mimic as well as possible the spectroscopic properties of the different iron complexes which are intermediates in the catalytic cycle of cytochrome P-450 (for reviews on this aspect, see COLLMAN et al., 1977; MANSUY, 1983a). The succesful synthesis and complete characterization of model iron-porphyrin complexes for all the stable intermediates have helped to provide a detailed picture of the variations of the coordination and spin state of the iron throughout the cytochrome P-450 catalytic cycle.

A second objective was to prepare model iron-porphyrin complexes containing iron ligands or exhibiting particular coordination structures that were expected in cytochrome P-450-iron-metabolite complexes. The formation of such cytochrome P-450-iron-metabolite complexes after biotransformation of various substrates had been detected by spectroscopic techniques. The possible structure of the bound metabolite and of the iron-metabolite bond had been deduced from various indirect studies on the enzymatic system itself. However, the isolation and complete characterization of iron-porphyrin model complexes containing identical ligands was of great help to confirm the proposed structures and to understand their specific properties. In many cases, the structures of the model complexes were established definitively by an X-ray analysis. Very often, the model complexes were prepared using reactions similar to those involved in the enzymatic system. Recent reviews have been devoted to this use of iron-porphyrins to understand the formation of cytochrome P-450-iron-ligand or -iron-metabolite complexes (MANSUY, 1981; MANSUY, 1983a; MANSUY and BATTIONI, 1985; BROTHERS and COLLMAN, 1986). These studies have greatly contributed to revealing the existence of an **important organometallic chemistry of cytochrome P-450** and also, in a more general manner, the **unique richness of the coordination chemistry of this cytochrome**. During its reactions with some substrates, the formation of more or less stable iron complexes containing σ-alkyl or σ-aryl Fe—C bonds, carbenic Fe=C bonds or nitrenic Fe=N bonds has been found. Moreover, after oxidation of substrates containing an amine function, iron-nitrosoalkane complexes are formed. The iron-nitrogen bonds in some of these are so stable even in vivo that cytochromes P-450 engaged in such complexes are completely inhibited. Thus, the preparation of "static" chemical models for cytochrome P-450-iron-metabolite complexes has led to a better understanding of the coordination chemistry and mechanisms of cytochrome P-450. It has also led to detailed molecular explanations for some implications of cytochromes P-450 in pharmacology and toxicology (for the various mechanisms of cytochrome P-450 inhibition, see a recent review by ORTIZ DE MONTELLANO and REICH, 1986).

A third objective of the use of iron-porphyrin models was to build up **catalytically active** chemical systems able to reproduce the different reactions catalyzed by cytochrome P-450-dependent monooxygenases. A success in this very difficult challenge would have at least three main consequences: (i) to develop homogeneous catalysts for the selective hydroxylation of alkanes or of aromatic hydrocarbons under mild conditions, the problem of these two reactions having no really good answer in chemistry, (ii) to design efficient model catalysts for regioselective and asymmetric oxidations in organic chemistry, (iii) to provide simple chemical systems for the preparation of the primary metabolites of a given drug or exogenous compound by its direct oxidation. In particular, this would lead to a simple preparation of oxidized derivatives of a drug that could be used either as authentic samples in the determination of the structures of metabolites formed in vitro or in vivo, or in large amounts, as oxidized metabolites necessary for the pharmacological and toxicological evaluation of a drug or of any other exogenous compound.

This review will be exclusively concerned with the third of these objectives of the use of metalloporphyrin models, which is to build up **catalytically** active chemical systems able to perform monooxygenation reactions (for recent reviews on metalloporphyrin-catalyzed oxidations, see MANSUY and BATTIONI, 1986 and MEUNIER, 1986).

2. Mechanisms of substrate oxidation by cytochrome P-450

In order to build up catalytically active models, it is necessary to understand as well as possible the detailed catalytic cycle of dioxygen activation and substrate oxidation by cytochrome P-450.

2.1. The catalytic cycle of dioxygen activation by cytochrome P-450

The cytochrome P-450-catalyzed insertion of an oxygen atom from dioxygen into a substrate requires a two-electron reduction of dioxygen in agreement with the following stoichiometric equation typical of monooxygenases:

$$RH + O_2 + 2e^- + 2H^+ \rightarrow ROH + H_2O$$

These two electrons coming from NADPH are transferred to the cytochrome P-450-heme via the two flavin cofactors of cytochrome P-450 reductase in microsomal monooxygenases. In fact, in the microsomal system, cytochrome P-450 can also receive electrons from NADH via another flavoprotein, cytochrome b_5 reductase, and cytochrome b_5 (PETERSON and PROUGH, 1986). In its resting state, two forms of cytochrome P-450 are in equilibrium: a hexacoordinate low-spin iron(III) complex, having a cysteinate and presumably an OH-containing entity as axial ligands, and a pentacoordinate high-spin iron(III) complex with the cysteinate as the only axial ligand. The position of this

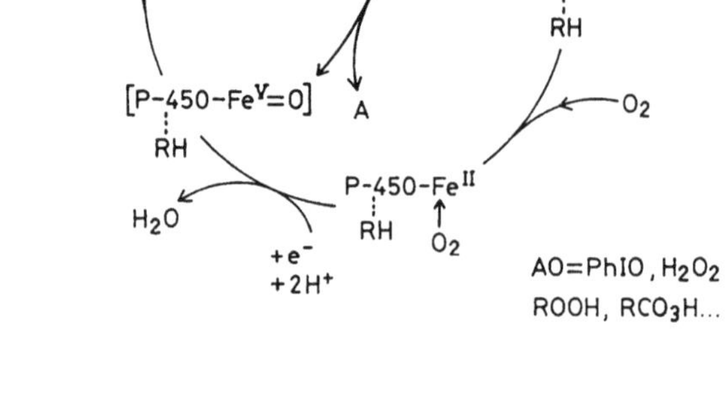

Fig. 1. Catalytic cycle of cytochrome P-450.

equilibrium is very much dependent on the nature of the cytochrome P-450 isoenzyme considered but also on several factors such as the temperature and the ionic strength. In general, the binding of a substrate to cytochrome P-450 which occurs on a protein hydrophobic active site close to the heme leads to a shift of this equilibrium towards the high-spin pentacoordinate state. The iron(III) coordinated to only one axial ligand in this enzyme-substrate complex is more easily reduced by one electron coming from NADPH via the reductase. This leads to the high-spin pentacoordinate cytochrome P-450-iron(II) complex, a stable intermediate in the catalytic cycle which has been studied by various spectroscopic methods. This ferrous complex has a vacant coordination position and readily binds many ligands such as CO, isocyanides, nitrogenous bases, amines, phosphines and O_2. The Fe(II)-O_2 intermediate derived from O_2 binding was studied by Mössbauer (Sharrock et al., 1976) and EXAFS (Dawson et al., 1986) spectroscopy in the case of the bacterial P-450_{cam}. It appears as a low-spin hexacoordinate complex with a formal Fe(III)$O_2^{\cdot -}$ structure analogous to that of oxyhemoglobin. The greater instability of this intermediate in microsomal cytochromes P-450 has so far precluded such spectroscopic studies. A one-electron reduction of the ferrous-dioxygen intermediate leads to the so-called active oxygen complex, the real oxidizing species. However, this key intermediate of the catalytic cycle has a lifetime so short that it could not be observed by any spectroscopic technique. Thus, contrary to those of the other intermediates of the catalytic cycle, the nature of the active oxygen complex is not yet known. Our knowledge of its possible structure is based only on studies of reactions of cytochrome P-450 with single oxygen atom donors (see above) and on comparisons with better known active oxygen complexes derived from other hemoproteins or from iron-porphyrins. From all these data, the most likely mechanism for dioxygen activation by cytochrome P-450 involves (i) a heterolytic cleavage of the O—O bond of a possible Fe(III)—O—OH intermediate formed by one-electron reduction of the ferrous-dioxygen complex and protonation of the terminal oxygen atom, (ii) the formation of a high-valent iron-oxo complex derived formally from the ferric state by removal of two electrons, (iii) the transfer of the oxygen atom of this iron-oxo complex into the substrate. Different structures are possible for this iron-oxo complex: a Fe(V)=O structure which has never yet been observed in hemoproteins or iron-porphyrins, a (porphyrin$^{+\cdot}$)—Fe(IV)=O structure which is found in compound I of horseradish peroxidase (Roberts et al., 1981) or in iron-porphyrin model complexes (see above), and an Fe(IV)=O structure with a protein-centered radical as in compound I of cytochrome c peroxidase (Hoffman et al., 1981) (Fig. 2). All these

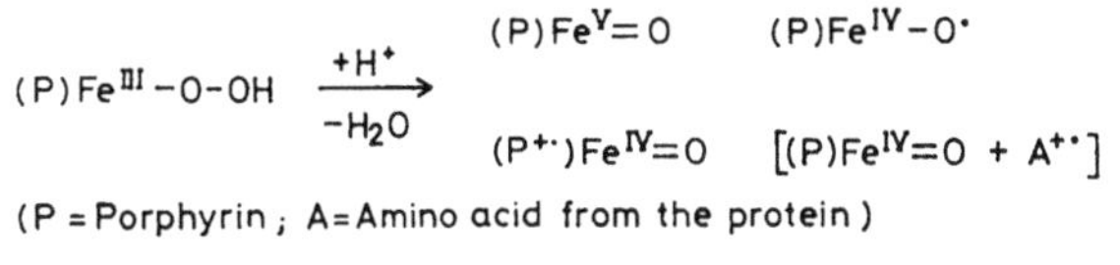

Fig. 2. Possible structures for the active oxygen complex of cytochrome P-450.

structures derive formally from a two-electron oxidation of the starting ferric state and the binding of an oxygen atom from O_2 to the iron. It is not known presently whether the cysteinate axial ligand of the iron is still bound in the active oxygen complex, but its presence should greatly influence the nature and intrinsic reactivity of this complex. This point has been discussed by several authors (see for instance, ORTIZ DE MONTELLANO, 1986a), but as no high-valent porphyrin (or hemoprotein)-iron-oxo complex bearing a thiolate ligand has been so far obtained, the question of the possible involvement of this ligand remains unanswered. Concerning the last step in the catalytic cycle, the transfer of the oxygen atom from the FeO species into the substrate, which leads to C—H bond hydroxylation, alkene epoxidation, aromatic ring hydroxylation or heteroatom oxidation, concerted and non-concerted mechanisms are theoretically possible. However, a great deal of indirect evidence suggests that non-concerted mechanisms are most often found (for a review, see ORTIZ DE MONTELLANO, 1986a). Most experimental results can be rationalized if one admits that the active oxygen complex has a free radical nature and reactivity. By considering its possible mesomeric form Fe(IV)—O˙, it is easily understandable that it could abstract hydrogen atoms from C—H bonds or add to double bonds of alkenes or aromatic rings. These mechanisms will be discussed in more detail in the following sections by comparison to those of model systems.

2.2. Oxidations catalyzed by cytochrome P-450 using oxygen atom donors different from dioxygen

During the last ten years, it has been shown that cytochrome P-450-dependent oxidations can be performed by using oxygen atom-donors AO instead of O_2 and NADPH. With such oxidants, the catalytic cycle of substrate oxidation is considerably simplified, as it consists of only two steps, the transfer of the oxygen atom from AO to the iron and the transfer of this oxygen atom from the FeO species to the substrate (Fig. 1). Single oxygen atom donors such as $NaIO_4$ and $NaClO_2$ (HRYCAY et al., 1975; DANIELSSON and WIKVALL, 1976; GUSTAFSSON and BERGMAN, 1976a) and later iodosylbenzene (LICHTENBERGER et al., 1976; GUSTAFSSON et al., 1976, 1979) and aryldimethylamine N-oxides (HEIMBROOK et al., 1984) have been successfully used for the cytochrome P-450-dependent oxidation of various substrates. Potential oxygen-atom donors involving an O—O bond such as alkylhydroperoxides or H_2O_2 have also been used for similar purposes (KADLUBAR et al., 1973; RAHIMTULA and O'BRIEN, 1974; NORDBLOM et al., 1976). However, with these oxidants, a further problem has been encountered because of the two possible modes of cleavage of their O—O bond. A heterolytic cleavage of the O—O bond is required for the formation of the high-valent Fe(V)=O species (Fig. 3), whereas a homolytic cleavage leads to an alkoxy radical and an Fe(IV) species. The relative importance of these two modes of cleavage of the O—O bond in cytochrome P-450-catalyzed oxidation of substrates by alkylhydroperoxides, peracids or H_2O_2 has been discussed in a recent review (ORTIZ DE MONTELLANO, 1986a).

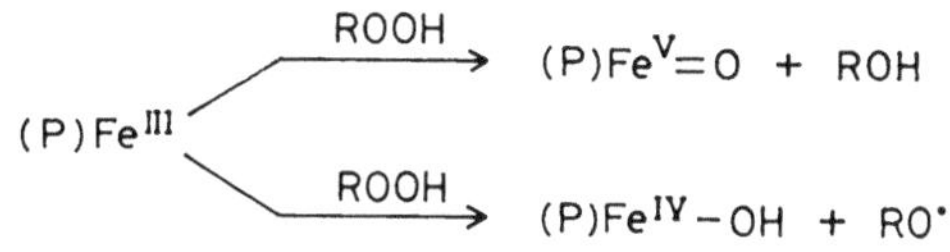

Fig. 3. The two possible modes of cleavage of oxidants containing a O–O bond by cytochrome P-450.

2.3. How to build up catalytically active models for cytochrome P-450?

In the following, all the model systems described will be based on a metalloporphyrin catalyst which mimics the cytochrome P-450 heme. An ideal model system would associate with the metalloporphyrin (preferably an iron-porphyrin) a thiolate ligand to mimic the axial cysteinate ligand of cytochrome P-450, a reducing agent and a proton donor necessary for a monooxygenation reaction (see section 1.1), and O_2 itself as the oxygen atom donor. However, because of the high oxidizing activity of the active oxygen metal-oxo intermediate in such systems, the direct use of thiolate ligands in chemical models seems very difficult: These thiolate ligands are very sensitive towards oxidizing agents. The cysteinate ligand of cytochrome P-450 is not too rapidly oxidized presumably because it is held by the protein far from the oxidizing oxygen atom of the active oxygen species. In fact, it is separated from this oxygen atom by the heme plane. It is still difficult to achieve such a clear separation in chemical systems and, so far, most model systems described which were catalytically active for alkane hydroxylation have not involved thiolate ligands. The long catalytic cycle of substrate oxidation by cytochrome P-450, which uses O_2 as oxygen atom donor and NADPH, appears difficult to mimic for two main reasons. The multienzyme monooxygenase system allows a complete separation of the iron-oxo hydroxylating species from the reducing agent (NADPH) in excess, since the electrons from NADPH are transferred to the heme via an electron transfer chain. This separation prevents the direct reaction of the FeO species with NADPH. In chemical model systems, it seems difficult to separate the metal-oxo active species from the reducing agent in excess, explaining the very low yields of substrate oxidation based on the reducing agent observed so far with model systems using O_2 itself (see following sections). The other role of the protein in the cytochrome P-450 catalytic cycle is to regulate the sequence of the different steps. With a simple chemical system, it appears far easier to mimic the shortened catalytic cycle of cytochrome P-450 which uses single oxygen atom donors. However, the level of difficulty of generating a high-valent iron-oxo complex upon reaction of an iron-porphyrin with an oxygen atom donor depends on the nature of this oxidant. In that regard, it is very likely that oxidants containing only one transferable oxygen atom linked to a good leaving group, such as iodosylbenzene

PhIO, hypochlorite ClO^- or tertiary amine oxides, would transfer their oxygen atom to metalloporphyrins more easily than oxidants containing an O—O bond. For the latter oxidants, such as alkyhydroperoxides and H_2O_2, there are, as mentioned in section 2.2., two possible modes of cleavage of the O—O bond after reaction with a metallic redox center and only the heterolytic cleavage will lead to the expected high-valent metal-oxo complex (Fig. 3). These considerations explain the order of presentation of the model systems in the following sections. It corresponds to the increasing level of complexity of these model systems as a function of the nature of the used oxygen atom donor (Fig. 4).

$$\mathrm{Fe(III)} + \mathrm{AO} \xrightarrow{-\mathrm{A}} \mathrm{Fe(V){=}O} \xrightarrow{+\mathrm{RH}} \mathrm{ROH} + \mathrm{Fe(III)}$$

$$\mathrm{Fe(III)} + \mathrm{R'OOH} \longrightarrow \begin{array}{l}\mathrm{Fe(V){=}O} + \mathrm{R'OH} \\ \text{or } \mathrm{Fe(IV){-}OH} + \mathrm{R'O^{\bullet}}\end{array} \xrightarrow{+\mathrm{RH}} \mathrm{ROH} + \mathrm{Fe(III)}$$

$$\mathrm{Fe(III)} + O_2 + 2e^- + 2H^+ \longrightarrow \mathrm{Fe(III){-}OOH} \xrightarrow{+H^+} \begin{array}{l}\mathrm{Fe(V){=}O} + H_2O \\ \text{or } \mathrm{Fe(IV){=}O} + \mathrm{H\,O^{\bullet}}\end{array} \xrightarrow{\mathrm{RH}} \mathrm{ROH} + \mathrm{Fe(III)}$$

Fig. 4. Pathways possibly involved in the three kinds of model systems using an iron-porphyrin and different oxygen atom donors.

3. Model systems using oxidants containing a single oxygen atom

3.1. Iodosylbenzene dependent systems

Since the first report by Groves et al. (1979a) showing that iodosylbenzene (PhIO) in the presence of catalytic amounts of iron-*meso*-tetraarylporphyrins was able to perform two important reactions catalyzed by cytochrome P-450, the hydroxylation of alkanes and the epoxidation of alkenes, a great deal of work has been devoted to the study of systems involving PhIO and various metalloporphyrin catalysts.

3.1.1. The various metalloporphyrins used as catalysts

The efficacy of a model PhIO-metalloporphyrin system can be evaluated by different parameters such as the catalytic activity of the metalloporphyrin, in moles of product formed per mole of catalyst (turnovers) per min for instance, and the selectivity of the oxidation reaction. But another important characteristic is the ability of the catalyst to remain intact in such a highly oxidizing medium. In fact, metalloporphyrins, like other macrocycles, tend to be oxidatively destroyed by the highly active oxygen species generated in model systems with not only PhIO but also the other oxidants that have been used so far. In order to evaluate the robustness of the catalysts in the oxidizing medium, a possible parameter is the total number of turnovers of the catalyst before its complete oxidative destruction. Natural iron-porphyrins such as iron protoporphyrin-IX are very sensitive towards oxidative degradation and have only rarely been used as catalysts in model systems (Tatsuno et al., 1986). Iron-porphyrins with aryl substituents on the four meso positions are more resistant and have been extensively used (Fig. 5). Iron-mesotetraphenyl-

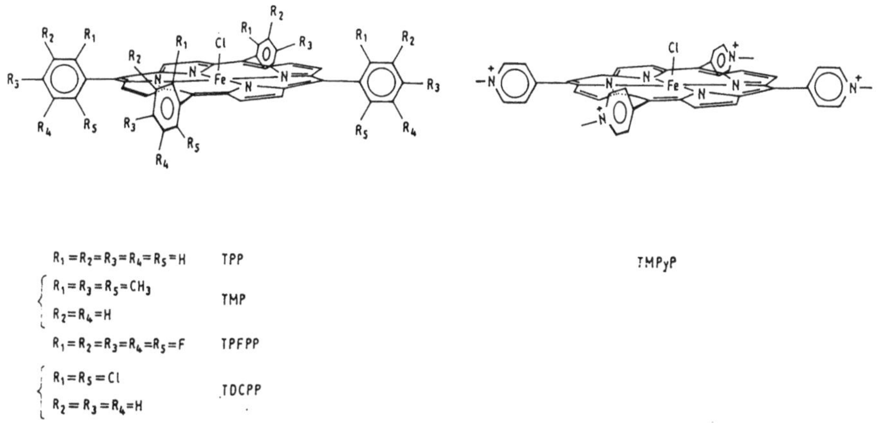

Fig. 5. Structure of different iron-porphyrins used as catalysts in model systems (abbreviations used in the text for the porphyrins are indicated).

porphyrin-chloride (Fe(TPP)(Cl)), which can be easily obtained from pyrrole and benzaldehyde, gives satisfactory results and has been used by most authors. Replacement of the chloride ligand of Fe(TPP)(Cl) by other axial ligands or counterions leads to qualitatively similar results (Nappa and Tolman, 1985). The use of iron-meso-tetramesityl-porphyrin, (Fe(TMP), which contains bulky methyl substituents in all the ortho positions of the *meso*-phenyl rings led to an important stabilization of the intermediate iron-oxo complexes and allowed their characterization by various spectroscopic techniques (Groves et al., 1981; Boso et al., 1983; Penner-Hahn et al., 1983, 1986). Major improvements in the catalytic activities and of the catalyst halftimes were obtained thanks to the use of tetraaryl-porphyrins containing halogen-substituted aryl groups such as pentafluorophenyl (Chang and Ebina, 1981; Nappa and Tolman, 1985) or 2,6-dichlorophenyl (Traylor et al., 1984a) groups. With the iron-porphyrin bearing these electron-withdrawing chloro-substituents that also provide steric protection, Fe(TDCPP)(Cl) (see fig. 5), as a catalyst, and C_6F_5IO in a $CH_2Cl_2:CH_3OH:H_2O$ mixture, more than 10,000 epoxide moles were obtained per mole of catalyst, with an initial rate as high as 300 turnovers per second (Traylor et al., 1985). In a more general manner, the use of metal complexes of this porphyrin as catalysts associated with many other oxidants than PhIO has led to higher catalytic activities and longer lifetimes of the catalyst (see following sections). Several other porphyrins bearing more or less bulky substituents on the meso-aryl groups have been used, particularly in order to achieve regioselective oxidation. They will be described in details in section 6. All these metalloporphyrins are soluble in hydrophobic solvents and have been used in general in CH_2Cl_2, C_6H_6 or CH_3CN. However a polar tetra-cationic iron-porphyrin, Fe(tetra-4N-methylpyridyl-porphyrin=TMPyP) Cl_5, was found to catalyze alkene epoxidation by PhIO in CH_3OH (LINDSAY-SMITH and MORTIMER, 1985b).

The catalytic activity of porphyrin complexes of different metal ions has been evaluated. Fe(III)- and Mn(III)-porphyrins have been used most often and appear to make the best catalysts. However, chromium- (GROVES and KRUPER, 1979), ruthenium- (DOLPHIN et al., 1983; LEUNG et al., 1983 and osmium- (CHE and CHUNG, 1986) porphyrins were also found to catalyze olefin epoxidation and alkane hydroxylation with lower activities.

3.1.2. Nature of the active oxygen complexes involved in Fe- or Mn-porphyrin-PhIO systems

One of the major interests of the study of such systems was the isolation of the first clearly defined high-valent porphyrin-iron-oxo complexes. A porphyrin-Fe(IV)=O complex has been prepared by reaction of O_2 with Fe(II)(TMP) at low temperature and homolytic cleavage of the O—O bond of a (TMP)Fe(III)—O—O—Fe(III)(TMP) intermediate dimer in the presence of 1-methyl-imidazole (BALCH et al., 1984) (Fig. 6). This complex exhibits spectroscopic properties analogous to those of compounds II of peroxidases (ENGLISH et al., 1983) and is a poor oxidant as it is only able to transfer its oxygen atom to phosphines.

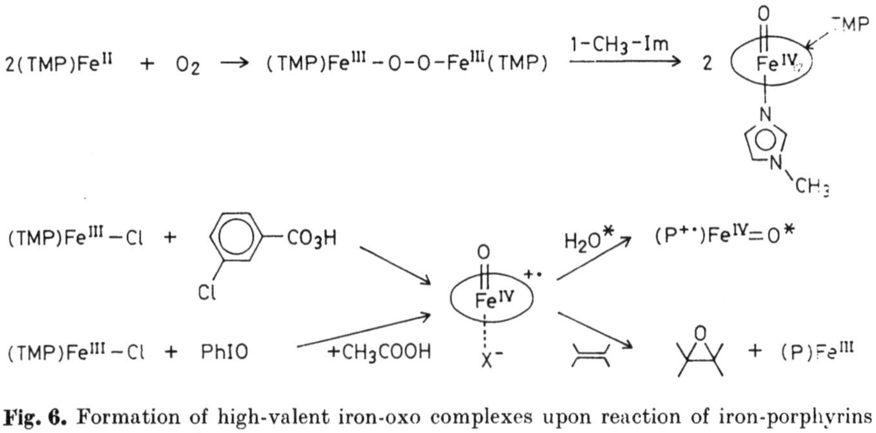

Fig. 6. Formation of high-valent iron-oxo complexes upon reaction of iron-porphyrins with oxygen-atom donors.

On the other hand, the complex derived from its one-electron oxidation is a powerful oxidizing agent. Such a complex having a formal Fe(V)=O structure has been also prepared by reaction of PhIO with Fe(TMP)(Cl) and treatment by CH_3COOH or by direct reaction of meta-chloroperbenzoic acid withFe (TMP)(Cl) (GROVES et al., 1981). This green complex is stable below −40 °C and was studied by ^{1}H-NMR, EPR, Mössbauer (Boso et al., 1983) and EXAFS (PENNER-HAHN et al., 1983; 1986) spectroscopy. All its characteristics are compatible with a (porphyrin-radical cation) Fe(IV)=O structure and similar to those of horseradish peroxidase compound I. It transfers its oxygen atom to alkenes almost quantitatively and exchanges its oxygen atom with water (GROVES et al., 1981) (Fig. 6). Recently, such $(TMP^{+\cdot})$Fe(IV)=O complexes have been prepared by chemical (BALCH et al., 1985) and electrochemical (LEE et al., 1985a; CALDERWOOD and BRUICE, 1986; YUAN et al., 1985a; GROVES and GILBERT, 1986) oxidation of (TMP)Fe(IV)=O. Porphyrin-Fe(IV)=O complexes have been also prepared by electrochemical oxidation of porphyrin-Fe(III)-OH complexes or by reduction of porphyrin-Fe(II)-O_2 complexes (SCHAPPACHER et al., 1985).

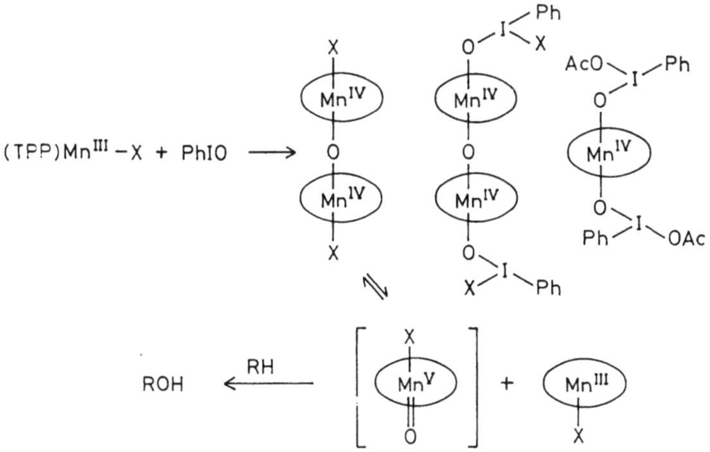

Fig. 7. Complexes isolated upon reaction of Mn(III)-porphyrins with PhIO and their reaction with hydrocarbons.

Reaction of Mn(III)-porphyrins with PhIO leads to various complexes the structure of which varies as a function of the nature of the solvent and of the counterion. The complexes characterized so far by X-ray analysis exhibit no Mn=O bond. They are either μ-oxo Mn(IV)-O-Mn(IV) dimers or Mn(IV) monomers as shown in Fig. 7 (SCHARDT et al., 1981, 1982; CAMENZIND et al., 1982; SMEGAL et al., 1983; SMEGAL and HILL, 1983a, 1983b). These Mn(IV) complexes are believed to be in equilibrium with a very reactive Mn(V)=O complex which could be responsible for their reactions with alkanes and alkenes (Fig. 7) (SMEGAL and HILL, 1983a).

3.1.3. Different reactions performed by model metalloporphyrin-PhIO systems

It is now clear that these model systems are able to perform all the kinds of oxidation catalyzed by cytochromes P-450 under similar mild conditions, i. e. the hydroxylation of alkanes, the epoxidation of alkenes, the hydroxylation of aromatic rings and the oxidation of molecules containing a heteroatom (O, N or S). The characteristics of these different reactions will be considered by comparison with those of the corresponding cytochrome P-450-dependent reactions.

3.1.3.1. Alkane hydroxylation

The hydroxylation of several alkanes such as cyclohexane, adamantane, decalin and n-heptane, is performed with satisfactory yields by using PhIO in the presence of Fe(III)- or Mn(III)-porphyrins. Alcohols are the major products and, in general, only minor amounts of the corresponding ketones are observed. Important characteristics of the hydroxylation of alkanes by these model systems which are similar to those found in the case of cytochrome P-450 are

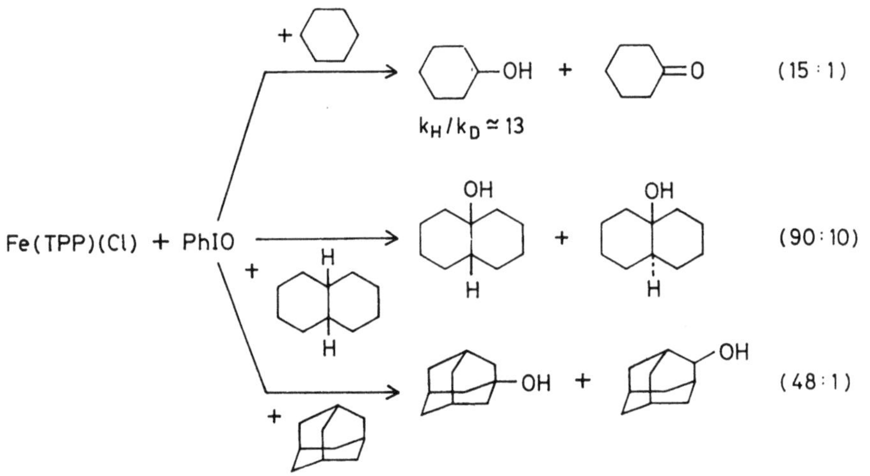

Fig. 8. Some characteristics of alkane hydroxylation by PhIO catalyzed by iron-porphyrins.

illustrated in Fig. 8: (i) a very high isotopic effect k_H/k_D of 12.9 ± 1 for cyclohexane hydroxylation (Groves et al., 1983c), (ii) a high retention of configuration in the hydroxylation of cis-decalin into 9-decalol (Groves and Nemo, 1983b; Lindsay-Smith and Sleath, 1983b), (iii) the preferential hydroxylation of tertiary C—H bonds in alkanes containing CH and CH_2 groups (relative reactivity of tertiary versus secondary C—H bonds in adamantane of about

50) (Groves and Nemo, 1983b). It is noteworthy that the Mn(TPP)(Cl)—PhIO system leads very often to higher yields of hydroxylated products of alkanes than the Fe(TPP)(Cl)—PhIO system, but also to higher ketone: alcohol ratios (Groves et al., 1980; Hill and Schardt, 1980; Fontecave and Mansuy, 1984a). Alkane activation by the Mn(TPP)(X)—PhIO system leads, in addition to the expected alcohols, to products deriving from the substitution of a hydrogen atom by a halogen X, coming from the catalyst or from the solvent, or by N_3 in the presence of NaN_3 in excess (Groves et al., 1980; Hill and Schardt, 1980; Hill and Smegal, 1982; Hill et al., 1983). These results have been interpreted by the involvement of free radicals derived from the alkane which can either abstract a halogen atom from the solvent or be oxidized by the Mn(IV)—X intermediate with transfer of a halogen or N_3 ligand. Intermediate formation of an alkane-derived free radical in these reactions was confirmed by a study of the oxidation of norcarane, in which products derived from an isomerisation of a norcaranyl radical were observed (Groves et al., 1980) (Fig. 9).

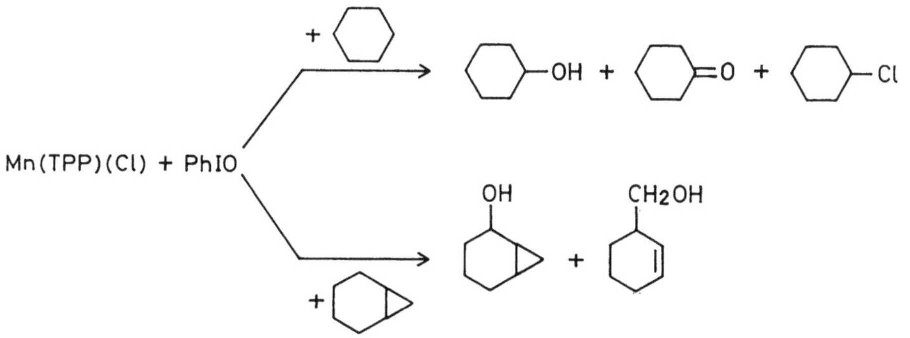

Fig. 9. Some characteristics of alkane oxidation by PhIO catalyzed by Mn-porphyrins.

3.1.3.2. Alkene oxidation

The four reactions observed upon cytochrome P-450-dependent oxidation of alkenes (i. e. the epoxidation of the double bond, hydroxylation of allylic C—H bonds, minor formation of aldehydes RCH_2CHO from $RCH{=}CH_2$ and the slow transformation of the iron-porphyrin catalyst into an N-alkyl porphyrin) have been easily reproduced by the iron-porphyrin—PhIO system. This system epoxidizes olefins with high yields based on the starting oxidant especially when Fe(TPFPP)(Cl) (Chang and Ebina, 1981) or Fe(TDCPP)(Cl) (Traylor et al., 1984) (Fig. 5) are used as catalysts. This reaction is totally stereospecific and corresponds to a *syn* addition of the oxygen atom to the double bond (Groves and Nemo, 1983a; Lindsay-Smith and Sleath, 1982). It is noteworthy that the *cis*-1,2-disubstituted olefins are much more reactive than their *trans*-isomers (Groves et al., 1983a). This has been explained by an approach of the double bond plane of the olefin almost perpendicular to the porphyrin plane. Theoretical calculations have indicated an approach with the double bond plane making a 45° angle with that of the porphyrin (SEVIN and FONTECAVE, 1986). Kinetic analyses of these epoxidation reactions led to important advances in our understanding of their mechanisms. In fact, two different results were obtained depending upon the conditions used. With the Fe(TDCPP)(Cl)—C_6F_5IO system in $CH_2Cl_2:CH_3OH:H_2O$ (80:18:2), the reaction order is one for the oxidant and for the catalyst and the reaction rate is not dependent on the alkene concentration (TRAYLOR et al., 1985). This indicates that the formation of the reactive iron-oxo complex is the rate-limiting step under these conditions. With Fe(TPFPP)(Cl) and C_6F_5IO in CH_2Cl_2, the epoxidation rate was independent of the olefin concentration but different alkenes were epoxidized at different rates and one alkene acted as a competitive inhibitor for the epoxidation of another (COLLMAN et al., 1985b). This result was interpreted by the kinetic scheme shown in Fig. 10, where the rate-determining step is the transfer of the oxygen atom from the iron-oxo complex to the alkene and where a preequilibrium between the iron-oxo complex, the alkene and an intermediate complex containing the elements of the iron-oxo complex and the alkene, precedes the oxygen atom transfer.

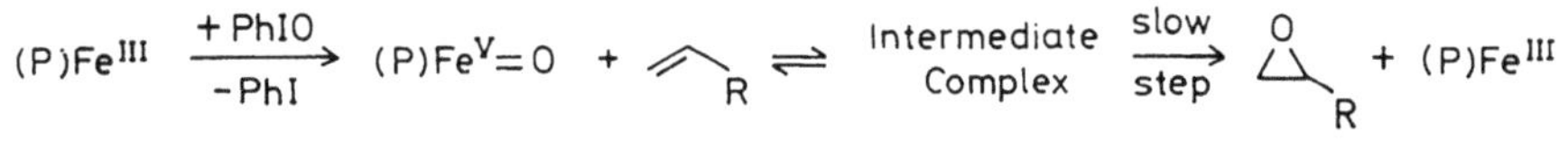

Fig. 10. Kinetic scheme proposed for alkene epoxidation by PhIO catalyzed by (FeTPP)(Cl).

During the epoxidation of monosubstituted alkenes by Fe(TPP)(Cl) (MANSUY et al., 1985) or Fe(TDCPP)(Cl) (MASHIKO et al., 1985; COLLMAN et al., 1986b) and PhIO, a slow transformation of the catalyst occurs with formation of a green pigment. Acidic demetallation of the pigments formed upon oxidation of $RCH{=}CH_2$ alkenes catalyzed by Fe(TDCPP)(Cl) leads to *N*-alkylporphyrins which exhibit an $N{-}CH_2CHOHR$ structure. An identical structure has been found for the green pigments derived from cytochrome P-450 which accumulate in the liver after oxidative metabolism of such $RCH{=}CH_2$ alkenes (ORTIZ DE MONTELLANO and REICH, 1986). In the model system as in cytochrome P-450, the stereochemistry found for the final *N*-alkylporphyrin corresponds to a *syn* addition of the oxygen atom and of a pyrrole nitrogen atom on the alkene double bond (COLLMAN et al., 1986b) (Fig. 11). The *N*-alkylporphy-

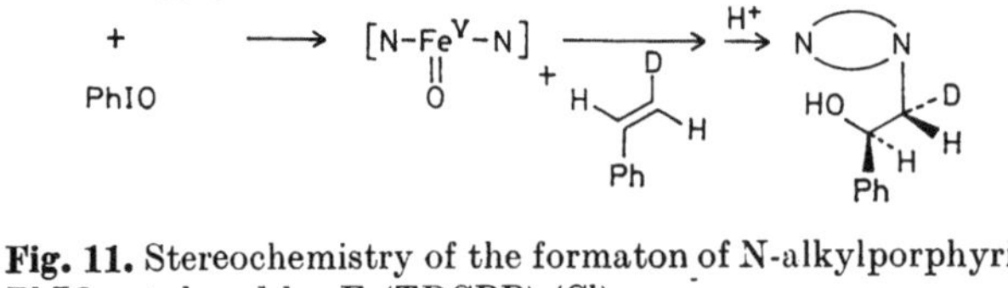

Fig. 11. Stereochemistry of the formaton of N-alkylporphyrins upon alkene oxidation by PhIO catalyzed by Fe(TDCPP) (Cl).

rins with an N—CH_2CHOHR structure, which have been isolated from oxidation of RCH=CH_2 alkenes either by Fe(TDCPP)(Cl) and PhIO or by cytochrome P-450, derive formally from addition of the pyrrole nitrogen atom to the less substituted carbon of the alkene double bond. *N*-alkylporphyrins deriving from addition of the pyrrole nitrogen atom to the more substituted carbon atom of the alkene double bond have recently been isolated from the oxidation of alkenes RCH=CH_2 by PhIO in the presence of less hindered iron-porphyrins such as Fe(TPP)(Cl) (ARTAUD et al., 1987) (see Fig. 12).

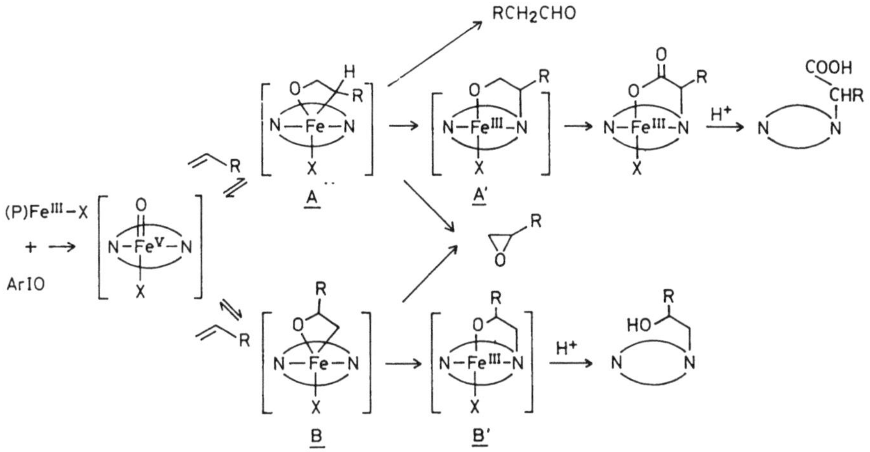

Fig. 12. Possible mechanisms involved in the oxidation of alkene double bonds by PhIO and iron-porphyrins.

Besides the epoxides which are always the major products of alkene oxidation by these model systems, and the *N*-alkylporphyrins, another class of products deriving from the oxidation of the double bond of some RCH=CH_2 alkenes, the aldehydes RCH_2CHO, have been detected in minor amounts (MANSUY et al., 1984d; GROVES and NEMO, 1983a; COLLMAN et al., 1986a). This aldehyde formation, which was also observed in cytochrome P-450-dependent reactions (MANSUY et al., 1984d), is particularly important in oxidation of styrenes. The aldehydes are not derived from an isomerisation of the corresponding epoxides but from the migration of a hydrogen atom during the reaction. Figure 12 gives a possible mechanism which allows one to explain all the aforementioned results concerning the oxidation of alkene double bonds by iodosoarenes in the presence of ironporphyrins. In this mechanism, the intermediate formed by reaction of the iron-oxo species with the alkene, for which kinetic (COLLMAN et al., 1985b) and spectroscopic evidence (GROVES and WATANABE, 1986a) has been obtained, is formulated as a four-membered metallocycle as suggested by several authors (MANSUY et al., 1984a; COLLMAN et al., 1985b; GROVES and WATANABE, 1986a). There is still no direct evidence for it, and other possible structures have been proposed (Groves and Watanabe, 1986a; Traylor et al., 1986a, b). However, the various final observed products can easily be explained by its different possible reactions. In fact, depending upon the sense of addition of the Fe=O species to the alkene double bond, two four-membered metallocycles may be formed. Both of them may undergo reductive elimination of their oxygen and σ-alkyl *cis* ligands of the iron to give the epoxide. Metallocycle *A* has another possible fate: a rearrangement with a 1,2-shift of a hydrogen atom in α position relative to the oxygen which leads to the observed aldehyde RCH_2CHO. Both metallocycles may also undergo a migration of their σ-alkyl ligand from the iron to a pyrrole nitrogen atom. Such Fe → N migrations have been previously observed for high-valent σ-alkyl-iron-porphyrin complexes formed as intermediates in the oxidation of σ-alkyl Fe(III)—R complexes (Mansuy et al., 1982a; Lançon et al., 1984; Balch and Renner, 1986). This isomerisation of metallocycles *A* and *B* leads to five-membered Fe—O—C—C—N metallocycles which, upon acidic demetallation, give the isolated *N*-alkylporphyrins. From the available data, it seems that the structure of the final *N*-alkylporphyrins depends greatly on the structure of the starting iron-porphyrin catalyst. With Fe(TDCPP)(Cl), $N—CH_2CHOHR$ structures have been found (Mashiko et al., 1985; Collman et al., 1986b) whereas with Fe(TPP)(Cl) or Fe(TpClPP)(Cl), *N*-alkylporphyrins with an N—CH(COOH)R structure have been isolated (Artaud et al., 1987). The latter derive presumably from a further oxidation of metallocycle *A*' in the reaction medium (Fig. 12).

During olefin oxidation by the model iron-porphyrin-ArIO systems, allylic alcohols are formed as well as epoxides (Groves and Subramanian, 1984; Mansuy et al., 1984a). Allylic alcohols coming from deuterated cyclohexene are mainly derived from the insertion of an oxygen atom into an allylic C—H bond but also from an allylic rearrangement of a possible free radical intermediate (Groves and Subramanian, 1984). The allylic alcohol:epoxide ratio depends greatly on the nature of the porphyrin substituents and on the nature of the metal (Mansuy et al., 1984a).

3.1.3.3. Aromatic hydrocarbon oxidations

Much less work has been devoted to the study of the oxidation of aromatic hydrocarbons by the metalloporphyrin-PhIO systems. In fact, yields of phenols in such reactions are generally considerably lower than the yields of alkane hydroxylation and alkene epoxidation. For instance, with the very efficient Fe(TPFPP)(Cl) catalyst and PhIO, cyclohexane is hydroxylated with a 71% yield and cyclohexene is epoxidized with a 95% yield whereas anisole is hydroxylated at the *ortho* and *para* positions with a total yield of only 11% (Chang and Ebina, 1981). Hydroxylation of para-deutero-anisole by this system is performed with 72% retention of deuterium in the product *para*-methoxyphenol. Such a shift of a deuterium atom from the carbon to be hydroxylated to the adjacent one is classical in cytochrome P-450-catalyzed hydroxylation of aromatic rings (JERINA and DALY, 1974). The level of deuterium retention in the *para*-hydroxylation of *para*-deutero-anisole by PhIO catalyzed by different metalloporphyrins depends on the nature of the catalyst: 72% for Fe(TPFPP)(Cl), 60% for Fe(TPP)(Cl) and Mn(TPFPP)(Cl) and only 26% for Mn(TPP)(Cl) (CHANG and EBINA, 1981). 1-Naphthol is also formed in low yields with 68% deuterium retention from the oxidation of labelled naphthalene by Fe(TPP)(Cl) and PhIO (LINDSAY-SMITH and SLEATH, 1982b) (Fig. 13).

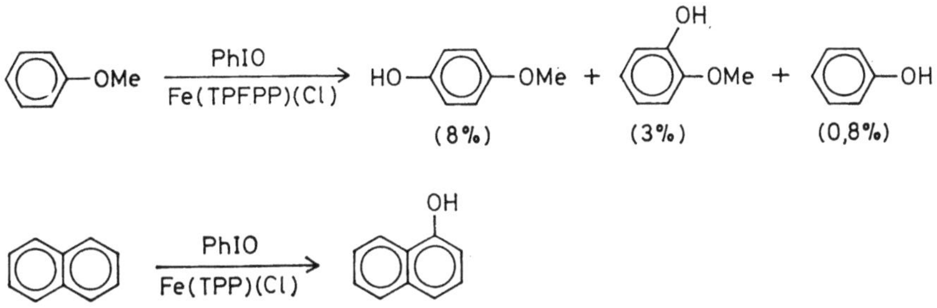

Fig. 13. Some examples of aromatic compound hydroxylation by PhIO and iron-porphyrins

3.1.3.4. Oxidation of compounds containing a heteroatom

Ethers containing an OCH_3 group are O-demethylated with low yields by the model Fe(TPP)(Cl)-PhIO system. An isotope effect $k_H:k_D$ of 9 is reported for the *O*-demethylation of anisole by this system (LINDSAY-SMITH et al., 1982; LINDSAY-SMITH and SLEATH, 1983a). This large isotope effect is close to that found for anisole-*O*-demethylation by rat liver microsomal cytochrome P-450 (LINDSAY-SMITH and SLEATH, 1983a). Tertiary amines are also *N*-dealkylated by these model systems but with an isotope effect $k_H:k_D$ of 1.3, this oxidation being initiated by a one-electron transfer from the amine to the iron-oxo complex and not by the abstraction of a hydrogen atom in α position to the nitrogen (LINDSAY-SMITH and MORTIMER, 1985a). Thioethers are oxidized with good yields by PhIO-metalloporphyrin systems. Sulfoxides are formed in a first step but a further oxidation leads to sulfones (ANDO et al., 1982). As the Fe- or Mn-porphyrin-PhIO systems were shown to reproduce quite well all the kinds of reaction catalyzed by cytochrome P-450, they have been used to mimic some oxidations of compounds which have important implications in pharmacology and toxicology. For instance, the oxidations of *N*-nitrosoamines (LINDSAY-SMITH et al., 1984) and of cyclopropylamines (GUENGERICH et al., 1984) by model systems have been found similar to the enzymatic oxidations.

3.1.4. The use of nitrogen and carbon analogues of iodosylbenzene

In the presence of Fe- or Mn-porphyrins, PhI=N-tosyl (a nitrogen analogue of PhIO) is able to transfer its *N*-tosyl-nitrene moiety to alkenes and alkanes, leading respectively to *N*-tosyl-aziridines and *N*-tosylamines (Breslow and Gellman, 1982, 1983; Mansuy et al., 1984b; Mansuy and Battioni, 1986). This subject is beyond the scope of this article and will not be discussed in detail (for a recent review, see Mansuy and Battioni, 1986). It is only mentioned to show that a chemistry of transfer of nitrogen into hydrocarbons exists as well as the oxygen transfer into the same substrates (Fig. 14). Carbon analogues of PhIO (the iodonium ylids PhI=CRR') also easily react with iron porphyrins with transfer of their carbene moiety to form stable metallocyclic iron complexes (Fig. 15) (Battioni et al., 1986a). The structures of these complexes are similar to those found in alkene oxidation by PhIO catalyzed by iron porphyrins (Fig. 12 and 15).

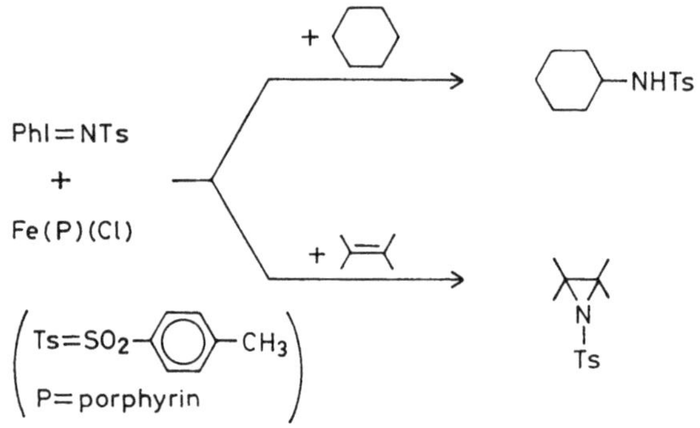

Fig. 14. Insertion of the N-tosyl moiety of PhINTs into hydrocarbons catalyzed by iron-porphyrins

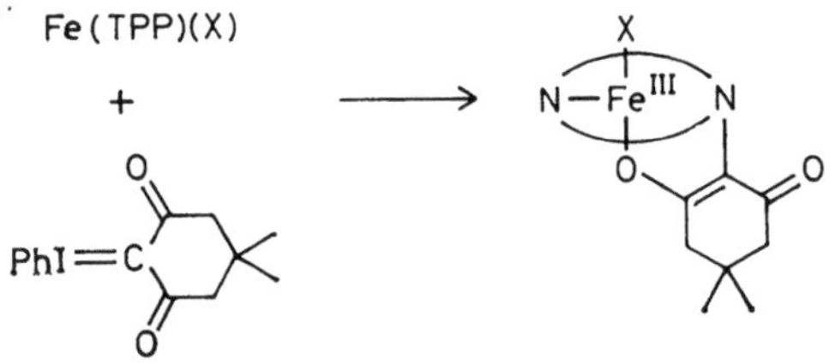

Fig. 15. Formation of Fe—O—C—C—N metallocycles upon reaction of a carbon equivalent of PhIO with ferric porphyrins.

3.2. Hypochlorite-dependent systems

The use of diluted solutions of NaOCl in water at pH around 12, in the presence of phase transfer agents, allows the oxidation of substrates dissolved in organic solvents in the presence of an Mn-porphyrin catalyst (Tabushi and Koga, 1979a; Guilmet and Meunier, 1980). With Mn(TPP) (Cl) as catalyst, benzyl alcohol is oxidized to benzaldehyde and alkanes are oxidized into the corresponding alkyl chlorides (Tabushi and Koga, 1979a). Styrene is epoxidized in good yield by such systems using NaOCl and Mn(TPP)(Cl) but alkenes containing allylic hydrogens lead to only low yields of epoxidation (GUILMET and MEUNIER, 1980). The system is considerably improved by the addition of nitrogen bases that can act as axial ligands to the catalyst, such as pyridine (GUILMET and MEUNIER, 1982) or imidazoles substituted by an aryl group (COLLMAN et al., 1983). With such ligands, the initial rate of styrene epoxidation can reach 11 turnovers per second (COLLMAN et al., 1985a) and the system becomes applicable to a large series of alkenes (MEUNIER et al., 1984). The use of pyridine leads also to an increase in the stereoselectivity of the epoxidation of stilbenes (GUILMET and MEUNIER, 1982). As with iodosoarenes, the use of Mn-tetraarylporphyrins containing bulky substituents in *ortho* position of the *meso*-phenyl rings leads to an increase in the epoxidation rates (BORTOLINI et al., 1983; COLLMAN et al., 1985a) and in the stereoselectivity of the epoxidation (BORTOLINI and MEUNIER, 1984). Moreover, the use of Mn-tetraarylporphyrins bearing halogen substituents at the *ortho* position of the *meso*-phenyl groups even leads to an efficient epoxidation of poorly reactive monosubstituted alkenes such as oct-1-ene (60—80% yield) (DE POORTER and MEUNIER, 1984). Olefin epoxidations can also be performed with increased rates by using systems in which the Mn-porphyrin is bound to a polymer (VAN DER MADE et al., 1983) or where the pH value of the NaOCl solution is decreased to 9.5 (MONTANARI et al., 1985). Kinetic and spectroscopic studies indicate that epoxidations by the hypochlorite-Mn-porphyrin systems have mechanisms very similar to those of the PhIO-Mn-porphyrin sytems. They involve a high-valent Mn-oxo complex (BORTOLINI et al., 1986) as the epoxidizing species and the formation of an intermediate complex between this Mn-oxo species and the alkene which could be a four-membered metallocycle (COLLMAN et al., 1984, 1985a; NOLTE et al., 1986) analogous to that described in the preceding section in PhIO-dependent systems (Fig. 16).

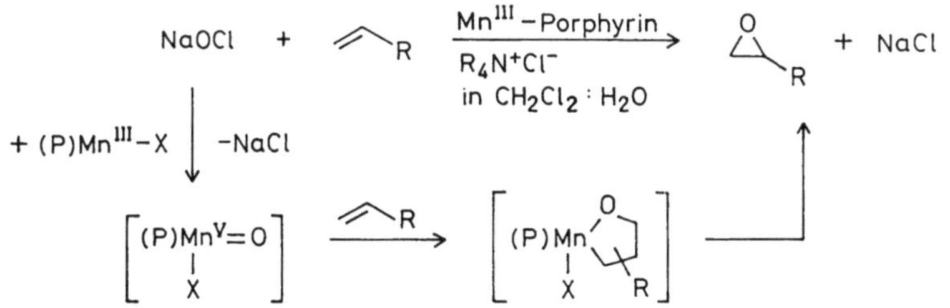

Fig. 16. Alkene epoxidations using NaOCl and Mn-porphyrins. A possible mechanism.

3.3. Systems dependent on other single oxygen atom donors

3.3.1. Amine-oxides

Dimethylaniline-*N*-oxides can act as oxygen atom donors for Fe- and Mn-porphyrins (SHANNON and BRUICE, 1981; NEE and BRUICE, 1982) as well as for cytochromes P-450 (HEIMBROOK et al., 1984). They have been used for alkene epoxidation. However, yields of alkene epoxidation and especially of alkane hydroxylation remain low when compared to those obtained with PhIO- or ClO^--dependent systems. This is due to competition between the substrate and dimethylaniline formed after transfer of the oxygen atom, for reaction with the metal-oxo species. A very detailed kinetic study of the different steps involved in the reactions performed by these *N*-oxides in the presence of Fe- (DICKEN et al., 1985) or Mn- (POWELL et al., 1984) porphyrins has been reported. An oxaziridine has also been used as an oxygen atom donor for the epoxidation of alkenes in the presence of Mn-porphyrins (YUAN and BRUICE, 1985a).

3.3.2. Potassium hydrogen persulfate

This water-soluble oxidant, $KHSO_5$, has been used in a biphasic system (CH_2Cl_2-buffer, pH8) in the presence of catalytic amounts of Mn(III)-porphyrins for alkene epoxidation and alkane hydroxylation (DE POORTER et al., 1985; DE POORTER and MEUNIER, 1985a, b). Its efficacy in olefin epoxidation is considerably increased by the presence of pyridine, like the previously described hypochlorite system. It is much superior to the latter system for alkane hydroxylation as it leads to good conversion of cyclohexane and adamantane in mild conditions.

3.3.3. Periodates

A periodate soluble in organic solvents, NBu_4IO_4, has been used as oxygen donor in the presence of catalytic amounts of an iron-porphyrin for the oxidation of thioethers into sulfoxides (TAKATA and ANDO, 1983). This system leads also to alkene epoxidation but less efficiently than PhIO or ClO^-.

4. Model systems using hydroperoxides as oxidants

4.1. Alkylhydroperoxides as oxidants

Alkylhydroperoxides ROOH (R = cumene or t-butyl), in the presence of catalytic amounts of Fe(TPP)(Cl), oxidize alkanes to the corresponding alcohols and ketones with good yields (80% for cyclohexane) (MANSUY et al., 1980). However, the Fe(TPP)(Cl)—ROOH system is completely unable to epoxidize alkenes (MANSUY et al., 1982b). A comparison of the properties of this system and of the Fe(TPP)(Cl)—PhIO system shows that different active oxygen species are formed. Reaction of Fe(TPP)(Cl) with cumenehydroperoxide fails to lead to the Fe(V)=O species characteristic of the PhIO system. In fact, this reaction leads presumably to a homolytic cleavage of the alkylhydroperoxide O—O bond with formation of an alkoxy radical as reactive species (MANSUY et al., 1982b) (Fig. 17). This radical can abstract hydrogen from alkanes, explaining why alkanes are hydroxylated eventually, while it cannot perform alkene epoxidation. However, cytochrome P-450 catalyzes alkene epoxidation by cumenehydroperoxide. This suggests that the strong electron-donating cysteinate ligand of cytochrome P-450-iron could play a key role in allowing heterolytic cleavage of the alkylhydroperoxide O—O bond (Mansuy and Battioni, 1986). Imidazole seems to play a similar role since its addition to the Fe(TPP)(Cl)—ROOH or Mn(TPP)(Cl)—ROOH (R = cumene or t-butyl) systems allows them to epoxidize alkenes such as cyclohexene, *cis*-stilbene and 2-methylhept-2-ene with yields between 20 and 50% (Mansuy et al., 1984c) (Fig. 17).

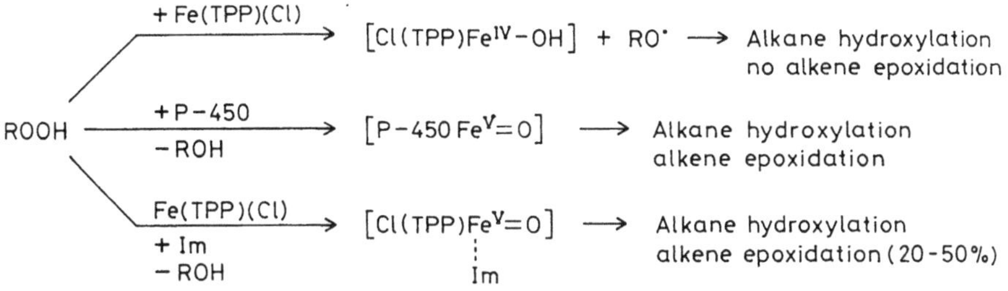

Fig. 17. Reactions of cumenehydroperoxide with P-450 or with iron porphyrins with or without imidazole. Possible active intermediates.

Kinetic studies on the heterolytic cleavage of the O—O bond of various peroxide compounds including peracids and alkylhydroperoxides by Fe(III)- and Mn(III)-porphyrins show that this type of cleavage and the formation of a high-valent metal-oxo species occurs more easily with acidic ROOH compounds and is facilitated by the presence of nitrogenous bases such as imidazole (Traylor et al., 1984b; Yuan and Bruice, 1985b, 1986; Lee and Bruice, 1985).

4.2. Hydrogen peroxide as oxidant

This oxidant appears especially interesting for two main reasons: (i) its reaction with Fe(III)-porphyrins should lead to an intermediate very similar to that formed in cytochrome P-450 reactions after a two-electron reduction of dioxygen, (ii) when diluted in water, it is a cheap and simple oxidant especially attractive for preparative oxygenation of substrates as the secondary product is water. It has recently been reported that H_2O_2 can be used in the presence of Mn(III)-porphyrins and imidazole for very efficient epoxidation of alkenes (Renaud et al., 1985) and very efficient hydroxylation of alkanes (Battioni et al., 1986b). For both reactions, imidazole is absolutely required. When Mn(TDCPP)(Cl) is used as a catalyst, quantitative conversions of various alkenes, including the poorly reactive non-1-ene, are obtained with very high yields (90—100%) within one hour at 20 °C. Epoxidation of 1,2-dialkyl (or diaryl)-alkenes is totally stereospecific and corresponds to a *syn* addition of the oxygen atom to the double bond. The Mn(TDCPP)(Cl)-imidazole-H_2O_2 system hydroxylates alkanes such as cyclohexane, cyclooctane, adamantan and ethylbenzene with yields between 40 and 85%. Alcohols and ketones are formed in a 1:1 ratio, except for adamantan which leads to adamantan-1-ol as a major product (Fig. 18) (Battioni et al., 1986b).

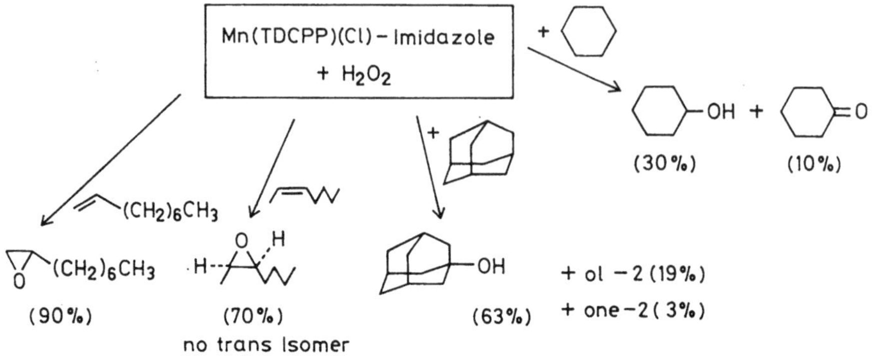

Fig. 18. Oxidations by H_2O_2 catalyzed by Mn-porphyrins and imidazole (yields based on the starting hydrocarbon).

In the presence of the water-soluble porphyrin Fe(TMPyP)$(Cl)_5$, H_2O_2 hydroxylates phenylalanine in tyrosine and hydroxytyrosine (Shimidzu et al., 1981). These aromatic hydroxylations occur with a 70% yield based on phenylalanine but at a low rate (7 turnovers per hour). Oxidation of thioethers into sulfoxides is also performed by H_2O_2 in the presence of Fe(TPP)(Cl). Also in this case, the addition of imidazole improves the rates and yields (Oae et al., 1982).

5. Model systems using O_2 in the presence of a reducing agent

5.1. Different systems reported

Borohydrides have been used as reducing agents in O_2-dependent oxidations of hydrocarbons catalyzed by Fe- or Mn-porphyrins. With $NaBH_4$ and Mn(TPP)(Cl), cyclohexene is oxidized to cyclohexanol as a major product with cyclohexen-3-ol as a minor product (Tabushi and Koga, 1979b). Cyclohexanol is believed to derive from the reduction of cyclohexene-oxide by $NaBH_4$ in excess, whereas cyclohexen-3-ol comes from allylic hydroxylation of cyclohexene. Similar systems using $NaBH_4$ in benzene-methanol mixtures and iron-porphyrin catalysts either soluble in the medium (Santa et al., 1985) or supported on polypeptides (Mori et al., 1985) have been also reported. By using tetra-*N*-butyl-ammonium borohydride, which is soluble in CH_2Cl_2, and Mn(TPP)(Cl), several terminal olefins such as styrene are oxidized to a mixture of the corresponding methyl ketone and of the alcohol derived from its reduction by the borohydride in excess (PERREE-FAUVET and GAUDEMER, 1981) (Fig. 19).

Sodium ascorbate can also be used as a reducing agent in a biphasic system where the Mn(TPP)(Cl) catalyst and the hydrocarbon substrate are present in benzene whereas the reducing agent is present in an aqueous phase buffered at pH 8.5 (Fig. 20). If a catalytic amount of a phase transfer agent is added to that system, it becomes able to oxidize alkenes very selectively into epoxides,

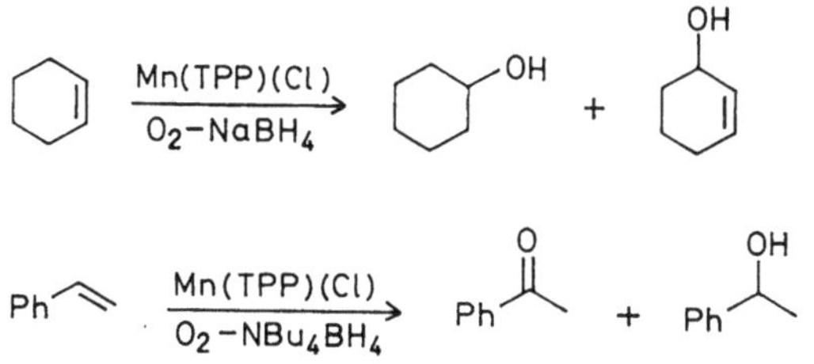

Fig. 19. Examples of alkene oxidations by O_2 in the presence of borohydrides catalyzed by Mn-porphyrins.

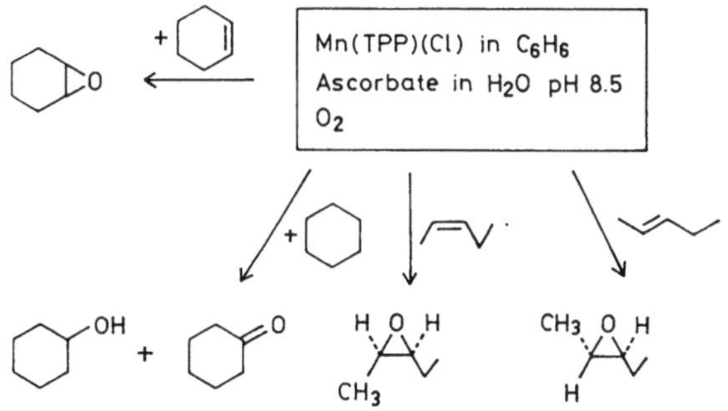

Fig. 20. Some reactions performed by the biphasic Mn(TPP)(Cl)-ascorbate-O_2 system.

and alkanes into alcohols and ketones (MANSUY et al., 1983). 1,2-dialkylethylenes are epoxidized by this system in a completely stereospecific manner whereas stilbenes are not. In fact, *cis*-stilbene leads to a 37:63 mixture of *cis* and *trans* epoxides (FONTECAVE and MANSUY, 1984a). This proportion is very similar to those reported for other systems using Mn(TPP)(Cl) as a catalyst and other oxidants such as PhIO or NaOCl. However, when compared to the Mn(TPP)(Cl)—PhIO system, the ascorbate—O_2—Mn(TPP)(Cl) system exhibits two major different characteristics: (i) it oxidizes cyclohexene exclusively into cyclohexene-oxide with no formation of allylic alcohol, (ii) it oxidizes alkanes with major formation of the corresponding ketones, whereas the Mn(TPP)(Cl)—PhIO system leads predominantly to the alcohols. This suggests that different active oxygen intermediates are involved in these two systems (FONTECAVE and MANSUY, 1984a).

Hydrogen itself has been used successfully as a reducing agent for monooxygenations by O_2 catalyzed by metalloporphyrins. In the presence of colloidal platinum and Mn(TPP)(Cl) as catalysts, cyclohexene is oxidized by O_2 and H_2 into the expected epoxide and minor amounts of cyclohexen-3-one (Tabushi and Yazaki, 1981). This Mn(TPP)(Cl)—O_2—H_2—Pt system also epoxidizes several other olefins with an order of reactivity of these olefins very similar to that found for the Mn(TPP)(Cl)—PhIO system. Addition of imidazole improved the epoxidation rates. The system also hydroxylates adamantan with predominant formation of the tertiary alcohol. The Mn(TPP)($OCOCH_3$)—O_2—H_2—Pt system was also used for the oxidation of more complex molecules such as geranyl acetate, and was found to epoxidize the double bond not carrying the

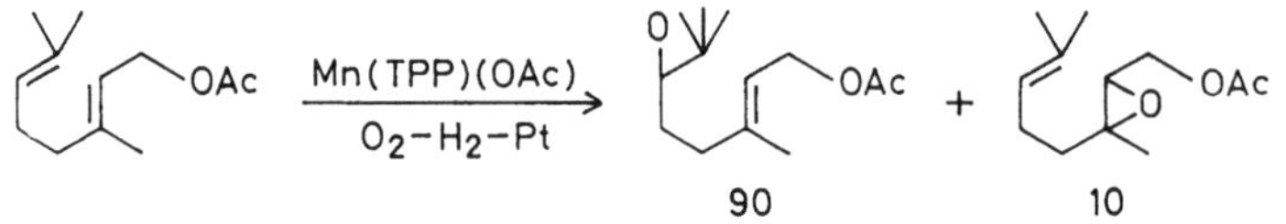

Fig. 21. Epoxidation of geranyl acetate by the Mn(TPP)(OAc)—O_2—H_2—Pt system.

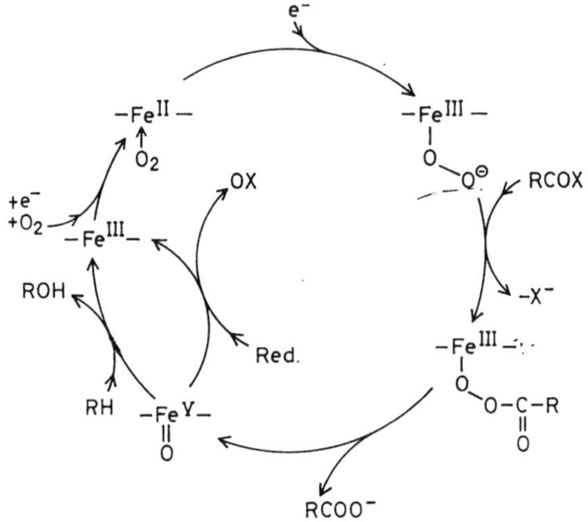

Fig. 22. The different steps involved in hydrocarbon oxidations by the Fe(TpivPP)(Cl)—O_2—H_2—Pt—$(PhCO)_2O$ system.

CH_2OCOCH_3 function in a regioselective manner (Tabushi and Mirimitsu, 1984) (Fig. 21). Iron-porphyrins act also as catalysts in O_2-dependent monooxygenation of hydrocarbons using H_2 and colloidal platinum. With the iron-"picket-fence"-porphyrin, Fe(TpivPP)(Cl), as a catalyst, the various steps of alkene epoxidation by such an O_2—H_2—Pt system have been studied (Tabushi et al., 1985). A crucial step is the heterolytic cleavage of the O—O bond of an Fe(III)—O—O^- intermediate (Fig. 22). The presence of protons, but also more importantly of an acid anhydride, greatly facilitates this step. This easy heterolytic cleavage of O—O bonds in porphyrin—Fe(III)—O—O—CO—R complexes has recently been demonstrated (Groves and Watanabe, 1986b).

It involves an extraordinarily low enthalpy of activation of 16.7 kJ per mol and is acid-catalyzed. In fact, this easier cleavage of the O—O bond of Fe(or Mn)-superoxo intermediates after acylation of the terminal oxygen atom by acid chlorides or acid anhydrides added to the medium was successfully used in systems based on Mn-porphyrins (GROVES et al., 1983) or Fe-porphyrins (KHENKIN and SHTEINMAN, 1984) and $O_2^{\cdot -}$. This heterolytic cleavage leads to the expected high-valent iron-oxo complex which can epoxidize alkenes and hydroxylate alkanes. Accordingly, the Fe(TpivPP)(Cl)—O_2—H_2—Pt—$(PhCO)_2O$ system epoxidizes cyclohexene at a rate of 36 turnovers per hour (TABUSHI et al., 1985). In the system described by KHENKIN and SHTEINMAN, the electrons necessary for the monooxygenations come from an electrode, while Fe(TPP)(Cl) or Fe(TMP)(Cl) is used as catalyst and benzoyl chloride as a stoichiometric acylating agent. This system demethylates anisole with a $k_H:k_D$ isotope effect of 7.

All these O_2-dependent systems are able to reproduce qualitatively the main reactions of cytochrome P-450 and particularly the hydroxylation of inert C—H bonds of alkanes under mild conditions. However, their efficacy remains inferior to that of the enzymatic system especially in terms of catalytic activity (number of turnovers per min) and of yields based on the reducing agent. In the Fe(TpivPP)(Cl)—O_2—H_2—Pt system, it was shown that the rate-limiting step is the arrival of the second electron in the catalytic cycle (TABUSHI et al., 1985). The low yields based on the reducing agent obtained with these model systems (between 0.1 and 5% for the borohydride-, ascorbate- and H_2—Pt-dependent systems as a function of the nature of the substrate) are due to competition at the level of the high-valent metal-oxo intermediate between the substrate and the reducing agent in excess (Fig. 22). As there is no separation between the active oxygen species and the reducing agent in excess in the model systems, unlike in the enzymatic system, the reduction of the metal-oxo species by the reducing agent is faster than its reaction with the hydrocarbons (FONTECAVE and MANSUY, 1984a; TABUSHI et al., 1985). The best yield based on the reducing agent reported so far, 56%, was obtained for cyclooctene epoxidation by an electrochemical system using Mn(TPP)(Cl) as a catalyst and $(PhCO)_2O$ as a stoichiometric acylating agent (CREAGER et al., 1986). However, because of the slow arrival of the electrons in that system, the rate remained low (about 2 turnovers per hour). Very recently, two systems were reported to give good yields based on the reducing agent (up to 50%) and higher rates (up to 3—9 turnovers per min) which are not too far from those of cytochrome P-450. The first one employs a dihydropyridine as a reducing agent in the presence of a flavin mononucleotide, O_2 as the oxygen atom donor, and a water-soluble anionic Mn-porphyrin and *N*-methyl-imidazole as catalysts. Nerol and cyclohexene are epoxidized by this system with turnover frequencies of 9 and 3.6 moles of epoxide formed per mole of catalyst per min, and yields based on the reducing agent up to 33% (TABUSHI and KODERA, 1986). The second system uses Zn powder as a reducing agent, O_2 as the oxygen atom donor, acetic acid as a source of protons and Mn(TPP)(Cl) and *N*-methylimidazole as catalysts (BATTIONI et al., 1987) (Fig. 23). Cyclooctene is epoxidized by this system with a turnover frequency of 3.3 moles of products formed per mole of catalyst per min and with yields based on the reducing agent of 50%. Other alkenes including cyclohexene and non-1-ene are epoxidized with

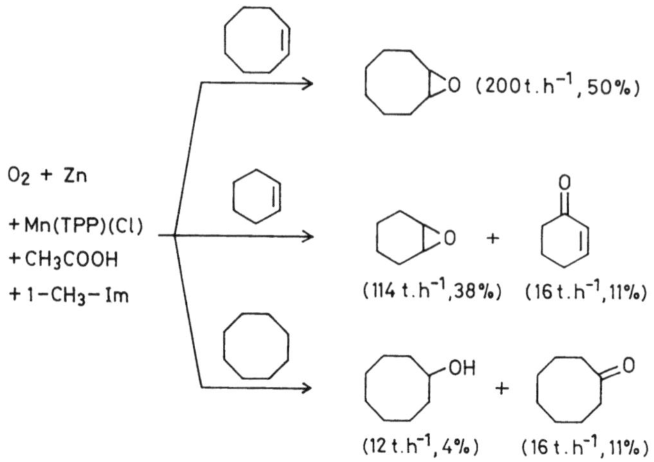

Fig. 23. Some characteristics of reactions performed by the Zn—O_2—Mn(TPP)(Cl)—CH_3COOH—N-methylimidazole system. ($t \cdot h^{-1}$ = turnovers per h.; yields based on the reducing agent %)

turnover frequencies between 0.4 and 3 min^{-1}. The system hydroxylates alkane) such as cyclooctane and adamantan with rates (28 and 32 turnovers per hours and yields based on the reducing agent (15 and 12%) which remain satisfactory. Interestingly, this system epoxidizes cyclooctene with a 95% conversion and 90% yield of cyclooctene-oxide within 1 h if a three-fold excess of Zn relative to cyclooctene is used. It is noteworthy that either *N*-methylimidazole or imidazole is absolutely required for monooxygenation to occur.

5.2. The possible roles of the porphyrin and axial ligand of iron in cytochrome P-450 and model systems

Very few studies have been devoted to comparisons between porphyrin- and non-porphyrin-Fe (or Mn) complexes under identical conditions. A comparison of the characteristics of the oxidation of hydrocarbons by PhIO catalyzed by either Fe(TPP)(Cl) or non-porphyrin-Fe(III) salts such as $FeCl_3$ or Fe(acetylacetonate)$_3$ showed that the porphyrin ligand played at least two roles: (i) to control more efficiently the substrate-derived intermediate free radicals by the iron complexes, and (ii) to trigger the chemoselectivity of the active oxygen iron complexes (FONTECAVE and MANSUY, 1984b). As far as the role of the axial iron ligand is concerned, the results obtained with the PhIO-iron-porphyrin system, which does not involve any axial thiolate ligand and which reproduces quite well the reactions of cytochrome P-450, indicated that the cysteinate ligand of iron in cytochrome P-450 is not necessary for the transfer of the oxygen atom of PhIO into the substrate. In systems using oxidants containing an O—O bond such as alkylhydroperoxides or H_2O_2, monooxygenation of hydrocarbons is observed only in the presence of an imidazole axial ligand of the Fe- or Mn-porphyrins (section 4.). It seems that the presence of this good electron-donor ligand allows the heterolytic cleavage of the O—O bond of the oxidant which is necessary for the formation of a formal Fe(V)=O active intermediate. Thus, it is tempting to speculate that a possible role of the good electron-donor cysteinate ligand in cytochrome P-450 is **to permit also the heterolytic cleavage of the O—O bond of hydroperoxides.** In oxidations by O_2 itself, a key or beneficial role of imidazole was most often observed in model systems (see above). This suggests that a good electron-donating axial ligand of the iron (imidazole in model systems or cysteinate in cytochrome P-450) greatly facilitates the heterolytic cleavage of the O—O bond of the possible Fe(III)- (or Mn(III))—OOH intermediate derived from a two-electron reduction of O_2. Besides this very important role in the O—O bond cleavage, the axial iron ligands should modulate the chemoselectivity and regioselectivity of the reactive metal-oxo intermediate.

6. Metalloporphyrin catalysts for regioselective and asymmetric oxidations

Interesting applications of model systems using appropriate iron macrocyclic complexes concern the selective epoxidation of the *cis* double bond of the *trans-trans-cis*-1,5,9-cyclododecatriene (Groves and Nemo, 1983a) by PhIO in the presence of Fe(TMP)(Cl), and of the double bond at the junction of rings A and B of a steroid which is a precursor of a contraceptive drug (Rohde et al., 1985), by PhIO in the presence of an iron-phthalocyanin (Fig. 24).

Several results show that a change in the nature of the *ortho*-substituents of the *meso*-phenyl groups in Fe(or Mn)-tetraarylporphyrin catalysts leads to marked variations in the regioselectivity of n-alkane hydroxylation (Mansuy et al., 1982b; Khenkin and Shteinman, 1984; Nappa and Tolman, 1985; Suslick et al., 1985; Cook et al., 1986) or of alkene oxidation (epoxidation versus allylic hydroxylation) (Mansuy et al., 1984a; De Carvalho and Meunier, 1986). This is presumably due, at least in part, to an approach of the substrate which is restricted by the more or less bulky *ortho* substituents of the *meso*-phenyl groups. The most spectacular results were obtained with an Mn(III)-tetraarylporphyrin (TTPPP) for which the *meso*-aryl groups are 2,4,6-triphenyl-phenyl groups. The very bulky *ortho*-phenyl substituents lead to such a restricted approach of linear alkanes that their hydroxylation occurs mainly in ω and $\omega-1$ positions (Cook et al., 1986). For instance, n-hexane is hydroxylated at positions 1, 2 and 3 with a 2:38:60 ratio by the Mn(TPP)

$(OCOCH_3)-PhIO$ system and with a 19:62:18 ratio by the Mn(TTPPP) $(OCOCH_3)-PhIO$ system. It should be possible by increasing the steric hindrance around the metal in the porphyrin even more to reach the very high regioselectivity that is characteristic of cytochrome P-450-dependent ω-hydroxylases able to hydroxylate linear alkanes on their terminal CH_3. Another interesting use of modified porphyrin catalysts concerns the asymmetric epoxidation of alkenes. For that purpose, Fe-porphyrins bearing

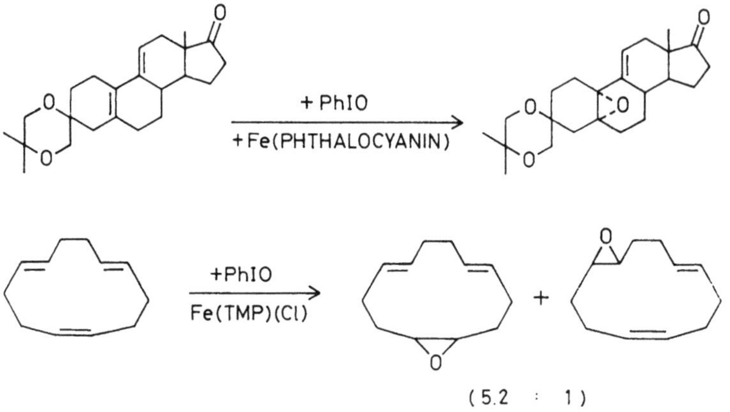

Fig. 24. Examples of regioselective epoxidations of polyolefins by PhIO catalyzed by iron-macrocycles.

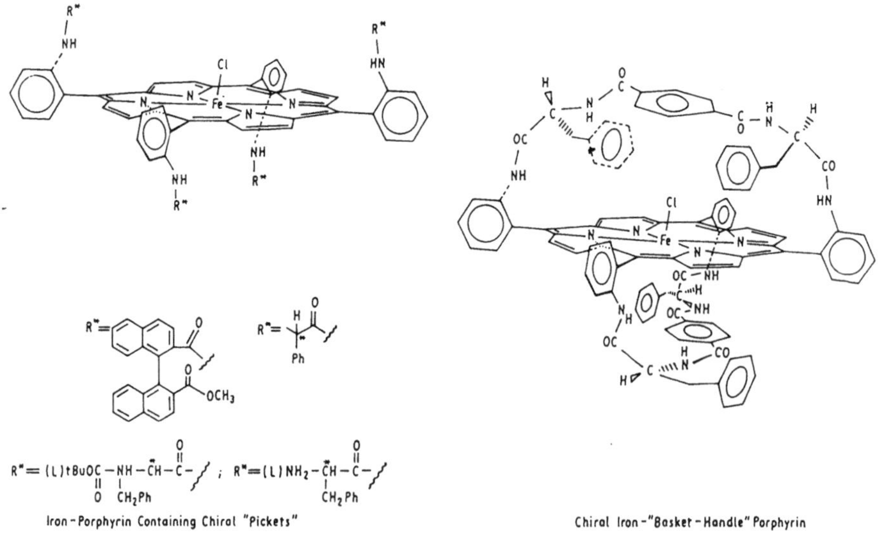

Fig. 25. Structure of chiral iron-porphyrins used as catalysts for asymmetric epoxidation of alkenes.

pickets containing chiral binaphthyl groups (Groves and Myers, 1983b) or amino acid residues (Mansuy et al., 1985b) or "basket-handles" containing chiral phenylalanine residues (Mansuy et al., 1985b), on both sides of the porphyrin ring, have been prepared (Fig. 25). They catalyze the asymmetric epoxidation of substituted styrenes with enantiomeric excesses between 12 and 50%.

7. Conclusion

A great deal of work has been devoted over the last seven years to the development of catalytically active synthetic model systems for cytochrome P-450. Several systems based on an Fe- or Mn-porphyrin and single oxygen atom donors such as PhIO, ClO^-, amine-oxides, alkylhydroperoxides and H_2O_2 have been found able to perform all the types of oxidation catalyzed by cytochrome P-450 including alkane hydroxylation, alkene epoxidation, aromatic ring hydroxylation and monooxygenation of thioethers. The reactions catalyzed by these model systems exhibit characteristics very similar to those of the corresponding reactions catalyzed by cytochrome P-450. This includes (i) a preference for the hydroxylation of tertiary C—H bonds over secondary and primary C—H bonds, (ii) high values of $k_H:k_D$ for the isotope effects observed during C—H bond hydroxylation, (iii) "NIH" shifts during hydroxylation of aromatic compounds. The ability of these model systems to mimic cytochrome P-450 closely is well illustrated by the study of olefin oxidation. The four reactions observed upon olefin oxidation by cytochrome P-450 — i. e. the epoxidation of the double bond, the hydroxylation of allylic C—H bonds, and, for the monosubstituted alkenes, the minor formation of aldehydes and the slow destruction of the iron-porphyrin catalyst with formation of *N*-alkylporphyrins — are all performed by the model systems with similar proportions.

Major improvements concerning the efficacy of the model systems have been obtained thanks to the use of Fe- or Mn-tetraarylporphyrins with aryl groups bearing electron-attracting substituents at *ortho* positions, such as TDCPP. These compounds have led to catalytic activities as high as 10—100 turnovers per second for olefin epoxidation. Moreover, they are very resistant towards oxidative degradation and can be used for at least 100,000 turnovers. In that regard, they are more potent and more resistant than cytochrome P-450 in the presence of the same oxidants. The use of strong electron-donating ligands for the Fe- or Mn-porphyrins, such as imidazole, has led to major progress. With such ligands, model systems using the readily available and simple oxidant H_2O_2 are able to perform alkene epoxidation and alkane hydroxylation with high conversions of the substrate and good yield of oxidized products. These model systems are now efficient enough to be used in preparative organic chemistry. The results obtained on the role of these imidazole ligands have suggested a possible role for the endogenous cysteinate ligand of cytochrome

P-450, which could be to permit the heterolytic cleavage of the O—O bond of H_2O_2 required for the formation of an Fe(V)=O species.

Several metalloporphyrin-based model system are now available for hydrocarbon oxidation by O_2 in the presence of a reducing agent. Catalytic activities and yields based on the reducing agent are still lower than those found for the corresponding cytochrome P-450-dependent reactions, though they now reach values of 9 turnovers of the catalyst per min and 50% (electronic yield) which are not too far from those reported for cytochrome P-450. Major advances with such O_2-dependent model systems as well as in the development of new catalysts for selective oxidations should be seen in the near future. First results have been obtained on asymmetric epoxidation of alkenes and regioselective hydroxylation of linear alkanes, but new catalysts for asymmetric C—H bond hydroxylation and regioselective oxidation of complex molecules remain to be prepared. The present model systems should already be very useful to study the oxidative metabolism of a drug and to isolate large amounts of its oxidized metabolites. Such progress would allow one to have specific catalysts for the selective preparation of each of these metabolites.

8. References

Ando, W., R. Tajima, and T. Takata, (1982), Tetrahedron Lett., **23**, 1685—1688.

Artaud, I., L. Devocelle, J. P. Battioni, J. P. Girault, and D. Mansuy, (1987), J. Am. Chem. Soc., **109**, 3782—3783.

Balch, A. L., Y. W. Chan, R. J. Cheng, G. N. La Mar, L. Latos-Grazynski, and M. W. Renner, (1984), J. Am. Chem. Soc., **106**, 7779—7785.

Balch, A. L., L. Latos-Grazynski, and M. W. Renner, (1985), J. Am. Chem. Soc., **107**, 2983—2985.

Balch, A. L. and M. W. Renner, (1986), J. Am. Chem. Soc., **108**, 2603—2608.

Battioni, J. P., I. Artaud, D. Dupre, P. Leduc, I. Akhrem, D. Mansuy, J. Fisher, R. Weiss, and I. Morgenstern-Badarau, (1986a), J. Am. Chem. Soc., **108**, 5598 to 5607.

Battioni, P., J. P. Renaud, J. F. Bartoli, and D. Mansuy, (1986b), J. Chem. Soc., Chem. Comm., 341—343.

Battioni, P., J. F. Bartoli, P. Leduc, M. Fontecave, and D. Mansuy, (1987), J. Chem. Soc., Chem. Comm., 791—792.

Bortolini, O. and B. Meunier, (1983), J. Chem. Soc., Chem. Comm., 1364—1366.

Bortolini, O. and B. Meunier, (1984), J. Chem. Soc. Perkin Trans II, 1967—1970.

Bortolini, O., M. Ricci, B. Meunier, P. Friant, I. Ascone, and J. Goulon, (1986), Nouv. J. Chim., **10**, 39—49.

Boso, B., G. Lang, T. J. McMurry, and J. T. Groves, (1983), J. Chem. Phys., **79**, 1122—1126.

Breslow, R. and S. H. Gellman, (1982), J. Chem. Soc., Chem. Comm., 1400—1401.

Breslow, R. and S. H. Gellman, (1983), J. Am. Chem. Soc., **105**, 6728—6729.

Brothers, P. J. and J. P. Collman, (1986), Acc. Chem. Res., **19**, 209—215.

Calderwood, T. S. and T. C. Bruice, (1986), Inorg. Chem., **25**, 3722—3724.

Camenzind, M. J., F. J. Hollander, and C. L. Hill, (1982), Inorg. Chem., **21**, 4301 to 4308.

Chang, C. K. and F. Ebina, (1981), J. Chem. Soc., Chem. Comm., 778—779.

Che, C. M. and W. C. Chung, (1986), J. Chem. Soc., Chem. Comm., 386—388.

COLLMAN, J. P. and T. N. SORRELL, (1977), in: Drug Metabolism Concepts, (D. M. JERINA, ed.), ACS Symposium series 44, Amer. Chem. Soc., Washington D. C., 27—45.
COLLMAN, J. P., T. KODADEK, S. A. RAYBUCK, and B. MEUNIER, (1983), Proc. Natl. Acad. Sci. USA, **80**, 7039—7041.
COLLMAN, J. P., J. I. BRAUMAN, B. MEUNIER, S. A. RAYBUCK, and T. KODADEK, (1984), Proc. Natl. Acad. Sci. USA, **81**, 3245—3248.
COLLMAN, J. P., J. I. BRAUMAN, B. MEUNIER, T. HAYASHI, T. KODADEK, and S. A. RAYBUCK, (1985a), J. Am. Chem. Soc., **107**, 2000—2005.
COLLMAN, J. P., T. KODADEK, S. A. RAYBUCK, J. I. BRAUMAN, and L. M. PAPAZIAN, (1985b), J. Am. Chem. Soc., **107**, 4343—4345.
COLLMAN, J. P., T. KODADEK, and J. I. BRAUMAN, (1986a), J. Am. Chem. Soc., **108**, 2588—2594.
COLLMAN, J. P., P. D. HAMPTON, and J. I. BRAUMAN, (1986b), J. Am. Chem. Soc., **108**, 7861—7862.
COOK, B. R., T. J. REINERT, and K. S. SUSLICK, (1986), J. Am. Chem. Soc., **108**, 7281 to 7286.
CREAGER, S. E., S. A. RAYBUCK, and R. W. MURRAY, (1986), J. Am. Chem. Soc., **108**, 4225—4227.
DANIELSSON, H. and K. WIKVALL, (1976), FEBS Lett., **66**, 299—302.
DAWSON, J. H., L. S. KAU, J. E. PENNER-HAHN, M. SONO, K. S. EBLE, G. S. BRUCE, L. P. HAGER, and K. O. HODGSON, (1986), J. Am. Chem. Soc., **108**, 8114—8116.
DE CARVALHO, M. E. and B. MEUNIER, (1986), Nouv. J. Chim., **10**, 223—227.
DE POORTER, B. and B. MEUNIER, (1984), Tetrahedron Lett., **25**, 1895—1896.
DE POORTER, B. and B. MEUNIER, (1985a), Nouv. J. Chim., **9**, 393—394.
DE POORTER, B., M. RICCI, and B. MEUNIER, (1985), Tetrahedron Lett., **26**, 4459—4462.
DE POORTER, B. and B. MEUNIER, (1985b), J. Chem. Soc. Perkin Trans II, 1735—1740.
DICKEN, C. M., F. L. LU, M. W. NEE, and T. C. BRUICE, (1985), J. Am. Chem. Soc., **107**, 5776—5789.
DOLPHIN, D., B. R. JAMES, and T. LEUNG, (1983), Inorg. Chim. Acta., **79**, 25—27.
ENGLISH, D. R., D. N. HENDRICKSON, and K. S. SUSLICK, (1983), Inorg. Chem., **22**, 367—368.
FONTECAVE, M. and D. MANSUY, (1984a), Tetrahedron, **40**, 4297—4311.
FONTECAVE, M. and D. MANSUY, (1984b), J. Chem. Soc., Chem. Comm., 879—881.
GROVES, J. T., T. E. NEMO, and R. S. MYERS, (1979), J. Am. Chem. Soc., **101**, 1032 to 1033.
GROVES, J. T. and W. J. KRUPER, (1979), J. Am. Chem. Soc., **101**, 7613—7615.
GROVES, J. T., W. J. KRUPER, and R. C. HAUSHALTER, (1980), J. Am. Chem. Soc., **102**, 6375—6377.
GROVES, J. T., R. C. HAUSHALTER, M. NAKAMURA, T. E. NEMO, and B. J. EVANS, (1981), J. Am. Chem. Soc., **103**, 2884—2886.
GROVES, J. T. and R. S. MYERS, (1983), J. Am. Chem. Soc., **105**, 5791—5796.
GROVES, J. T. and T. E. NEMO, (1983a), J. Am. Chem. Soc., **105**, 5786—5791.
GROVES, J. T. and T. E. NEMO, (1983b), J. Am. Chem. Soc., **105**, 6243—6248.
GROVES, J. T., Y. WATANABE, and T. J. MCMURRY, (1983), J. Am. Chem. Soc., **105**, 4489—4490.
GROVES, J. T. and D. V. SUBRAMANIAN, (1984), J. Am. Chem. Soc., **106**, 2177—2181.
GROVES, J. T. and J. A. GILBERT, (1986), Inorg. Chem., **25**, 123—125.
GROVES, J. T. and Y. WATANABE, (1986a), J. Am. Chem. Soc., **108**, 507—508.
GROVES, J. T. and Y. WATANABE, (1986b), J. Am. Chem. Soc., **108**, 7834—7836.
GUENGERICH, F. P. and T. L. MACDONALD, (1984), Acc. Chem. Res., **17**, 9—16.
GUENGERICH, F. P., R. J. WILLARD, J. P. SHEA, L. E. RICHARDS, and T. L. MACDONALD, (1984), J. Am. Chem. Soc., **106**, 6446—6447.
GUILMET, E. and B. MEUNIER, (1980), Tetrahedron Lett., **21**, 4449—4450.
GUILMET, E. and B. MEUNIER, (1982), Nouv. J. Chim. **6**, 511—513.
GUSTAFSSON, J. A. and J. BERGMAN, (1976), FEBS Lett., **70**, 276—280.

GUSTAFSSON, J. A., E. G. HRYCAY, and L. ERNSTER (1976), Arch. Biochem. Biophys., **174**, 440–453.
GUSTAFSSON, J. A., L. RONDAHL, and J. BERGMAN, (1979), Biochemistry, **18**, 865–870.
HEIMBROOK, D. C., R. I. MURRAY, K. D. EGEBERG, S. G. SLIGAR, M. W. NEE, and T. C. BRUICE, (1984), J. Am. Chem. Soc., **106**, 1514–1515.
HILL, C. L. and B. C. SCHARDT, (1980), J. Am. Chem. Soc., **102**, 6374–6375.
HILL, C. L. and J. A. SMEGAL, (1982), Nouv. J. Chim., **6**, 287–289.
HILL, C. L., J. A. SMEGAL, and T. J. HENLY, (1983), J. Org. Chem., **48**, 3277–3281.
HOFFMAN, B. M., J. E. ROBERTS, C. H. KANG, and E. MARGOLIASH, (1981), J. Biol. Chem., **256**, 6556–6564.
HRYCAY, E. G., J. A. GUSTAFSSON, M. INGELMAN-SUNDBERG, and L. ERNSTER, (1975), Biochem. Biophys. Res. Commun. **66**, 209–216.
JERINA, D. M. and J. W. DALY, (1974), Science, **185**, 573–581.
KADLUBAR, F. F., K. C. MORTON, and D. M. ZIEGLER, (1973), Biochem. Biophys. Res. Commun., **54**, 1255–1261.
KHENKIN, A. M. and A. A. SHTEINMAN, (1984), J. Chem. Soc., Chem. Comm., 1219 to 1220.
LANCON, D., P. COCOLIOS, R. GUILARD, and K. M. KADISH, (1984), J. Am. Chem. Soc., **106**, 4472–4478.
LEE, W. A., T. S. CALDERWOOD, and T. C. BRUICE, (1985), Proc. Natl. Acad. Sci., USA, **82**, 4301–4305.
LEE, W. A. and T. C. BRUICE, (1985), J. Am. Chem. Soc., **107**, 513–514.
LEUNG, T., B. R. JAMES, and E. DOLPHIN, (1983), Inorg. Chim. Acta, **79**, 180–181.
LICHTENBERGER, F., W. NASTAINCZYK, and V. ULLRICH, (1976), Biochem. Biophys. Res. Commun., **70**, 939–946.
LINDSAY-SMITH, J. R., R. E. PIGGOTT, and P. R. SLEATH, (1982), J. Chem. Soc., Chem. Comm., 55–56.
LINDSAY-SMITH, J. R. and P. R. SLEATH, (1982), J. Chem. Soc. Perkin Trans II, 1009 to 1015.
LINDSAY-SMITH, J. R. and P. R. SLEATH, (1983a), J. Chem. Soc. Perkin Trans II, 621–628.
LINDSAY-SMITH, J. R. and P. R. SLEATH, (1983b), J. Chem. Soc. Perkin Trans II, 1165–1169.
LINDSAY-SMITH, J. R., M. W. NEE, J. B. NOAR, and T. C. BRUICE, (1984), J. Chem. Soc., Perkin Trans II, 255–260.
LINDSAY-SMITH, J. R. and D. N. MORTIMER, (1985a), J. Chem. Soc., Chem. Comm., 64–65.
LINDSAY-SMITH, J. R. and D. N. MORTIMER, (1985b), J. Chem. Soc., Chem. Comm., 410–411.
MANSUY, D., J. F. BARTOLI, J. C. CHOTTARD, and M. LANGE, (1980), Angew. Chem. Int. Ed. Engl., **198**, 909–910.
MANSUY, D., (1981), in: Reviews in Biochemical Toxicology, (E. HODGSON, J. R. BEND, R. M. PHILPOT, eds.), Elsevier, New York, vol. 3, 283–320.
MANSUY, D., J. P. BATTIONI, D. DUPRE, E. SARTORI, and G. CHOTTARD, (1982a), J. Am. Chem. Soc., **104**, 6159–6161.
MANSUY, D., J. F. BARTOLI, and M. MOMENTEAU, (1982b), Tetrahedron Lett., **23**, 2781–2784.
MANSUY, D., (1983), in: The Coordination Chemistry of Metalloenzymes, (I. BERTINI, R. S. DRAGO, and C. LUCHINAT, eds.), D. Reidel Publishing Company, 343–361.
MANSUY, D., M. FONTECAVE, and J. F. BARTOLI, (1983), J. Chem. Soc., Chem. Comm., 253–254.
MANSUY, D., J. LECLAIRE, M. FONTECAVE, and P. DANSETTE, (1984a), Tetrahedron, **40**, 2847–2857.
MANSUY, D., J. P. MAHY, A. DUREAULT, G. BEDI, and P. BATTIONI, (1984b), J. Chem. Soc., Chem. Comm., 1161–1163.

Mansuy, D., P. Battioni, and J. P. Renaud, (1984c), J. Chem. Soc., Chem. Comm., 1255–1257.
Mansuy, D., J. Leclaire, M. Fontecave, and M. Momenteau, (1984d), Biochem. Biophys. Res. Commun., **119**, 319–325.
Mansuy, D. and P. Battioni, (1985), in: Drug Metabolism, (G. Siest, ed.), Pergamon Press, Oxford, New York, 195–203.
Mansuy, D., L. Devocelle, I. Artaud, and J. P. Battioni, (1985a), Nouv. J. Chim., **9**, 711–716.
Mansuy, D., P. Battioni, J. P. Renaud, and P. Guerin, (1985b), J. Chem. Soc., Chem. Comm., 155–156.
Mansuy, D. and P. Battioni, (1986), Bull. Soc. Chim. Belg., **95**, 959–971.
Mashiko, T., D. Dolphin, T. Nakano, and T. G. Traylor, (1985), J. Am. Chem. Soc., **107**, 3735–3736.
Meunier, B., E. Guilmet, M. E. de Carvalho, and R. Poilblanc, (1984), J. Am. Chem. Soc., **106**, 6668–6675.
Meunier, B., (1986), Bull. Soc. Chim. France **4**, 578–594.
Miwa, G. T. and A. Y. H. Lu, (1986), in: Cytochrome P-450, Structure, Mechanism, and Biochemistry, (P. R. Ortiz de Montellano, ed.), Plenum Press, New York and London, 77–88.
Montanari, F., M. Penso, S. Quici, and P. Vigano, (1985), J. Org. Chem., **50**, 4888 to 4893.
Mori, T., T. Santa, and M. Hirobe, (1985), Tetrahedron Lett., **26**, 5555–5558.
Nordblom, G. D., R. E. White, and M. J. Coon, (1976), Arch. Biochem. Biophys., **175**, 524–533.
Nappa, M. J. and C. A. Tolman, (1985), Inorg. Chem., **24**, 4711–4719.
Nee, M. W. and T. C. Bruice, (1982), J. Am. Chem. Soc., **104**, 6123–6125.
Nolte, R. J., J. A. Razenberg, and R. Schuurman, (1986), J. Am. Chem. Soc., **108**, 2751–2752.
Oae, S., Y. Watanabe, and K. Fujimori, (1982), Tetrahedron Lett., **23**, 1189–1192.
Ortiz de Montellano, P. R., (1986), in: Cytochrome P-450, Structure, Mechanism, and Biochemistry, (P. R. Ortiz de Montellano, ed.), Plenum Press, New York and London, 217–271.
Ortiz de Montellano, P. R. and N. O. Reich, (1986), in: Cytochrome P-450, Structure, Mechanism and Biochemistry, (P. R. Ortiz de Montellano, ed.) Plenum Press, New York and London, 273–314.
Penner-Hahn, J. E., T. J. McMurry, M. Renner, L. Latos-Grazynski, K. S. Eble, I. M. Davis, A. L. Balch, J. T. Groves, J. H. Dawson, and K. O. Hodgson, (1983), J. Biol. Chem., **258**, 12761–12764.
Penner-Hahn, J. E., K. S. Eble, T. J. McMurry, M. Renner, A. L. Balch, J. T. Groves, J. H. Dawson, and K. O. Hodgson, (1986), J. Am. Chem. Soc., **108**, 7819 to 7825.
Perree-Fauvet, M. and A. Gaudemer, (1981), J. Chem. Soc., Chem. Comm., 874–875.
Peterson, J. A. and R. A. Prough, (1986), in: Cytochrome P-450, Structure, Mechanism and Biochemistry, (P. R. Ortiz de Montellano, ed.), Plenum Press, New York and London, 89–117.
Poulos, T. L., B. C. Finzel, I. C. Gunsalus, G. C. Wagner, and J. Kraut, (1985), J. Biol. Chem., **260**, 16122–16130.
Powell, M. F., E. F. Pai, and T. C. Bruice, (1984), J. Am. Chem. Soc., **106**, 3277 to 3285.
Rahimtula, A. D. and P. J. O'Brien, (1974), Biochem. Biophys. Res. Commun., **60**, 440–447.
Rein, H., C. Jung, O. Ristau, and J. Friedrich, (1984), in: Cytochrome P-450, (K. Ruckpaul and H. Rein, eds.), Akademie-Verlag, Berlin, 163–249.
Renaud, J. P., P. Battioni, J. F. Bartoli, and D. Mansuy, (1985), J. Chem. Soc., Chem. Comm., 888–889.

ROBERTS, J. E., B. M. HOFFMAN, R. RUTTER, and L. P. HAGER, (1981), J. Biol. Chem., **256**, 2118–2121.
ROHDE, R., G. NEEF, G. SAUER, and R. WIECHERT, (1985), Tetrahedron Lett., **26**, 2069–2072.
SANTA, T., T. MORI, and M. HIROBE, (1985), Chem. Pharm. Bull., **33**, 2175–2178.
SCHAPPACHER, M., R. WEISS, R. MONTIEL-MONTOYA, A. TRAUTWEIN, and A. TABARD, (1985), J. Am. Chem. Soc., **107**, 3736–3738.
SCHARDT, B. C., F. J. HOLLANDER, and C. L. HILL, (1981), J. Chem. Soc., Chem. Comm., 765–766.
SCHARDT, B. C., F. J. HOLLANDER, and C. L. HILL, (1982), J. Am. Chem. Soc., **104**, 3964–3972.
SEVIN, A. and M. FONTECAVE, (1986), J. Am. Chem. Soc., **108**, 3266–3272.
SHANNON, P. and T. C. BRUICE, (1981), J. Am. Chem. Soc., **103**, 4580–4582.
SHARROCK, M., P. G. DEBRUNNER, C. SCHULZ, J. D. LIPSCOMB, V. MARSHALL, and I. C. GUNSALUS, (1976), Biochem. Biophys. Acta, **420**, 8–26.
SHIMIDZU, T., T. IYODA, and N. KANDA, (1981), J. Chem. Soc., Chem. Comm., 1206 to 1207.
SMEGAL, J. A., B. C. SHARDT, and C. L. HILL, (1983), J. Am. Chem. Soc., **105**, 3510 to 3515.
SMEGAL, J. A. and C. L. HILL, (1983a), J. Am. Chem. Soc., **105**, 3515–3521.
SMEGAL, J. A. and C. L. HILL, (1983b), J. Am. Chem. Soc., **105**, 2920–2922.
SUSLICK, K., B. COOK, and M. FOX, (1985), J. Chem. Soc., Chem. Comm., 580–582.
TABUSHI, I. and N. KOGA, (1979a), Tetrahedron Lett., **38**, 3681–3684.
TABUSHI, I. and N. KOGA, (1979b), J. Am. Chem. Soc., **101**, 6456–6458.
TABUSHI, I. and A. YAZAKI, (1981), J. Am. Chem. Soc., **103**, 7371–7373.
TABUSHI, I. and K. MIRIMITSU, (1984), J. Am. Chem. Soc., **106**, 6871–6873.
TABUSHI, I., M. KODERA, and M. YOKAYAMA, (1985), J. Am. Chem. Soc., **107**, 4466 to 4473.
TABUSHI, I. and M. KODERA, (1986), J. Am. Chem. Soc., **108**, 1101–1103.
TAKATA, T. and W. ANDO, (1983), Tetrahedron Lett., **34**, 3631–3634.
TATSUNO, Y., A. SEKIYA, K. TANI, and T. SAITO, (1986), Chemistry Lett., 889–892.
TRAYLOR, P. S., D. DOLPHIN, and T. G. TRAYLOR, (1984a), J. Chem. Soc., Chem. Comm., 279–280.
TRAYLOR, T. G., W. A. LEE, and D. V. STYNES, (1984b), J. Am. Chem. Soc., **106**, 755 to 764.
TRAYLOR, T. G., J. C. MARSTERS, T. NAKANO, and B. E. DUNLAP, (1985), J. Am. Chem. Soc., **107**, 5537–5539.
TRAYLOR, T. G., T. NAKANO, B. E. DUNLAP, P. S. TRAYLOR, and D. DOLPHIN, (1986a), J. Am. Chem. Soc., **108**, 2782–2784.
TRAYLOR, F. G., Y. IAMAMOTO, and T. NAKANO, (1986b), J. Am. Chem. Soc., **108**, 3529–3531.
VAN DER MADE, A. W., J. W. M. SMEETS, R. J. M. NOLTE, and W. DRENTH, (1983), J. Chem. Soc., Chem. Comm., 1204–1206.
WHITE, R. E. and M. J. COON, (1980), Ann. Rev. Biochem. **49**, 315–356.
YUAN, L. C., T. S. CALDERWOOD, and T. C. BRUICE, (1985), J. Am. Chem. Soc., **107**, 8273–8274.
YUAN, L. C. and T. C. BRUICE, (1985a), J. Chem. Soc., Chem. Comm., 868–869.
YUAN, L. C. and T. C. BRUICE, (1985b), Inorg. Chem. **24**, 986–987.
YUAN, L. C. and T. C. BRUICE, (1986), J. Am. Chem. Soc., **108**, 1643–1650.

Chapter 3

Structural Multiplicity and Functional Versatility of Cytochrome P-450

F. Peter GUENGERICH

1. Introduction

About forty years ago some of the first mixed-function oxidase reactions were observed. Cytochrome P-450 (P-450) was found to catalyze many of these transformations. The interest in P-450 reactions has increased because of the implications of these transformations in a number of fields such as pharmacology, drug metabolism, chemical carcinogenesis, endocrinology and toxicology.

Today we recognize that many forms of P-450 exist and have made some advances in associating individual forms with certain reactions. We also understand some of the factors involved in catalysis of the reactions and the regulatory factors which influence the amounts and activities of the P-450s in mammalian systems.

2. Individual forms of P-450

2.1. General

In the past decade many individual forms of P-450 have been characterized, probably more than most workers previously thought would exist. With technical advances in protein chemistry and the introduction of recombinant DNA technology, the primary sequences of these P-450s (or at least mRNAs related to them) are rapidly becoming available (Black and Coon, 1986). The different forms of microsomal P-450 are tabulated here (Tables 1—5).

Several points should be made here. The term "forms" is used here instead of "isozymes". While the different forms of P-450 discussed here are probably all distinct gene products (as opposed to being the result of post-translational modification), the term "isozyme" implies that all of the P-450s catalyze the same reaction. Today it is apparent that different P-450s catalyze different reactions, as shown by substrate selectivity and regio- and stereoselective oxidation of a given substrate by different P-450s. Tables 1—5 list the P-450s characterized in rat, rabbit, and mouse liver microsomes, extrahepatic microsomal P-450s, and human microsomal P-450s. No attempt has been made to include mitochondrial P-450s (which have been treated elsewhere and have relatively distinct functions and electron transfer mechanisms, Lambeth et al., 1982), P-450s from other animal or plant sources, or P-450s in prokaryotes and yeasts. Only P-450s are listed which either have been purified to apparent electrophoretic homogeneity, have been characterized by cDNA cloning and sequencing, or, in a few cases, show extensive immunochemical similarity to known forms. When a P-450 form has been isolated in the author's own laboratory, the nomenclature applied there is used. Attempts have been made to list all other preparations with the same properties. The reader should consider the possibility that some of the P-450 forms may prove to be separable into more than one distinct protein in the future, and, in the lack of available evidence, the effects of any post-translational modification are not really considered here.

The classification manner in which the P-450s will be discussed is somewhat arbitrary. The scheme does not necessarily imply sequence similarity or even orthology of function. Efforts are being made elsewhere to classify the P-450 genes on the basis of sequence information (NEBERT et al., 1987)

2.2. Barbiturate-inducible P-450s

Some of P-450s in this category (rat P-450_{PB-B}, rabbit P-450 2) are present in livers of untreated animals only at very low levels and the levels can be raised as much as 100-fold by administration of inducers such as the prototype phenobarbital (THOMAS et al., 1979; PICKETT et al., 1981; GUENGERICH et al., 1982b). The sequences of these two particular P-450s are very similar to each other and an apparent human ortholog has been identified (P-450 pHB1, PHILLIPS et al., 1985). This is a multi-gene family and at least one related gene is expressed to give P-450_{PB-D}, which differs in only 13 residues but has roughly an order of magnitude less catalytic activity towards several substrates (YUAN et al., 1983). Other strain variants may exist. P-450_{PB-B} and P-450 2 are present in lungs of untreated rats and rabbits, respectively, and are not inducible in that tissue.

Other forms of P-450 (e.g., rat P-450_{PB-C}) are present in substantial amounts in untreated animals and are induced only about two-fold by administration of phenobarbital.

Several forms of P-450 have been isolated from phenobarbital-treated mice but the inducibility of these proteins has not been established (HUANG et al., 1976).

2.3. Polycyclic aromatic hydrocarbon-inducible P-450s

The mammalian species examined to date contain two genes in this family. Rat P-$450_{\beta NF-B}$, mouse P_1-450, and rabbit P-450 6 appear to be orthologs as judged by sequence similarity, spectral properties (ferrous carbonyl λ_{max} at 447 nm), and similar substrate specificities (however, note the discrepancies with regard to warfarin hydroxylation — KAMINSKY et al., 1980, 1984). The coding sequence has been inserted into a yeast and catalytic activity can be expressed (SAKAKI et al., 1985, 1986). The human ortholog — human P_1-450 — has not been isolated but its existence, at least in extrahepatic tissues, seems very likely from the results of immunochemical studies and a cDNA clone has been sequenced to deduce the primary amino acid sequence (JAISWAL et al., 1985b).

The product of the other gene (rat P-450_{ISF-G}, rabbit P-450 4, mouse P_3-450) has been isolated from several species. It also has a ferrous carbonyl λ_{max} at 447 nm but is isolated largely in the high-spin form (although conversion of

P-450 iron from the low- to high-spin state has been postulated to raise the oxidation-reduction potential, it should be noted that the potentials of some of the high-spin proteins are quite negative — GUENGERICH, 1983; HUANG et al., 1986—and this does not appear to be the general case). The P-450s isolated from the different species have considerable sequence similarity and all appear to be active in catalyzing acetanilide 4-hydroxylation. The human ortholog is probably $P\text{-}450_{PA}$, as we have tentatively concluded from a series of immunochemical studies (GUENGERICH et al., 1987).

The induction patterns of the two P-450s are similar. The suggestion has been made that induction is coordinate (HARDWICK et al., 1985). However, the induction of the two P-450s shows a temporal dissociation (FAGAN et al., 1986), $P\text{-}450_{ISF-G}$ (unlike $P\text{-}450_{\beta NF-B}$) is not induced in extrahepatic tissues (GOLDSTEIN and LINKO, 1984) or cultured cells (STEWARD et al., 1985), and the suggestion has been made that induction of P_3-450 does not involve a role for the so-called Ah receptor (COOK and HODGSON, 1986).

2.4. Steroid-inducible P-450s

The first family considered here is inducible by the classic inducer pregnenolone 16α-carbonitrile (PCN) or by dexamethasone or several other steroids (ELSHOURBAGY and GUZELIAN, 1980). Phenobarbital induces in vivo but not in cell culture. Triacetyloleandomycin induces through a mechanism which probably involves stabilization of mRNA or protein as opposed to enhanced rates of transcription (SCHUETZ et al., 1986; WATKINS et al., 1986). The family includes rat $P\text{-}450_{PCN-E}$, rabbit P-450 3c, and human $P\text{-}450_{NF}$. At least two mRNAs are present in the rat and the relationship to protein structure and catalytic activity is unclear (GONZALEZ et al., 1986b). Another nagging problem is the low level of catalytic activity in purified preparations (ELSHOURBAGY and GUZELIAN, 1980; GUENGERICH et al., 1986a). Antibodies inhibit microsomal activities thought to be associated with this system, but the possibility exists that a related, inactive P-450 has been purified. Alternatively, the active protein may be inactivated during purification. WAXMAN et al. (1985b) have postulated that dephosphorylation may be involved in deactivation but we have been unable to substantiate this finding in the case of human $P\text{-}450_{NF}$ in our own laboratory (BORK, R. W. and GUENGERICH, F. P., unpublished).

Immunochemical and catalytic activity measurements suggest that $P\text{-}450_{PCN-E}$ is present throughout life in male rats; testosterone is required for neonatal imprinting and for maintenance in adulthood (WAXMAN et al., 1985a; DANNAN et al., 1986). Females have the enzyme in the neonatal period but lose it during puberty. The enzyme is inducible by PCN, dexamethasone, phenobarbital, or triacetyloleandomycin in either sex. Testosterone treatment shows an effect only in gonadectomized male rats. Whether all of these phenomena can be explained by changes in a single protein is still a matter for investigation (GONZALEZ et al., 1986b).

Several other rat P-450s are influenced by steroid hormones. Rat P-450_{UT-A} appears during puberty in male rats and disappears during old age; it is never present in females (WAXMAN et al., 1985a; KAMATAKI et al., 1985). Levels of the enzyme are dependent upon testosterone (both for neonatal imprinting and for maintenance (WAXMAN et al., 1985a; DANNAN et al., 1986)). The orthologs of rat P-450_{UT-A}, based on sequence similarity, are rabbit P-450 1 and human P-450_{MP-1} and P-450_{MP-2}; however, these P-450s are apparently not regulated by hormones. Rabbit P-450 1 and human P-450_{MP-1} and P-450_{MP-2} are coded for by multi-gene families and the extent of multiplicity in these families is as yet unclear.

Rat P-450_{UT-I} is female-specific and does not appear in males, except those which are very old (KAMATAKI et al., 1985). It appears in females during puberty and is partially suppressed by neonatal ovariectomy (WAXMAN et al., 1985a). The enzyme shows immunochemical cross-reaction with male-specific rat P-450_{UT-A}, although specific antibodies for each protein can be prepared by cross-adsorption.

This general subject is discussed in further detail in chapter 5 by MORGAN and GUSTAFSSON.

2.5. Ethanol-inducible P-450s

Rat P-450j and rabbit P-450 3a belong to this family. The structural human P-450 ortholog is termed human P-450j, which has been sequenced but not isolated in a functionally-active form. This appears to be a relatively simple gene family in rats and humans and may only code for a single mRNA (SONG et al., 1987).

2.6. Clofibrate-inducible P-450s

GIBSON et al. (1982) have purified a form of rat P-450 induced by administration of certain hypolipidemic agents which also cause peroxisomal proliferation. The coding sequence of a cDNA clone has recently been published (HARDWICK et al., 1987). The cDNA has been inserted in a yeast and activity is expressed.

2.7. Other P-450s (experimental animals)

Other P-450s which have been characterized are listed in the tables and their properties will not be reiterated. Some of these are induced to small extents by certain compounds. In particular, P-450_{UT-F} (whose sequence has recently been established — NEBERT et al., 1987) is induced somewhat by barbiturates or polycyclic aromatic hydrocarbons and its level is somewhat higher in females (particularly prepubertal females) than males.

Table 1. Rat Liver P-450s

P-450	Other similar preparations[a]	Inducibility	Characteristic catalytic reactions	Other distinguishing features
1. P-450_{UT-A} (GUENGERICH et al., 1982a)	P-450h; P-450_{PB-2c}; P-450male; P-450M-1; P-450RLM5; P-450A; P-450 16α; P-450CC-25	Elevated by testosterone in castrated males (not females) (WAXMAN et al., 1985a; DANNAN et al., 1986); "typical" P-450 inducers suppress this P-450 (GUENGERICH et al., 1982a; DANNAN et al., 1983)	Testosterone 2α,16α-hydroxylation (WAXMAN et al., 1985a), estradiol 2- and 4-hydroxylation (DANNAN et al., 1986), ethylmorphine N-demethylation (GUENGERICH et al., 1982a), nifedipine oxidation (GUENGERICH et al., 1986a)	Male-specific, shows developmental pattern (WAXMAN et al., 1985a); allelozymic forms exist (RAMPERSAUD et al., 1985) [b] YOSHIOKA et al., 1987
2. P-$450_{\beta NF-B}$ (GUENGERICH 1977, 1978; RYAN et al., 1979; GUENGERICH et al., 1982a)	P-450c; P-448_2; P-448IId; PcB-448L; MC-P-448; MC-1; P-450_{MC}; Form 5; P-446; P-448; MC-P_{448}; P-447; P-$450B_2$BNF; P-450_{MC-I} and P-450_{MC-II}	Highly inducible by polycyclic aromatic hydrocarbons (i. e., 3-methylcholanthrene, β-naphthoflavone), co-planar polyhalogenated biphenyls, isosafrole, phenothiazine (THOMAS et al., 1983)	7-Ethoxyresorufin O-deethylation, benzo[a]-pyrene hydroxylation (-3, -9) and epoxidation (-7,8)	Ferrous-carbonyl λ_{max} at 447 nm [b] YABUSAKI et al., 1984
3. P-450_{PB-B} (GUENGERICH, 1977, 1978; GUENGERICH et al., 1982a; RYAN et al., 1979)	P-450b; P-450PB-4; PB P-450; P-450PB—1; P-450 "C"; P-450_{PB}; PB P_{450}; P-450_4; PB-1; P-450; P-450(L), P-450(M), and P-450(H); P-450a, P-450b, and P-450c; P-$450B_2$PB; P-$450_{b_{LE}}$	Highly inducible by phenobarbital and other barbiturates, *trans*-stilbene oxide, 1,1-di(p-chlorophenyl)-2,2-dichloroethylene (DDE) (YOSHIOKA et al., 1984), *o*-substituted polyhalogenated biphenyls (DANNAN et al., 1983)	Benzphetamine N-demethylation, aminopyrine N-demethylation (GUENGERICH et al., 1982a), 7-pentoxyresorufin O-dealkylation (LUBET et al., 1985)	Related forms may be expressed in liver (RYAN et al., 1982b) [b] MIZUKAMI et al., 1983; YUAN et al., 1983; FUJII-KURIYAMA et al., 1982

4. P-450$_{PB-C}$ (GUENGERICH et al., 1982a)	P-450PB-1; P-450k	Phenobarbital-increase is only about 2-fold (cf. P-450$_{PB-B}$); complex polyhalogenated biphenyl mixtures do not induce (DANNAN et al., 1983)	R-Warfarin 7-hydroxylation (GUENGERICH et al., 1982a)	[b] GONZALEZ et al., 1986a
5. P-450$_{PB-D}$ (GUENGERICH 1977, 1978; RYAN et al., 1982a)	P-450e; P-450PB-5	Highly inducible by phenobarbital, other barbiturates, and *o*-substituted polyhalogenated biphenyls—induction pattern similar to P-450$_{PB-B}$	Specificity for substrates similar to P-450$_{PB-B}$ but activity towards all substrates is much lower	[b] MIZUKAMA et al., 1983; YUAN et al., 1983
6. P-450$_{PCN-E}$ (ELSHOURBAGY and GUZELIAN, 1980; GUENGERICH et al., 1982a)	P-450p; P-450PB-2a; P-450$_{PCN}$; PCN-P-450	PCN, dexamethasone and other steroids, phenobarbital, triacetyloleandomycin (SCHUETZ et al., 1986)	Testosterone 6β-hydroxylation, nifedipine oxidation, S-mephenytoin 4-hydroxylation (SHIMADA and GUENGERICH, 1985), warfarin 10-hydroxylation (GUENGERICH et al., 1982a)	Sex-specific (for males) in adults; females express enzyme at birth but then stop at puberty; inducible in both sexes (WAXMAN et al., 1985a) [b] GONZALEZ et al., 1985
7. P-450$_{PCN-2}$ (GONZALEZ et al., 1986b)		PCN, dexamethasone, triacetyloleandomycin (?)		Evidence for its existence based on isolation of second cDNA clone related to P-450$_{PCN}$ [b] GONZALEZ et al., 1986b

Table 1. — Continued

P-450	Other similar preparations[a]	Inducibility	Characteristic catalytic reactions	Other distinguishing features
8. P-450_{UT-F} (RYAN et al., 1979; GUENGERICH et al., 1982a	P-450a; P-450PB-3; Form 1	Weak induction by barbituates or polycyclic hydrocarbons	Testosterone 7α-hydroxylation	Ferrous carbonyl λ_{max} at 452 nm [b] NEBERT et al., 1987
9. P-450_{ISF-G} (RYAN et al., 1980; GUENGERICH et al., 1982a)	P-450d; P-448IIa; PCB P-450H; MC-2; Form 4; P-450; P-450_{HCB}	Similar to P-$450_{\beta NF-B}$ (vide supra)	Estradiol 2-hydroxylation, acetanilide 4-hydroxylation (SUNDHEIMER et al., 1983), N-hydroxylation of several aromatic amines (HAMMONS et al., 1985; KAMATAKI et al., 1983a)	Ferrous carbonyl λ_{max} at 447 nm; distinct spectral complexes form with isosafrole metabolites, isolated in high-spin state [b] KAWAJIRI et al., 1984
10. P-450_{UT-II} (LARREY et al., 1984)		Not apparently inducible (LARREY et al., 1984)	Debrisoquine 4-hydroxylation, sparteine Δ^2- hydroxylation, bufuralol 1'-hydroxylation (LARREY et al., 1984)	
11. P-450_{UT-I} (KAMATAKI et al., 1983b; WAXMAN et al., 1985a)	P-450i; P-450 2d; P-450-female; P-450 15α	Not apparently inducible	5α-Androstane-3α,17β-diol 3,17-disulfate 15β-hydroxylation (MACGEOCH et al., 1984)	Female-specific; appears in very old males (KAMATAKI et al., 1985)

12. P-450j (Ryan et al., 1985)	P-450et, $P\text{-}450_{acc}$ (Yang et al., 1985)	Ethanol, isoniazide, several ketones	Oxidation of several alcohols to aldehydes, N,N-dimethylnitrosamine N-demethylation, CCl_4 reduction	Isolated in partially high-spin state [b] Song et al., 1987
13. P-452 (Gibson et al., 1982)	$P\text{-}450_{\omega}$	Clofibrate, other hypolipidemic agents	ω- and ω-1 hydroxylation of fatty acids, azo dye reduction	Ferrous carbonyl λ_{max} at 452 nm [b] Hardwick et al., 1987
14. P-450f (Ryan et al., 1984)	P-450RLM 3			
15. P-450g (Ryan et al., 1984)				
16. P-450RLM 2 (Jansson et al., 1985a)			Testosterone 6β-, 15α-, 7β-hydroxylation	
17. P-450RLM 5a (Jansson et al., 1985b)				
18. P-450PB-6 (Waxman, 1986)				More acidic than $P\text{-}450_{PB-B}$, $P\text{-}450_{PB-D}$

Table 1. — Continued

P-450	Other similar preparations[a]	Inducibility	Characteristic catalytic reactions	Other distinguishing features
19. Taurodeoxycholate 7α-hydroxylase (Murakami and Okuda, 1981)			Taurodeoxycholate 7α-hydroxylation	
20. P-450, P-451 (Agosin et al., 1979)			1-[1-(4'-ethyl)-phenoxy]-3,7-dimethyl-6,7-epoxy-*trans*-2-octene hydroxylation	
21. $P\text{-}450_{7\alpha}$ (Andersson et al., 1985; Tang and Chiang, 1986)			Cholesterol 7α-hydroxylation	
22. $P\text{-}450_{14\alpha\text{-DM}}$ (Trzaskos et al., 1986)		Cholestyramine?	Lanosterol 14α-demethylation	

[a] Because of restrictions on the number of references, literature citations are not given here. See Guengerich (1987a) for further treatment.
[b] Coding sequence has been reported (either protein or a related cDNA). See indicated reference.

Table 2. Rabbit Liver P-450s[a]

P-450	Other similar preparations[b]	Inducibility	Characteristic catalytic reactions	Other distinguishing features
1. P-450 1 (DIETER et al., 1982)	P-450 LM1 (HAUGEN et al., 1975)	Apparently not inducible; polymorphism exists in rabbits	Pregnenolone, progesterone 21-hydroxylase, estradiol 2-hydroxylase, benzo(a)-pyrene 3-hydroxylation	[c] TUKEY et al., 1985
2. P-450 2 (HAUGEN et al., 1975)	Purified in several laboratories (see SCHWAB and JOHNSON, 1987)	Highly inducible by pheno-barbital and other barbi-turates	Benzphetamine N-demethylation	[c] TARR et al., 1983
3. P-450 2 (HEINEMANN and OZOLS, 1982, 1983)		Highly inducible by pheno-barbital and other barbi-turates		Sequence similar to P-450 2 of TARR et al. (1983) but 13 residues differ; existence of >1 mRNA in this family has been described by SATO and by PHILPOT (un-published results) but exact relationship of pro-teins and mRNAs unclear
4. P-450 3a (KOOP et al., 1982)		Ethanol, imidazole, iso-niazid, trichloroethylene, benzene, pyrazole	Oxidation of several alco-hols to aldehydes, N,N-dimethylnitrosamine N-demethylation	High-spin ferric enzyme

Table 2. — Continued

P-450	Other similar preparations[b]	Inducibility	Characteristic catalytic reactions	Other distinguishing features
5. P-450 3b (KOOP and COON, 1979; JOHNSON, 1980)		Apparently not inducible	Progesterone 6β- and 16α-hydroxylation	[c] OZOLS et al., 1985
6. P-450 3b (subform) (SCHWAB and JOHNSON, 1985b)				A lower activity variant allelic to the normal P-450 3b is found in some individual animals
7. P-450 3c (INGELMAN-SUNDBERG et al., 1979; MIKI et al., 1981)	P450LM3; P-450B$_1$	PCN and some other steroids, triacetyloleandomycin, phenobarbital	Androstenedione 6β-hydroxylation	Stimulated by cytochrome b_5 [c] DALET et al., 1986
8. P-450 4 (HAUGEN et al., 1975)	P-448$_1$; P-448; P-450 LM4b	Polycyclic aromatic hydrocarbons	Acetanilide 4-hydroxylation, 2-acetylaminofluorene N-hydroxylation, acetaminophen oxidation	Partially high-spin ferric, ferrous carbonyl λ_{max} at 448 nm [c] OKINO et al., 1985; OZOLS, 1986
9. P-450 4 (BOSTRÖM and WIKVALL, 1982)		Cholestyramine	Cholesterol 7α-hydroxylation	Similar to other P-450 4 but possibly distinguished by several criteria

10. P-450 5 (Wolf et al., 1978)	P-450$_{II}$ (Wolf et al., 1978)	Phenobarbital	N-Hydroxylation of 2-aminoanthracene and 2-aminofluorene	
11. P-450 6 (Johnson and Muller-Eberhard, 1977; Ueng and Alvares, 1982)	P-450b; P-450$_{III}$	Polycyclic aromatic hydrocarbons	Benzo[a]pyrene hydroxylation	[c] Okino et al., 1985
12. P-450 7 (Haugen et al., 1975)				
13. P-450 CN (Miki et al., 1982)				Cyanide-sensitive
14. P-450$_2$ (Komori et al., 1984)			High affinity for 7-alkoxycoumarins	
15. P-450$_4$, P-450$_6$, P-450$_7$, and P-450$_8$ (Aoyama et al., 1982)				

[a] See Schwab and Johnson (1987) for further treatment.
[b] These proteins have also been referred to as LM 1, LM 2, ... (Haugen et al., 1975).
[c] Coding sequence has been reported. See indicated reference.

Table 3. Mouse Liver P-450s[a]

P-450	Inducibility	Characteristic catalytic reactions	Other distinguishing features
1. P_1-450 (NEGISHI and NEBERT, 1979)	Polycyclic aromatic hydrocarbons	Benzo[a]pyrene hydroxylation (-9) and epoxidation (-7,8)	[b] KIMURA et al., 1984
2. P_3-450 (NEGISHI and NEBERT, 1979)	Polycyclic aromatic hydrocarbons	Acetanilide 4-hydroxylation	[b] KIMURA et al., 1984
3. P_2-450 (OHYAMA et al., 1984)	Polycyclic aromatic hydrocarbons	Isosafrole metabolism to form spectral complexes	Reported to be a strain variant of P_3-450 differing in only one substitution (KIMURA and NEBERT, 1986)[b]
4. P-450 A_2 (HUANG et al., 1976)	Phenobarbital (?)		
5. P-450 C_2 (HUANG et al., 1976)	Phenobarbital (?)	Ethylmorphine N-demethylation	
6. P-$450_{15\alpha}$ (HARADA and NEGISHI, 1984a)		Testosterone 15α-hydroxylation	Female specific

7. P-$450_{16\alpha}$ (HARADA and NEGISHI, 1984b)	Testosterone 16α-hydroxylation

[a] Mice contain several other orthologs of rat P-450s, which have been demonstrated by immunoelectrophoresis and immunoinhibition of specific warfarin hydroxylase activities but have not been further characterized (KAMINSKY et al., 1984).
[b] Coding sequence is reported from cDNA studies. See indicated reference.

Table 4. Extrahepatic Microsomal P-450s

P-450	Animal species	Tissue source	Other similar preparations	Characteristic catalytic reactions	Other distinguishing features (including inducibility)
1. P-450 2 (SERABJIT-SINGH et al., 1983)	Rabbit	Lung, kidney (DEES et al., 1982), aorta, bladder (SERABJIT-SINGH et al., 1985)	P-450_{LM-2} (GUENGERICH, 1977), P-450a (OGITA et al., 1983), P-450p-1 (YAMAMOTO et al., 1984)	Benzphetamine N-demethylation	Not inducible in lung — levels can be decreased by administration of certain compounds
2. P-450 5 (SERABJIT-SINGH et al., 1983)	Rabbit	Lung, bladder (SERABJIT-SINGH et al., 1985)			Not inducible in lung
3. P-450 6 (SERABJIT-SINGH et al., 1983)	Rabbit	Lung, kidney (LIEM et al., 1980), aorta, bladder (SERABJIT-SINGH et al., 1985)			Inducible in lung by polycyclic aromatic hydrocarbons

Table 4. — Continued

P-450	**Animal species**	**Tissue source**	**Other similar preparations**	**Characteristic catalytic reactions**	**Other distinguishing features (including inducibility)**
4. P-$450_{PG-\omega}$ (WILLIAMS et al., 1984)	Rabbit	Lung	P-450p-2 (YAMAMOTO et al., 1984)		Also in liver — elevated during pregnancy
5. P-450 3a	Rabbit	Kidney, nasal mucosa (DING et al., 1986)			
6. P-$450_{17\alpha}$ (ZUBER et al., 1986)	Several	Adrenals		Progesterone 17α-hydroxylation	Sequence deduced from cDNA (see ref.)
7. P-450_{C-21} (JOHN et al., 1986)	Several	Adrenals		Progesterone 21-hydroxylation	Sequence deduced from cDNA (see ref.)
8. Aromatase	Several — purified from humans (EVANS et al., 1986)	Placenta		Aromatization of testosterone to 17β-estradiol	Also in breast, ovary, and other tissues

9. "P_1-450" (rat P-$450_{\beta NF-B}$)	Several	Placenta, lymphocytes, monocytes, lung, kidney, others (Fujino et al., 1982; Jaiswal et al., 1985a, 1985b)		Benzo[a]pyrene hydroxylation	Identified by catalytic activity and immunochemical techniques
10. P-450_{PB-B}	Rat	Lung, kidney (Guengerich et al., 1982b)		Same as rat liver P-450_{PB-B}	Shown to be P-450_{PB-B} as opposed to P-450_{PB-D} (Omniecinski, 1986); levels present constitutively and not inducible in lung (cf. rabbits)
11. P-450b (Ogita et al., 1983)	Rabbit	Kidney		Prostaglandin A_1 ω-hydroxylation, myristate ω- and ω-1 hydroxylation	
12. (P-450) (Ichihara et al., 1981)	Rabbit	Intestine		Hexadecane hydroxylation	
13. P-450a (Ichihara et al., 1983)	Rabbit	Intestinal mucosa			
14. P-450b (Ichihara et al., 1983)	Rabbit	Intestinal mucosa			

Table 4. — Continued

P-450	Animal species	Tissue source	Other similar preparations	Characteristic catalytic reactions	Other distinguishing features (including inducibility)
15. P-450ia (Kaku et al., 1984)	Rabbit	Intestinal mucosa		Prostaglandin A_1 ω-hydroxylation	
16. P-450ca (Kaku et al., 1985)	Rabbit	Colon mucosa		Prostaglandin A_1 ω-hydroxylation	
17. P-450cb (Kaku et al., 1985)	Rabbit	Colon mucosa			
18, P-448c (Kaku et al., 1985)	Rabbit	Colon mucosa			

Table 5. Human P-450s[a]

P-450	Other similar preparations	Major catalytic activities	Other notes
1. P-450$_{DB}$ (Distlerath et al., 1985)	P-450$_{NT}$ (Bigersson et al., 1986), P-450$_{BufI}$ (Gut et al., 1984, 1986)	Debrisoquine 4-hydroxylation (Distlerath et al., 1985), sparteine Δ^2- and Δ^5 oxidation (Distlerath et al., 1985; Gut et al., 1986), bufuralol 1′-hydroxylation (Distlerath et al., 1985), propranolol 4-hydroxylation (Distlerath et al., 1985), encainide O-demethylation (Distlerath et al., 1985), metoprolol hydroxylation (Gut et al., 1986), nortryptilline 10-hydroxylation (Bigersson et al., 1986)	Related to polymorphism of debrisoquine 4-hydroxylation; ortholog of rat P-450$_{UT-H}$ (Distlerath and Guengerich, 1984)
2. P-450$_{PA}$ (Distlerath et al., 1985)	P-450d (Wrighton et al., 1986a), P-450 4 (Quattrochi et al., 1986), P_3-450 (Jaiswal et al., 1986)	Phenacetin O-deethylase (low K_m) (Distlerath et al., 1985)	Apparent ortholog of mouse P_3-450, rat P-450$_{ISF-G}$, and rabbit P-450 4 (P-450d preparation is inactive and P-450 4 and P_3-450 are derived sequences) (Quattrochi et al., 1986; Jaiswal et al., 1986) (protein was not isolated)

Table 5. — Continued

P-450	Other similar preparations	Major catalytic activities	Other notes
3. P_1-450 (JAISWAL et al., 1985b; KAWAJIRI et al., 1986)		Benzo[a]pyrene hydroxylase (?)	cDNA derived from human hepatoma library; several studies with cross-reacting antibodies suggest this protein (never isolated) is present and inducible in extrahepatic tissues including placenta, lymphocytes, and monocytes (JAISWAL et al., 1985b; FUJINO et al., 1982)
4. P-450_{MP-1} (WANG et al., 1980, 1983; SHIMADA et al., 1986)	P-450 meph(?) (GUT et al., 1987)	S-Mephenytoin 4-hydroxylation (also some N-demethylation), phenytoin 4-hydroxylation, tolbutamide methyl hydroxylation, hexobarbital 3' hydroxylation (SHIMADA et al., 1986; R. G. KNODELL, S. D. HALL, G. R. WILKINSON, and F. P. GUENGERICH, submitted)	Related to S-mephenytoin 4-hydroxylase polymorphism; ortholog of rat P-450_{UT-A} and rabbit P-450 1; sequence of a cDNA clone has been reported (UMBENHAUER et al., 1987)
5. P-450_{MP-2} (SHIMADA et al., 1986)		Same as P-450_{MP-1}	Same as P-450_{MP-1}, other related forms may exist
6. P-450 BufII (GUT et al., 1986)		Bufuralol 1'-hydroxylation (GUT et al., 1986)	Very similar to P-450_{DB} but altered catalytic activity

7. P-450_{NF} (Wang et al., 1983; Guengerich et al., 1986a)	Nifedipine oxidation (Guengerich et al., 1986a), oxidation of other dihydropyridines (Böcker and Guengerich, 1986), quinidine N- and 3-hydroxylations (Guengerich et al., 1986a), erythromycin N-demethylation (Watkins et al., 1985), benzphetamine N-demethylation (Guengerich et al., 1986a), aldrin epoxidation (Guengerich et al., 1986a), testosterone 6β-hydroxylation (Guengerich et al., 1986a), cortisol 6β hydroxylation (Ged et al., 1985), estradiol 2- and 4-hydroxylation (Guengerich et al., 1986a)	Related to apparent polymorphism of nifedipine oxidation (Kleinbloesem et al., 1984); ortholog of rat P-450_{PCN-E} and rabbit P-450 3c; probably inducible by barbiturates, some steroids, and rifampicin (Watkins et al., 1986); cDNA cloning indicates existence of related mRNAs
8. P-450 HLp (Watkins et al., 1985)	Activities listed under P-450_{NF} (?)	Very closely related to P-450_{NF} (Beaune et al., 1986)
9. P-450 9 (Beaune et al., 1985)	Unknown	
10. P-450HFLa (Kitada et al., 1985)	Unknown	Only expressed in fetal liver

Table 5. — Continued

P-450	Other similar preparations	Major catalytic activities	Other notes
11. P-450j (WRIGHTON et al., 1986a; SONG et al., 1987)		Ethanol oxidation, N,N-dimethyl nitrosamine N-demethylation, CCl_4 reduction (?)	Isolated protein is inactive, substrate selectivity suggested by inhibition experiments, sequence derived from cDNA isolated with ortholog (rat P-450j) (also orthologous to rabbit P-450 3a)
12. P-450 pHB_1 (PHILLIPS et al., 1985)		Unknown	Only partial cDNA clone sequence available; identified in library as ortholog of rat $P\text{-}450_{PB-B}$ (PHILLIPS et al., 1985), rat $P\text{-}450_{UT-F}$ (NEBERT et al., 1987)
13. $P\text{-}450_{C-21}$ (WHITE et al., 1984; HIGASHI et al., 1986)		Pregnenolone 21-hydroxylation	Protein specific to adrenals; involved in polymorphism related to [disease]; protein has not been isolated — sequence derived from cDNA clone

[a] See Table 6 for known coding sequences.

Table 6. Known Sequences of Human P-450s (RNA or protein)

		1				5					10					15					20					25					30					35
P-450$_{MP}$	1	M	D	S	L	V	V	L	V	L	C	L	S	C	L	L	L	L	S	L	W	R	Q	S	S	G	R	G	K	L	P	P	G	P	T	P
P$_1$-450	1	M	L	F	P	I	S	M	S	A	T	E	F	L	L	A	S	V	I	F	C	L	V	F	W	V	I	R	A	S	R	P	Q	V	P	K
P$_3$-450	1	M	A	L	S	Q	S	V	P	F	S	A	T	E	L	L	L	A	S	A	I	F	C	L	V	F	W	V	L	K	G	L	R	P	R	V
P-450$_{NF}$	1	M	A	L	I	P	D	L	A	M	E	T	W	L	L	L	A	V	S	L	V	L	L	Y	L	Y	G	T	H	S	H	G	L	F	K	K
P-450p	1	M	A	L	I	P	D	L	A	M	E	T	W	L	L	L	A	V	S	L	V	L	L	Y	L	Y	G	T	H	S	H	G	L	F	K	K
P-450pHB(1)	1																																			
P-450$_{c-21}$	1	M	L	L	L	G	L	L	L	L	P	L	L	A	G	A	R	L	L	W	N	W	W	K	L	R	S	L	H	L	P	P	L	A	P	G

		36				40					45					50					55					60					65					70
P-450$_{MP}$	36	L	P	V	I	G	N	I	L	Q	I	G	I	K	D	I	S	K	S	L	T	N	L	S	K	V	Y	G	P	V	F	T	L	Y	F	G
P$_1$-450	36	G	L	K	N	P	P	G	P	W	G	W	P	L	I	G	H	M	L	T	L	G	K	N	P	H	L	A	L	S	R	M	S	Q	Q	Y
P$_3$-450	36	P	K	G	L	K	S	P	P	E	P	W	G	W	P	L	L	G	H	V	L	T	L	G	K	N	P	H	L	A	L	S	R	M	S	Q
P-450$_{NF}$	36	L	G	I	P	G	P	T	P	L	P	F	L	G	N	I	L	S	Y	H	K	G	F	C	M	F	D	M	E	C	H	K	K	Y	G	K
P-450p	36	L	G	I	P	G	P	T	P	L	P	F	L	G	N	I	L	S	Y	H	K	G	F	C	M	F	D	M	E	C	H	K	K	Y	G	K
P-450pHB(1)	36																																			
P-450$_{c-21}$	36	F	L	H	L	L	Q	P	D	L	P	I	Y	L	L	G	L	T	Q	K	F	G	P	I	Y	R	L	H	L	G	L	Q	D	V	V	V

Table 6. — Continued

		71 75 80 85 90 95 100 105
P-450$_{MP}$	71	L K P I V V L H G Y E A V K E A L I D L G E E F S G R G I F P L A E R
P_1-450	71	G D V L Q I R I G S T P V V V L S G L D T I R Q A L V R Q G D D F K G
P_3-450	71	R Y G D V L Q I S I G S T P V L V L S R L D T I R Q A L V R Q G D D F
P-450$_{NF}$	71	V W G F Y D G Q Q P V L A I T D P D M I K T V L V K E C Y S V F T N R
P-450p	71	V W G F Y D G Q Q P V L A I T D P D M I K L V L V K E C Y S V F T N R
P-450pHB(1)	71	
P-450$_{c-21}$	71	L N S K R T I E E A M V K K W A D F A G R P E P L T Y K L V S K N Y P

		106 110 115 120 125 130 135 140
P-450$_{MP}$	106	A N R G F G I V F S N G K K W K E I R R F S L M T L R N F G M G K R S
P_1-450	106	R P D L Y T F T L I S N G Q S M S F S P D S G P V W A A R R R L A Q N
P_3-450	106	K G R P D L Y T S T L I T D G Q S L T F S T D S G P V W A A R R R L A
P-450$_{NF}$	106	R P F G P V G F M K S A I S I A E D E E W K R L R S L L S P T F T S G
P-450p	106	R P F G P V G F M K S A I S I A E D E E W K R L R S L L S P T F T S G
P-450pHB(1)	106	
P-450$_{c-21}$	106	D L S L G D Y S L L W K A H K K L T R S A L L L G I R D S M E P V V E

		141				145					150					155					160					165					170					175
P-450$_{MP}$	141	I	E	D	R	V	Q	E	E	A	R	C	L	V	E	E	L	R	K	T	K	A	S	P	C	D	P	T	F	I	L	G	C	A	P	C
P$_1$-450	141	G	L	K	S	F	S	I	A	S	D	P	A	S	S	T	S	C	Y	L	E	E	H	V	S	K	E	A	E	V	L	I	S	T	L	Q
P$_3$-450	141	Q	N	A	L	N	T	F	S	I	A	S	D	P	A	S	S	S	S	C	Y	L	E	E	H	V	S	K	E	A	K	A	L	I	S	R
P-450$_{NF}$	141	K	L	K	E	M	V	P	I	I	A	Q	Y	G	D	V	L	V	R	N	L	R	R	E	A	E	T	G	K	P	V	T	L	K	D	V
P-450p	141	K	L	K	E	M	V	P	I	I	A	Q	Y	G	D	V	L	V	R	N	L	R	R	E	R	E	T	G	K	P	V	T	L	K	D	V
P-450pHB(1)	141																												R	L	R	A	N	I	D	P
P-450$_{c-21}$	141	Q	L	T	Q	E	F	C	E	R	M	R	A	Q	P	G	T	P	V	A	I	E	E	E	F	S	L	L	T	C	S	I	I	C	Y	L

		176				180					185					190					195					200					205					210
P-450$_{MP}$	176	N	V	I	C	S	I	I	F	H	K	R	F	D	Y	K	D	Q	Q	F	L	N	L	M	E	K	L	N	E	N	I	K	I	L	S	S
P$_1$-450	176	E	L	M	A	G	P	G	H	F	N	P	Y	R	Y	V	V	V	S	V	T	N	V	I	C	A	I	C	F	G	R	R	Y	D	H	N
P$_3$-450	176	L	Q	E	L	M	A	G	P	G	H	F	D	P	Y	N	Q	V	V	V	S	V	A	N	V	I	G	A	M	C	F	G	Q	H	F	P
P-450$_{NF}$	176	F	G	A	Y	S	M	D	V	I	T	S	T	S	F	G	V	N	I	D	S	L	N	N	P	Q	D	P	F	V	E	N	T	K	K	L
P-450p	176	F	G	A	Y	S	M	D	V	I	T	S	S	S	F	G	V	N	V	D	S	L	N	N	P	Q	D	P	L	V	E	N	T	K	K	L
P-450pHB(1)	176	T	L	F	L	I	R	T	F	S	N	V	I	S	S	I	V	F	G	D	R	F	D	Y	K	D	R	E	L	L	S	L	F	R	I	M
P-450$_{c-21}$	176	T	F	G	D	K	I	K	D	D	N	L	M	P	A	Y	Y	K	C	I	Q	E	V	L	K	T	W	S	H	W	S	I	Q	I	V	D

Table 6. — Continued

		211 215 220 225 230 235 240 245
P-450$_{MP}$	211	P W I Q I C N N F S P I I D Y F P G T H N K L L K N V A F M K S Y I L
P$_1$-450	211	H Q E L L S L V N L N N N F G E V V G S G N P A D F I P I L R Y L P N
P$_3$-450	211	E S S D E M L S L V K N T H E F V E T A S S G N P L D F F P I L R Y L
P-450$_{NF}$	211	L R F D F L D P F F L S I T V F P F L I P I L E V L N I C V F P R E V
P-450p	211	L R F D F L D P F F L S I T V F P F L I P I L E V L N I C V F P R E V
P-450pHB(1)	211	L V I V P V H V N S T G Q L Y E M F S S V M K Q L P G P Q Q Q A F Q L
P-450$_{c-21}$	211	V I P F L R F F P N P G L R R L K Q A I E K R D H I V E M Q L R Q H K

		246 250 255 260 265 270 275 280
P-450$_{MP}$	246	E K V K E H Q E S M D M N N P Q D F I D C F L M K M E K E K H N Q P S
P$_1$-450	246	P S L N A F K D L N E K F Y S F M Q K M V K E H Y K T F E K G H I R D
P$_3$-450	246	P N P A L Q R F K A F N Q R F L W F L Q K T V Q E H Y Q D F D K N S V
P-450$_{NF}$	246	T N F L R K S V K R M K E S R L E D T Q K H R V D F L Q L M I D S Q N
P-450p	246	T N F L R K A V K R M K E S R L E D T Q K H R V D F L Q L M I D S H K
P-450pHB(1)	246	L Q G L E D F I A K K V E H N T P L D P N S P R D F I D S F L I R M Q
P-450$_{c-21}$	246	E S L V A G Q W R D M M D Y M L Q G V A Q P S M E E G S G Q L L E G H

		281 285 290 295 300 305 310 315
P-450$_{MP}$	281	E F T I E S L E N T A V D L F G A G T E T T S T T L R Y A L L L L L K
P$_1$-450	281	I T D S L I E H C Q E K Q L D E N A N V Q L S D E K I I N I V L D L F
P$_3$-450	281	R D I T G A L F K H S K K G P R A S G N L I P Q E K I V N L V N D I F
P-450$_{NF}$	281	S K E T E S H K A L S D L E L V A Q S I I F I F A G Y E T T S S V L S
P-450p	281	N S K E T E S H K A L S D L E L V A Q S I I F I F A G Y E T T S S V L
P-450pHB(1)	281	E E E K N P N T E F Y L E K L V M T S L N L F I G G T E T V S T T L R
P-450$_{c-21}$	281	V H M A A V D L L I G G T E T T A N T L S W A V V F L L H H P E I Q Q

		316 320 325 330 335 340 345 350
P-450$_{MP}$	316	H P E V T A K V Q E E I E R V I G R N R S P C M Q D R S H M P Y T D A
P$_1$-450	316	G A G F D T V T T A I S W S L M Y L V M N P R V Q R K I Q E E L D T V
P$_3$-450	316	G A G F D T F T T A I S W S L M Y L V T K P E I Q R K I Q K E L D T V
P-450$_{NF}$	316	F I M Y E L A T H P D V Q Q K L Q E E I D A V L P N K A P P T Y D T V
P-450p	316	S F I M Y E L A T H P D V Q Q K L Q E E I D A V L P N K A P P T Y D T
P-450pHB(1)	316	Y G F L L L I K H P G V E A K V H E E I D R V I G K N R Q P K F E D R
P-450$_{c-21}$	316	R L Q E E L D H E L G P G A S S S R V P Y K D R A R L P L L N A T I A

Table 6. — Continued

		351 355 360 365 370 375 380 385
P-450$_{MP}$	351	V V H E V Q R C I D L L P T S L P H A V T C D I K F R N Y L I P K G T
P$_1$-450	351	I G R S R R P R L S D R S H L P Y M E A F I L E T F R H S S F V P F T
P$_3$-450	351	I G R E R R P R L S D R P Q L P Y L E A F I L E T F R H S S F L P F T
P-450$_{NF}$	351	L Q M E Y L D M V V N E T L R L F P I A M R L E R V C K K D V E I N G
P-450p	351	V L Q M E Y L D M V V N E T L R L F P I A M R L E R V C K K D V E I N
P-450pHB(1)	351	A K M P Y M E A M I H Q I Q R F G D V I P M I W P G R V K K D T K F R
P-450$_{c-21}$	351	E V L R L R P V V P L A L P H R T T R P S S I S G Y D I P E G T V I I

		386 390 395 400 405 410 415 420
P-450$_{MP}$	386	T I L I S L T S V L H D N K E F P N P E M F D P H H F L D E G D N F K
P$_1$-450	386	I P H S T T R D T S L K G F Y I P K G R C V F V N Q W Q I N H D Q K L
P$_3$-450	386	I P H S T T R D T T L N G F Y I P K K C C V F V N Q W Q V N H D P E L
P-450$_{NF}$	386	M F I P K G W V V M I P S Y A L H R D P K Y W T E P E K F L P E R F S
P-450p	386	G M F I P K G W V V M I P S Y A L H R D P K Y W T E P E K F L P E R F
P-450pHB(1)	386	D F F L P K G T E V Y P M L G S V L R D P I F L S K P Q D F N P Q H F
P-450$_{c-21}$	386	P N L Q G A H L D E T V W E R P H E F W P D R F L E P G K N S R A L A

		421 425 430 435 440 445 450 455
$P\text{-}450_{MP}$	421	K S K Y F M P F S A G K R I C V G E A L A G M E L F L F L T S I L Q N
P_1-450	421	W V N P S E F L P E R F L T P D G A I D K V L S E K V I I F G M G K R
P_3-450	421	L E D P S E F R P E R F L T A D G T A I N K P L S E K M M M L V G M G
$P\text{-}450_{NF}$	421	K K N K D N I D P Y T Y T P F G S G P R N C I G M R F A L M N M K L A
P-450p	421	S K K N K D N I D P Y I Y T P F G S G P R N C I G M R F A L M N M K L
P-450pHB(1)	421	T E L E G A P K K S D A F V P F S I G Q P N C F G E G L A R M E L F L
$P\text{-}450_{c-21}$	421	F G C G A P V C L G E P L A R L E L F V V L T R L L Q A F T L L P S G

		456 460 465 470 475 480 485 490
$P\text{-}450_{MP}$	456	F N L K S L V D P K N L D T T P V V N G F A S V P P F Y Q L C F I P V
P_1-450	456	K C I G E T V A R W E V F L F L A I L L Q R V E F S V P L G V K V D M
P_3-450	456	K R R C I G E V L A K W E I F L F L A I L L Q Q L E F S V P P G V K V
$P\text{-}450_{NF}$	456	L I R V L Q N F S F K P C K E T Q I P L K L S L G G L L Q P E K P V V
P-450p	456	A L I R V L Q N F S F K P C K E T Q I P L K L S L G G L L Q P E K P V
P-450pHB(1)	456	F F T T V M Q N F R L K S S Q S P K D I D V S P K H V G F P R F P R N
$P\text{-}450_{c-21}$	456	D A L P S L Q P L P H C S V I L K M Q P F Q V R L Q P R G M G A H S P

Table 6 — Continued.

		491 495 500 505 510 515 520
P-450_{MP}	491	
P_1-450	491	T P I Y G L T M K H A C C E H F Q M Q L R S
P_3-450	491	D L T P I Y G L T M K H A R C E H V Q A R L P F S I N
P-450_{NF}	491	L K V E S R D G T V S G A
P-450p	491	V L K V E S R D G T V S G A
P-450pHB(1)	491	Y T M S F L P R
P-450_{c-21}	491	G Q N Q

Sources: P-450_{MP}, Umbenhauer et al., 1987; P_1-450, Jaiswal et al., 1985b; P_3-450, Jaiswal et al., 1986; P-450_{NF}, Beaune et al., 1986; P-450p, Molowa et al., 1986; P-450pHB(1), Phillips et al., 1985; P-450_{c-21}, Higashi et al., 1986. See Tables 1—5 for references to sequences derived in other laboratories. All of these sequences with the exception of P-450_{c-21} are available for computerized readout through the National Biomedical Research Foundation, Washington, D. C. (1986) **Protein Sequence Data Base of the Protein Identification Source, Release No. 8.0.** Sequences are not adjusted for maximum overlap; in the case of the partial clone P-450pHB(1) the sequence is adjusted to overlap in the region surrounding Cys 442. Note that the P-450p sequence has been derived from partial clones (Molowa et al., 1986), the deduced sequence is highly similar to that of P-450_{NF} but a frameshift is present. Boxes indicate the region of sequences which are aligned in Table 7.

2.8. Human P-450s

The known human P-450s are listed in Table 5. Some of these have already been mentioned in terms of their structural similarities to P-450 forms characterized in animals. The available coding sequences of human P-450s are given in Table 6.

Of the P-450s listed, only a few have been isolated in functionally catalytic form and demonstrated to possess distinct activities. Some are probably functional but characteristic substrates have not been identified (i.e., P-450_9, P-450 HFLa). P-450_{NF} is functionally active although a higher level of activity is expected (cf. the ortholog rat P-450_{PCN-E}) (GUENGERICH et al., 1986a). In two cases, a protein has been purified by using immunoaffinity chromatography under denaturing conditions (with an antibody raised to an ortholog) (i.e., P-450j, P-450d (WRIGHTON et al., 1986a, 1986b)). In several other cases, cDNA clones have been isolated and sequenced utilizing probes made with orthologs from experimental animals. In most of the cases examined to date genomic blotting has detected large regions of hybridizable DNA and strongly suggested relatively large multi-gene families and the possibility of transcription of several related mRNAs (and proteins). Assignment of protein sequences on the basis of cDNA sequences should be made cautiously. Another complication in such systems is that polymorphisms (often unrelated to catalytic function) occur frequently in humans—in most cases cDNA libraries were not constructed from the same individuals that protein comparisons are being made with. Considerably more effort will be needed to define which protein residues can vary without loss of function.

3. Functional versatility of P-450s

3.1. Extent of P-450 substrate multiplicity

One of the features of the P-450s which was first recognized was the diversity of substrates. At first the only recognized features of similarity were the hydrophobic nature of the materials and the tendency for oxidation to occur at the more electrophilic sites. Ultimately differences in the actions of microsomal preparations (from variously treated animals) on certain compounds led to the concept of P-450 multiplicity. Today we recognize that the broad specificity is due in part to the multiplicity of the P-450s. However, each of the P-450s can still oxidize (and in some cases reduce) a wide variety of chemicals. For instance, the ability of P-450_{PB-B} to oxidize more than 70 different substrates has been noted before (GUENGERICH, 1987a).

The total number of P-450 substrates easily extends into the thousands. Most tend to be non-polar, although highly water-soluble compounds such as ethanol, acetone, and N,N-dimethylnitrosamine are now recognized to be substrates (C. S. YANG et al., 1985). The smallest known substrates appear to be methanol (MW 32) (Morgan et al., 1982) and acetylene (MW 26) (Kunze et al., 1983). Methane has not been tested as a substrate. Some of the larger ones are avermectin B_{1a} (WM 872) (Miwa et al., 1982) and macrolide antibiotics such as erythromycin (MW 734) and triacetyloleandomycin (MW 814) (Wrighton et al., 1985), although the extent of insertion of the substrate into the enzyme is unknown. Even larger is the cyclic peptide cyclosporin A (MW 1325), an immunosuppressive agent which appears to be metabolized by P-450s (Burke and Whiting, 1986). Larger peptides and proteins do not appear to be substrates for P-450s. Thus, we see that P-450s can oxidize substrates containing only one carbon or as many as 66.

3.2. Common chemical mechanisms of catalysis

The chemical mechanisms for the oxidation of different compounds by P-450s are probably similar for the different proteins and have been discussed elsewhere in terms of general features (Guengerich and Macdonald, 1984; Guengerich and Liebler, 1985; Guengerich, 1987b). The chemistry all involves a formal $(FeO)^{3+}$ complex, formed by the elimination of H_2O from the iron site after two electrons have been added. Studies with chemical models suggest that the electron distribution may involve Fe^{IV} with the other formal charge delocalized in the porphyrin ring (Groves et al., 1985; Groves and Watanabe, 1986) (see chapter by Mansuy). Briefly, we can divide the oxidative reactions into the following categories:

(1) *C-Hydroxylation:* formation of alcohols at C atoms. The mechanism is thought to involve abstraction of a hydrogen atom and "oxygen rebound" to the radicaloid site then follows.

$$(FeO)^{3+} + -\underset{|}{\overset{|}{C}}-H \rightarrow (FeOH)^{3+} + -\underset{|}{\overset{|}{C}}\cdot \rightarrow Fe^{3+} + -\underset{|}{\overset{|}{C}}-OH$$

Key experiments in support of this hypothesis include measurement of kinetic hydrogen isotope effects, lack of stereospecificity (Groves et al., 1978), and allylic rearrangement of putative intermediates (Groves and Subramanian, 1984). With some low potential alkanes such as quadricyclane, the initial hydrogen abstraction may involve sequential transfer of an electron and proton (Stearns and Ortiz de Montellano, 1985).

(2) *Heteroatom oxygenation:* N, S, P, and I atoms are converted to heteroatom oxides. The mechanism is thought to involve abstraction of a non-bonded electron and subsequent oxygen rebound (Guengerich and Macdonald, 1984).

$$(FeO)^{3+} + -\underset{|}{\overset{|}{N}}: \rightarrow (FeO)^{2+} + -\underset{|}{\overset{|}{N}}{}^{\cdot}_{+} \rightarrow Fe^{3+} + -\underset{|}{\overset{|}{N}} \rightarrow O$$

(3) *Heteroatom release:* dealkylation at a site containing N, O, Cl, Br or I. With O, C, and Br this reaction may follow the pathway of C-hydroxylation as in (1). Recently we have found that O-dealkylation occurs with carboxylic acid esters, as well as ethers, to give oxidative hydrolysis (GUENGERICH, unpublished). With (amine) N, the mechanism starts with electron abstraction to form an aminium radical which rapidly rearranges to yield a carbon-centered radical. Oxygen rebound to this carbon site yields a carbinolamine which breaks down to a free amine and a carbonyl group (GUENGERICH and MACDONALD, 1984; GUENGERICH et al., 1984).

In the oxidation of dihydropyridine compounds, the same process is thought to occur. However, the end result is dehydrogenation (formation of a pyridine). With 4-aryl substituents (or only two hydrogens at the 4-position), dehydrogenation occurs with loss of a proton from the 4-position (BÖCKER and GUENGERICH, 1986). When an alkyl group is present in the 4-position, the (alkyl) group is released as a radical during pyridine formation (AUGUSTO et al., 1982; BÖCKER and GUENGERICH, 1986).

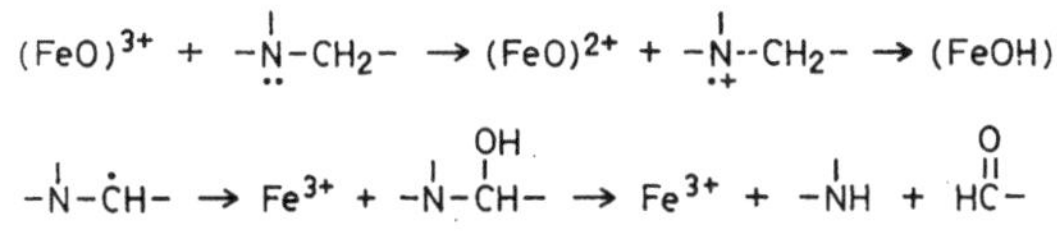

(4) *Epoxidation:* formation of an epoxide at an olefinic or aryl moiety.

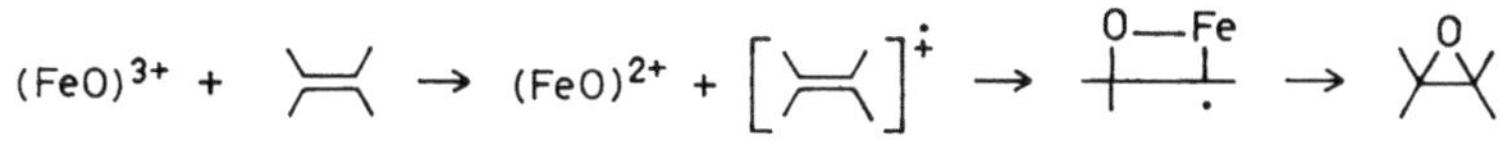

While few doubts are raised concerning the existence of a stepwise mechanism for epoxidation, there is some uncertainty over the precise electronic arrangements of the intermediates. Experiments support the roles of most of the potential intermediates shown above, and reactions may proceed from alternative forms as a function of the individual substrate.

The C-Fe-O-C intermediate is supported by the finding of GROVES et al. (1986) that a propylene olefinic proton exchanged with the medium during epoxidation.

(5) *Oxidative group migration:* with olefins oxidation can result in the formation of a carbonyl group and the migration of an anion to a neighboring position:

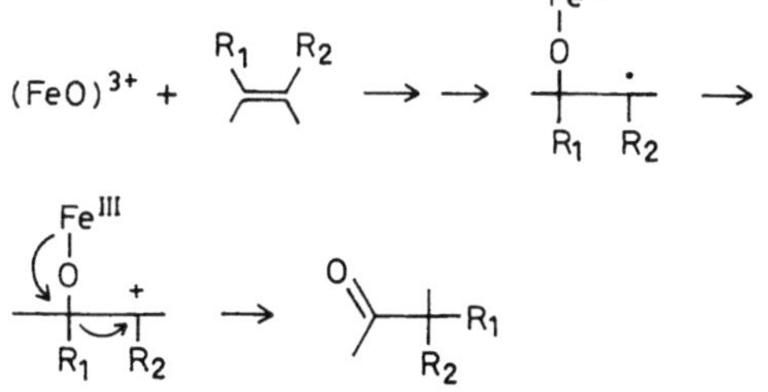

This reaction usually accompanies epoxidation but is not necessarily the result of epoxide rearrangement per se (LIEBLER and GUENGERICH, 1983; MILLER and GUENGERICH, 1982; GROVES and MYERS, 1983; GUENGERICH and MACDONALD, 1984; TRAYLOR et al., 1986; MANSUY et al., 1984). With arene molecules this rearrangement has been termed the "NIH shift" (DALY et al., 1972): epoxides may be intermediates (in the formation of phenols) but are not necessarily so. In addition to the intermediate shown here, structures listed under item (4) may also be applicable.

(6) *Inactivation of P-450 by heme alkylation:* this occurs during the oxidation of terminal olefins by attack of one of the intermediates shown under item (4) on one of the porphyrin nitrogens. The exact electronic structure of the attacking species is uncertain, although some precedent exists for involvement of the form shown:

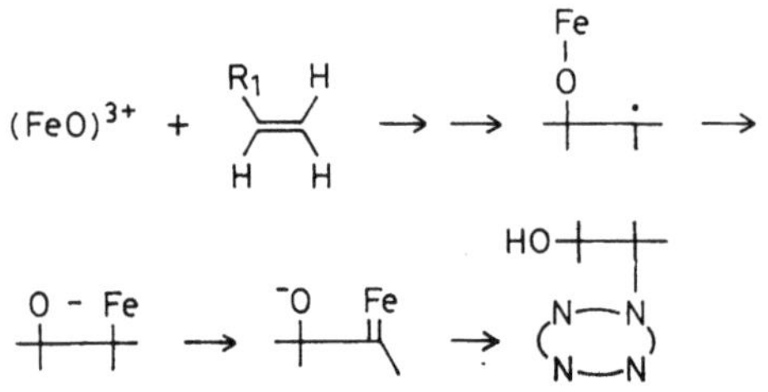

The chemistry is linked to the processes of epoxidation and oxidative group migration (see above). Similar reactions are observed during the oxidation of acetylenes and adducts have been characterized (ORTIZ de MONTELLANO and CORREIA, 1983).

During the oxidation of olefins and acetylenes, other types of mechanism-based inactivation can occur (GUENGERICH, 1986; HALPERT et al., 1983, 1986; GAN et al., 1984), although the chemistry has not been as well characterized. P-450 substrates other than olefins and acetylenes can also lead to mechanism-based inactivation (GUENGERICH et al., 1984; ORTIZ DE MONTELLANO and CORREIA, 1983) although the chemistry has not been elucidated yet.

In addition to the above mixed-function oxidative reactions, P-450s also catalyze a number of reductions. For instance, P-450s have been reported to reduce CCl_4, some azo dyes, arene oxides, tertiary amine N-oxides, and nitro compounds. These reductions involve sequential one-electron transfers; in some cases the products may be unstable and react with oxygen (i. e., CCl_4 and nitro compounds).

3.3. Specificity

Specificity of individual P-450s can be seen at several levels, as exemplified below in each case.

(1) *Preference for different structures:* As an example, aminopyrine N-de-

methylation is catalyzed by rat P-450$_{PB-B}$ in preference to other rat liver P-450s (GUENGERICH et al., 1982a). In contrast, rat P-450$_{ISF-G}$ is most active in the N-oxygenation of 2-naphthylamine (HAMMONS et al., 1985). In many cases, the substrate specificity of individual P-450 forms has been considered only in terms of V_{max} data (or in terms of single substrate concentration data which have not been shown to be saturating). In some cases comparisons have been made in terms of both V_{max} and K_m. This selectivity can sometimes be altered by what appear to be modest changes. For instance, 7-ethoxyresorufin O-deethylation is catalyzed primarily by rat liver P-450$_{\beta NF-B}$ but 7-pentoxyresorufin O-dealkylation is catalyzed mainly by rat P-450$_{PB-B}$ (LUBET et al., 1985).

(2) *Preference for one of a pair of enantiomers:* For example, only the S-isomer of mephenytoin is hydroxylated at the 4-position by human P-450$_{MP}$ (SHIMADA et al., 1986) and purified P-450$_{DB}$ preferentially hydroxylates the (+) enantiomer of bufuralol (GUT et al., 1986; DISTLERATH et al., 1985).

(3) *Regioselectivity of hydroxylation:* A single compound may be hydroxylated at different sites by different forms of P-450. Extensive investigations have shown the tendency of different P-450s to hydroxylate the individual sites of warfarin, testosterone, polycyclic hydrocarbons, and numerous other compounds (KAMINSKY et al., 1980) (Fig. 1). HARADA et al. (1984) have shown that

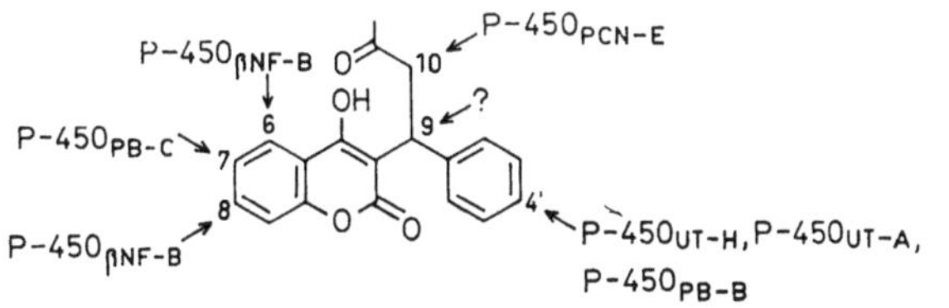

Fig. 1A

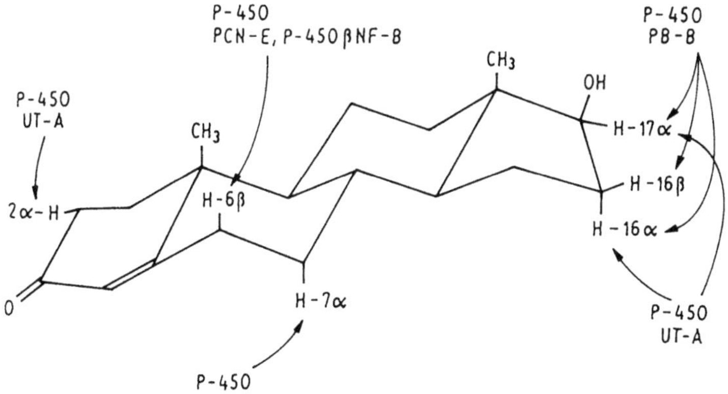

Fig. 1B

Fig. 1. Regioselectivity and stereoselectivity of P-450 hydroxylations. (A) Hydroxylation of warfarin at various carbon atoms by rat liver P-450s. (B) Hydroxylation of testosterone by various rat liver P-450s.

such regioselectivity can be altered to some extent with a single P-450 by hindering one reaction with deuterium substitution.

(4) *Stereoselectivity of oxidation at a single site within a molecule:* Certain polycyclic hydrocarbons form epoxides by insertion of oxygen from either face of the plane of the substrate, depending upon the individual form of P-450 involved (JERINA et al., 1982; S. K. YANG et al., 1985) (Fig. 2). Such stereoselectivity is also common in steroid hydroxylation (Fig. 1B).

As mentioned above, the chemistry carried out by the individual forms of P-450 is probably very similar and one cannot really assign the different types of reaction (i.e., C-hydroxylation, heteroatom oxygenation, etc.) to individual P-450s, for each can be shown to catalyze several types of reaction. The various aspects of specificity are probably due primarily to differences in protein structure that dictate the juxtaposition of compounds relative to the activated oxygen atom. One could consider some limited exceptions, such as the tendency of rat $P\text{-}450_{ISF-G}$ and rabbit P-450 4 to catalyze N-hydroxylation of aromatic amines (HAMMONS et al., 1985; KAMATAKI et al., 1983a). However, these same proteins also carry out other reactions efficiently as well (i.e., estradiol 2- and 4-hydroxylation, acetanilide 4-hydroxylation, phenacetin O-deethylation), and some N-hydroxylations are preferentially catalyzed by other P-450s (e.g., rat $P\text{-}450_{PB-B}$ is most active towards 4,4'-methylene-*bis*-(2-chloroaniline), F. F. KADLUBAR and F. P. GUENGERICH, unpublished results).

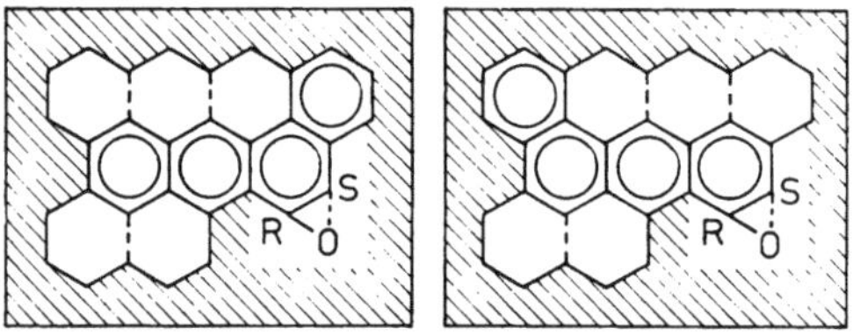

Fig. 2. A model for a substrate binding sites of a P-450 with rat $P\text{-}450_{\beta NF-B}$ (P-450c), the phenyl rings of polycyclic aromatic hydrocarbons fit into the hexagonal sites shown, with the $(FeO)^{3+}$ site being behind the plane of the page. See JERINA et al., (1982) and VAN BLADEREN et al. (1982). A) Benzanthracene 5,6 epoxidation. B) Benzanthracene 8,9 epoxidation.

In two cases attempts have been made to begin to define the geometry of P-450 substrate binding sites. The first involves $P\text{-}450_{\beta NF-B}$. JERINA has used information regarding the ability of this enzyme to catalyze the hydroxylation and epoxidation of polycyclic aromatic hydrocarbons to develop a site model based upon annulated benzene rings (JERINA et al., 1982; THAKKER et al., 1986) (Fig. 2). One uses this model by placing the new substrate under consideration into the pattern and observing (1) whether or not the substrate will fit at all, (2) what position(s) should be attacked by the oxygen, and (3) what side the molecule should be attacked from (what the stereochemistry of epoxidation should be). This model has been used to explain the oxidation of a number of polycyclic aromatic hydrocarbons. The model does not predict the catalytic rates, of course. P-$450_{\beta NF-B}$ oxidizes substrates other than polycyclic aromatic hydrocarbons, and some steroids and other substrates are hydroxylated in a highly specific manner (i.e., testoterone, 7-ethoxyresorufin) but these have not been fitted yet. S. K. Yang et al., (1985) have also pointed out that this model, as proposed, does not hold for 12-methylbenz[a,h]anthracene and 7,12-dimethylbenz[a,h]anthracene and certain reactions on benzo[a]-pyrene.

Another model involves substrates for the debrisoquine 4-hydroxylases, rat P-450_{UT-H} and human P-450_{DB}. Wolff et al., (1985) developed a working model in which the basic nitrogen, which seems to be present in all characteristic substrates (some of which have been assumed from clinical studies), is about 5 Å away from the site of hydroxylation (Fig. 3). A putative anionic charge on the protein (carboxylate?) is proposed to fix the site of the presumably charged amine. A hydrophobic region seems to be near the hydroxylation site, for many benzylic reactions occur and substrates which do not have aromatic rings have a number of methylene groups in this area (i.e., perhexiline, sparteine). Kronbach et al., (1985) have suggested that the distance should be increased from 5 Å to 7 Å, although such a change leaves out debrisoquine 4-hydroxylation, the classical reaction for this enzyme (Fig. 3B).

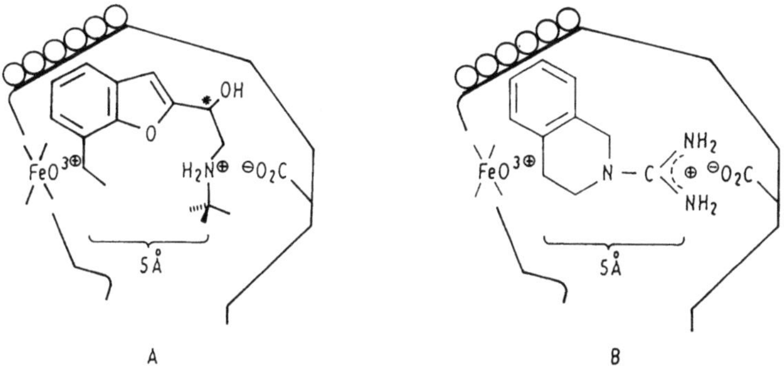

Fig. 3. A model for a binding site of a P-450: rat P-450_{UT-H} and human P-450_{DB}. A) Bufuralol 1' hydroxylation. B) Debrisoquine 4-hydroxylation. See Wolff et al., (1985); Kronbach et al., (1985); Heinrich (1986). Rat P-450_{UT-H} and human P-450_{DB}: a basic nitrogen group is set approximately 5 Å from the site of hydroxylation by $(FeO)^{3+}$. The putative carboxylate attached to the protein is offered as one possibility to explain the position of the nitrogen. A hydrophobic region is nearby the hydroxylation site.

This model still requires considerable refinement before use. In contrast to the P-$450_{\beta NF-B}$ model of Jerina which was based upon planar substrates, the P-450_{DB} model really needs a three-dimensional sense for use. Two other points should be made. First, a variety of strong inhibitors are known for P-450_{DB} and several of these have been reported to be competitive (Otton et al., 1984). It is of interest that the most effective inhibitor, quinidine, is not hydroxylated by P-450_{DB} (but is specifically hydroxylated by human P-450_{NF}) (GUENGERICH et al., 1986b). Therefore an active site model should position quinidine in such a way that all sites which would be hydroxylated are placed away from the formal $(FeO)^{3+}$ species. The second point is that, although the catalytic specificities of rat P-450_{UT-H} and human P-450_{DB} are similar and the proteins are immunochemically related (DISTLERATH et al., 1985; DISTLERATH and GUENGERICH, 1984), the two proteins are not identical in their reactions, as revealed by the inhibition studies of HEINRICH (1986).

3.4. Protein structure-activity relationships

Today many P-450 sequences are available and comparison of these sequences has provided insight into some of the regions which are related to function. Early comparisons of P-450 sequences suggested that the cysteine binding the heme should be situated either near residue 150 or 435 because of the extensive similarity of the sequences in those regions (BLACK and COON, 1986). The binding of the heme at the latter cysteinyl position is now generally accepted because (1) some mammalian P-450s lack the cysteinyl residue near position 150 (BLACK and COON, 1986) and (2) X-ray crystallography reveals that the heme in bacterial P-450_{cam} is bound to a sequence that resembles that near position 435 in the mammalian P-450s (POULOS et al., 1985).

While the position of the heme prosthetic group now seems clear (at least the axial ligand — the distal ligand remains unspecified but may be a hydroxyl), the portions of the mammalian P-450s which bind NADPH-P-450 reductase and organic substrates are unknown. These sequences may be revealed by comparison of sequences, labelling studies with photoaffinity labels, cross-linking reagents, and mechanism-based inactivators, other labelling/protection studies, and site-specific mutagenesis work. In principle these features would best be revealed by X-ray crystallography, as in the case of bacterial P-450_{cam}, but the precedents for crystallization of intrinsic membrane proteins are few at this time.

Another matter for consideration is the extent to which we can predict catalytic specificity from comparison of protein sequences. The known human coding sequences are elaborated in Table 6, and this is expanded in Table 7 for comparison of N-terminal and heme-binding regions. Inspection of the original references cited in Tables 1—6 reveals that a general correlation exists between sequence similarity and catalytic function. For instance, rat P-$450_{\beta NF-B}$, mouse P_1-450, and rabbit P-450 6 all metabolize benzo[a]pyrene in a similar manner, rat P-450_{PCN-E} and human P-450_{NF} catalyze testosterone 6β-hydroxylation, and rat P-450_{ISF-G} and rabbit P-450 4 N-hydroxylate several aromatic amines. Similar patterns of regioselective warfarin hydroxylation and immunochemical reactivity were revealed between rats and mice (KAMINSKY et al., 1984). However, the differences in activity between structurally similar proteins

Table 7. Comparison of N-Terminal Sequences and Heme-binding Region Sequences of Human P-450 Enzymes

P-450	Residue																			
	1				5					10					15					20
P-450$_{MP}$	M	D	S	L	V	V	L	V	L	C	L	S	C	L	L	L	L	S	L	W	R
P$_1$-450	M	L	F	P	I	S	M	S	A	T	E	F	L	L	A	S	V	I	F	C	L
P$_3$-450	M	A	L	S	Q	S	V	P	F	S	A	T	E	L	L	L	A	S	A	I	F
P-450$_{NF}$[a]	M	A	L	I	P	D	L	A	M	E	T	W	L	L	L	A	V	S	L	V	L
P-450$_{C-21}$	M	L	L	L	G	L	L	L	L	P	L	L	A	G	A	R	L	L	W	N	W
	435[b]																				
P-450$_{MP}$	F	S	A	G	L	R	I	C	V	G	E	A	L	A	G	M	E	L	F	L	F
P$_1$-450	F	G	M	G	K	R	K	C	I	G	E	T	V	A	R	W	E	V	F	L	F
P$_3$-450	V	G	M	G	K	R	R	C	I	G	E	V	L	A	K	W	E	I	F	L	F
P-450$_{NF}$[a]	F	G	S	G	P	R	N	C	I	G	M	R	F	A	L	M	N	M	K	L	A
pHP-450(1)	F	S	I	G	Q	P	N	C	F	G	E	G	L	A	K	M	E	L	F	L	F
P-450$_{C-21}$	F	G	C	G	A	P	V	C	L	G	E	P	L	A	R	L	E	L	F	V	V

[a] P-450$_{NF}$ and P-450p are identical in these regions.
[b] The cysteine position is that of P-450$_{MP}$ (Umbenhauer et al., 1987).

should also be kept in mind. For instance, although rat P-450$_{\beta NF-B}$ and mouse P$_1$-450 are very similar in sequence, the rat protein has high warfarin 6- and 8-hydroxylase activities but the mouse protein does not (Kaminsky et al., 1984). Rat P-450$_{PB-B}$ and P-450$_{PB-D}$ differ only in 13 residues but P-450$_{PB-D}$ has only sluggish activity toward many substrates. Human P-450$_{MP}$ and rabbit P-450 1 are about 80% similar as judged by comparison of related cDNA sequences (Umbenhauer et al., 1987); rabbit P-450 1 has characteristic estradiol 2-hydroxylase and pregnenolone 21-hydroxylase activities (Schwab and Johnson, 1985a) but human P-450$_{MP}$ has neither activity. The apparent rat ortholog of human P-450$_{MP}$ is P-450$_{UT-A}$ (80% sequence similarity, Umben-

HAUER et al., 1987; YOSHIOKA et al., 1987), a male-specific protein. However, S-mephenytoin 4-hydroxylation is catalyzed in humans by P-450_{MP} and in rats by P-450_{PCN-E}, whose sequence is not closely related to that of P-450_{UT-A} (SHIMADA and GUENGERICH, 1985; SHIMADA et al., 1986). These exceptions are not so unusual when we consider that single residue changes in proteins made through natural processes or directed mutagenesis are capable, in some cases, of causing dramatic changes in protein function.

4. Factors affecting the function of P-450s

4.1. General aspects of regulation of P-450s

An in-depth consideration of the molecular regulation of P-450s is beyond the scope of this review. Much of our information has come from studies with experimental animal models and cultured cell lines. For recent reviews the reader is referred to ADESNIK and ATCHISON (1986) and WHITLOCK (1986). This review will focus primarily on what is known about factors regulating P-450 activity in humans. A few comments will be made about basic aspects of regulation in preparation for consideration of human systems.

No evidence for amplification of P-450 genes has been found to date. Polymorphisms have been reported to occur at the DNA sequence level in experimental animals and yield proteins with altered activity (KIMURA and NEBERT, 1986).

A key mechanism for altering P-450 activities in experimental animals is through induction by administered chemicals; hormonal regulation also occurs. In these cases a fairly common phenomenon is an elevation of the specific mRNA coding for the protein which is elevated by the treatment. This increase is the result of an enhanced rate of mRNA synthesis in many cases. The exact events underlying increased rates of transcription are unknown. However, in the case of one P-450 (mouse P_1-450 and orthologs), evidence for a receptor exists and this putative *trans*-acting factor has been partially purified (POLAND and KNUTSON, 1982). Recent experiments involving chimeric gene constructs have provided evidence for the existence of a repressor protein as well (JONES et al., 1985; GONZALEZ and NEBERT, 1985; SOGAWA et al., 1986). In some cases, however, a correlation cannot be demonstrated between increased P-450 levels and either mRNA transcription rates or mRNA levels (SONG et al., 1987; SIMMONS et al., 1987). Some of these discrepancies may be due to lack of discrimination of antibodies and cDNA probes. However, post-translational factors may actually be involved, such as enhanced stability of specific mRNAs or changes in the stability of altered proteins. Some compounds that cause increased levels of individual P-450s appear to act in part by decreasing the degradation of these P-450 proteins (SCHUETZ et al., 1986; WATKINS et al., 1986; STEWARD et al., 1985).

The addition of the heme prosthetic group to P-450 can be considered a posttranslation modification. Heme can be exchanged among different P-450s (SADANO and OMURA, 1983, 1985) and in certain situations the heme pool can be depleted, leading to decreased P-450 activity (ORTIZ DE MONTELLANO and CORREIA, 1983). The only other post-translational modification which appears to influence catalytic activity is the phosphorylation of rat P-450 cholesterol 7α-hydroxylase (GOODWIN et al., 1982; SANGHVI et al., 1983; TANG and CHIANG, 1986), although even that situation is controversial (BERGLUND et al., 1986). The P-450s which have been examined to date do not appear to be heavily glycosylated (GUENGERICH et al., 1982a; ARMSTRONG et al., 1983).

Cofactor supply is also a consideration in the regulation of P-450 activity in vivo. The general subject, particularly as related to nutritional influences, has been reviewed elsewhere (THURMAN and KAUFMANN, 1980; GUENGERICH, 1984). The supply of NADPH may become limiting during starvation. Oxygen concentrations can be important, especially with compounds which can be reduced by P-450s (e.g., CCl_4), and oxygen gradients exist in the liver (JONES and MASON, 1978; JI et al., 1982).

Finally, it should be pointed out that regulation of P-450s is cell-specific. In complex tissues containing different types of cell, usually only one or a few contain P-450s. Even in liver, hepatocytes in different regions of the tissue differ in their concentrations of P-450s, and basal and induced patterns vary with the specific form under consideration (BARON et al., 1984). The molecular basis of these differences is not understood.

4.2. Regulation of P-450 activities in humans

Our understanding of the regulation of human P-450s is not as complete as in the case of experimental animals. Only very recently have these enzymes been purified and probes become available for studies at the molecular level. Much of our insight has been gained at the level of clinical pharmacokinetics. These studies have revealed some very interesting aspects of inter-individual variations and the importance of P-450s in humans. On the other hand, the variation among individuals has also increased the complexity of trying to understand phenomena such as enzyme induction.

In general, dramatic sex differences in P-450-related activities are not found in humans (GIUDICELLI and TILLEMENT, 1977; MACLEOD et al., 1979; RICHARDSON et al., 1985). The greatest difference between the sexes has been noted for the clearance of librium (chlorazepam) (ROBERTS et al., 1979). Even some steroid hormones such as testosterone (GUENGERICH et al., 1986a) and 17β-estradiol (FISHMAN et al., 1980) are handled in the same way in both sexes, in contrast to what has been reported in rodents (see chapter 5 by MORGAN and GUSTAFSSON).

Rodents also show pronounced developmental changes in levels of individual P-450s (WAXMAN et al., 1985a; CRESTEIL et al., 1986) as a function of hormonal status. Such striking changes have not been reported in humans. The matter of whether changes occur in the very aged is controversial (GILLETTE, 1979); even in rats the differences tend to be small (McMARTIN et al., 1980). Differences are observed, however, between adult (or adolescent) and fetal liver in humans. P-450_{MP} is not present in fetal liver (SHIMADA et al., 1986) but is present in adults — the time at which the enzyme is expressed is unknown (the mRNA is not present in fetal liver either — UMBENHAUER et al., 1987). P-450 HLFa is present in fetal human liver and not in adult liver (KITADA et al., 1985).

Human P-450s can be induced, although the evidence is indirect. Most of this evidence comes from enhanced rates of in vivo clearance of drugs after treatment with other drugs, and the subject has been reviewed elsewhere (DISTLERATH and GUENGERICH, 1987). Comparisons between smokers and non-smokers also show differences in clearance of certain drugs. The feeding of charcoal-cooked foods has also been shown to increase rates of antipyrine clearance (ANDERSON et al., 1982). Rates of excretion of some endogenous compounds have been used as indices of P-450 function. The basis of changes in glucaric acid excretion (LECAMWASAM et al., 1975) is unknown. 6β-hydroxycortisol also has been used as a marker (PARK, 1981); this metabolite now appears to be formed by P-450_{NF} (GED et al., 1985). In some cases efforts have also been made to compare rates of in vitro drug metabolism in samples obtained from individuals who have been given various drugs (DISTLERATH and GUENGERICH, 1987). For instance, liver microsomes prepared from an individual who received large doses of barbiturates, steroids, and macrolide antibiotics showed unusually high rates of oxidation of erythromycin (WATKINS et al., 1985) and nifedipine (GUENGERICH et al., 1986a). Subsequent studies showed that very high levels of P-450_{NF} (P-450p) were present in this same sample (WATKINS et al., 1985; GUENGERICH et al., 1986a).

Studies in isolated cells have been few. Many human cells of varying origin show enhanced aryl hydrocarbon hydroxylase activity in culture after treatment with polycyclic hydrocarbons (JONES et al., 1984; HANKINSON et al., 1985). In cultured human fetal hepatocytes, this induction is enhanced by the presence of dexamethasone, which does not induce by itself and may be acting to stabilize mRNA (MATHIS et al., 1986). The inducibility of human lymphocytes and monocytes by polycyclic hydrocarbons is well documented (WHITLOCK and GELBOIN, 1979), and attempts to correlate the inducibility with the susceptibility to smoking-related lung cancer have been made (KELLERMAN et al., 1973; JAISWAL et al., 1985a). Hepatocytes are difficult to culture, even in the short term. MOLOWA et al. (1986) have reported that in such cultures, however, P-450p (P-450_{NF}) is inducible by dexamethasone (mRNA levels were also increased). We have also found in our own laboratory that levels of this enzyme (and its mRNA) can be increased in culture (GUILLOUZO et al., 1985) by triacetyloleandomycin. Levels of P-450_{PA} were elevated by 3-methylcholanthrene treatment of such cultures. P-450_{MP} (and its mRNA) was refractory to induction by phenobarbital, 3-methylcholanthrene, β-naphthoflavone, isosafrole and triacetyloleandomycin in these studies (P. H. BEAUNE, R. W. BORK, D. R. UMBENHAUER, A. GUILLOUZO, M. V. MARTIN, and F. P. GUENGERICH, unpublished).

Another aspect of regulation in humans involves polymorphisms. The term "polymorphism" has been used broadly to refer to unimodal and polymodal variations in drug oxidation in humans. Several examples of such variation have been documented (KÜPFER and PREISING, 1983). With the debrisoquine/sparteine and mephenytoin polymorphisms, family studies have shown that the "poor metabolizer" phenotype is heritable (MAHGOUB et al., 1977). P-450s have been isolated which appear to be involved in some of these polymorphisms (i.e., P-450_{DB}, P-450_{MP}, P-450_{NF}). In the case of P-450_{NF} can a relationship be shown between immunochemically-determined amounts of the protein (P-450_{NF}) and in vitro levels of the catalytic activity (nifedipine oxidase) (GUENGERICH et al., 1986a; WATKINS et al., 1985). Levels of mRNA related to P-450_{NF} were correlated with P-450_{NF} in our own studies (BEAUNE et al., 1986) and in the work of MOLOWA et al. (1986). Levels of P-450_{MP} mRNA are not correlated to either P-450_{MP} levels or to *S*-mephenytoin 4-hydroxylase activity in various liver samples (UMBENHAUER et al., 1987).

In no case has the molecular basis of the polymorphism been elucidated. There are two major possibilities. The first is mutations in the structural genes coding for the P-450s involved in the reactions that result in altered catalytic activity. The other most likely hypothesis is that polymorphisms involve changes in the regulation of levels of P-450 proteins. Most of the available evidence argues against the latter possibility, although the lack of specificity of the probes used to date may be masking some differences in levels of mRNA and protein.

5. References

ADESNIK, M. and M. ATCHISON, (1986), CRC Crit. Rev. Biochem. **19**, 247–305.

AGOSIN, M., A. MORELLO, R. WHITE, R. REPETTO, and J. PEDEMONTE, (1979), J. Biol. Chem. **254**, 9915–9920.

ANDERSON, K. E., A. H. CONNEY, and A. KAPPAS, (1982), Nutrit. Rev. **40**, 161 to 171.

ANDERSSON, S., H. BOSTRÖM, H. DANIELSSON, and K. WIKVALL, (1985), Methods Enzymol. **111**, 364–377.

AOYAMA, T., Y. IMAI, and R. SATO, (1982), in: Microsomes, Drug Oxidations, and Drug Toxicity, (R. SATO and R. KATO, eds.), Japan Scientific Societies Press, Tokyo, 83 to 84.

ARMSTRONG, R. N., C. PINTO-COELLVO, D. E. RYAN, P. E. THOMAS, and W. LEVIN, (1983), J. Biol. Chem. **258**, 2106–2108.

AUGUSTO, O., H. S. BEILAN, and P. R. ORTIZ DE MONTELLANO, (1982), J. Biol. Chem. **257**, 11288–11295.

BARON, J., T. T. KAWABATA, S. A. KNAPP, J. M. VOIGT, J. A. REDICK, W. B. JAKOBY, and F. P. GUENGERICH, (1984), in: Foreign Compound Metabolism, (J. CALDWELL and G. D. PAULSON, eds.), Taylor and Francis, London, 17–36.

BEAUNE, P. H., J.-P. FLINOIS, L. KIFFEL, P. KREMERS, and J.-P. LEROUX, (1985), Biochim. Biophys. Acta **840**, 364–370.
BEAUNE, P. H., D. R. UMBENHAUER, R. W. BORK, R. S. LLOYD, and F. P. GUENGERICH, (1986), Proc. Natl. Acad. Sci. U.S.A. **83**, 8064–8068.
BERGLUND, L., I. BJÖRKHEM, B. ANGELIN, and K. EINARSSON, (1986), Acta Chem. Scand. B **40**, 457–461.
BIGERSSON, C., E. T. MORGAN, H. JÖRNVALL, and C. VON BAHR, (1986), Biochem. Pharmacol. **35**, 3165–3166.
BLACK, S. D. and M. J. COON, (1986), in: Cytochrome P-450: Structure, Mechanism, and Biochemistry, (P. R. ORTIZ DE MONTELLANO, ed.), Plenum Press, New York, 161–216.
BÖCKER, R. and F. P. GUENGERICH, (1986), J. Med. Chem. **29**, 1596–1603.
BÖSTROM, H. and K. WIKVALL, (1982), J. Biol. Chem. **257**, 11755–11759.
BURKE, M. D. and P. H. WHITING, (1986), Clin. Nephrol. **25**, Suppl. 1, 5111–5116.
COOK, J. C. and E. HODGSON, (1986), Chem.-Biol. Interact. **58**, 233–240.
CRESTEIL, T., P. BEAUNE, C. CELIER, J.-P. LEROUX, and F. P. GUENGERICH, (1986), J. Pharmacol. Exptl. Therap. **236**, 269–276.
DALET, C., J. M. BLANCHARD, P. S. GUZELIAN, J. BARWICK, H. HARTLE, and P. MAUREL, (1986), Nucleic Acids Res. **14**, 5999–6015.
DALY, J. W., D. M. JERINA, and B. WITKOP, (1972), Experientia, **28**, 1129–1149.
DANNAN, G. A., F. P. GUENGERICH, L. S. KAMINSKY, and S. D. AUST, (1983), J. Biol. Chem. **258**, 1282–1288.
DANNAN, G. A., D. J. WAXMAN, and F. P. GUENGERICH, (1986), J. Biol. Chem. **261**, 10728–10735.
DEES, J. H., B. S. S. MASTERS, U. MULLER-EBERHARD, and E. F. JOHNSON, (1982), Cancer Res. **42**, 1423–1432.
DIETER, H. H., U. MULLER-EBERHARD, and E. F. JOHNSON, (1982), Biochem. Biophys. Res. Commun. **105**, 515–520.
DING, X., D. R. KOOP, B. L. CRUMP, and M. J. COON, (1986), Mol. Pharmacol. **30**, 370–378.
DISTLERATH, L. M. and F. P. GUENGERICH, (1984), Proc. Natl. Acad. Sci. U.S.A. **81**, 7348–7352.
DISTLERATH, L. M., P. E. B. REILLY, M. V. MARTIN, G. G. DAVIS, G. R. WILKINSON, and F. P. GUENGERICH, (1985), J. Biol. Chem. **260**, 9057–9067.
DISTLERATH, L. M. and F. P. GUENGERICH, (1987), in: Mammalian Cytochromes P-450, (F. P. GUENGERICH, ed.), Vol. 1, CRC Press, Boca Raton, Florida, 133.
ELSHOURBAGY, N. A. and P. S. GUZELIAN, (1980), J. Biol. Chem. **255**, 1279–1282.
EVANS, C. T., D. B. LEDESMA, T. Z. SCHULZ, E. R. SIMPSON, and C. R. MENDELSON, (1986), Proc. Natl. Acad. Sci. U.S.A. **83**, 6387–6391.
FAGAN, J. B., J. V. PASTEWKA, S. R. CHALBERG, E. GOZUKARA, F. P. GUENGERICH, and H. V. GELBOIN, (1986), Arch. Biochem. Biophys. **244**, 261–272.
FISHMAN, J., H. L. BRADLOW, J. SCHNEIDER, K. E. ANDERSON, and A. KAPPAS, (1980), Proc. Natl. Acad. Sci. U.S.A. **77**, 4957–4960.
FUJII-KURIYAMA, Y., Y. MIZUKAMI, K. KAWAJIRI, K. SOGAWA, and M. MURAMATSU, (1982), Proc. Natl. Acad. Sci. U.S.A. **79**, 2793–2797.
FUJINO, T., S. S. PARK, D. WEST, and H. V. GELBOIN, (1982), Proc. Natl. Acad. Sci. U.S.A. **79**, 3682–3686.
GAN, L-S. L., A. L. ACEBO, and W. L. ALWORTH, (1984), Biochemistry **23**, 3827–3836.
GED, C., P. H. BEAUNE, I. DALET, P. MAUREL, and J-P. LEROUX, (1985), in: Abstracts, International Symposium on Clinical and Basic Aspects of Enzyme Induction and Inhibition, July 6–7, 1985, Essen, Federal Republic of Germany.
GIBSON, G. G., T. C. ORTON, and P. P. TAMBURINI, (1982), Biochem. J. **203**, 161–168.
GILLETTE, J. R., (1979), Fed. Proc. **38**, 1900–1909.
GIUDICELLI, J. F. and J. P. TILLEMENT, (1977), Clin. Pharmacokin. **2**, 157–166.
GOLDSTEIN, J. A. and P. LINKO, (1984), Mol. Pharmacol. **25**, 185–191.

Gonzalez, F. J. and D. W. Nebert, (1985), Nucleic Acids Res. **13**, 7269–7288.
Gonzalez, F. J., D. W. Nebert, J. P. Hardwick, and C. B. Kasper, (1985), J. Biol. Chem. **260**, 7435–7441.
Gonzalez, F. J., S. Kimura, B-J. Song, J. Pastewka, H. V. Gelboin, and J. P. Hardwick, (1986a), J. Biol. Chem. **261**, 10667–10672.
Gonzalez, F. J., B. J. Song, and J. P. Hardwick, (1986b), Mol. Cell Biol. **6**, 2969 to 2976.
Goodwin, C. D., B. W. Cooper, and S. Margolis, (1982), J. Biol. Chem. **257**, 4469 to 4472.
Groves, J. T., G. E. Avaria-Neisser, K. M. Fish, M. Imachi, and R. L. Kuczkowski, (1986), J. Am. Chem. Soc. **108**, 3837–3838.
Groves, J. T., G. A. McClusky, R. E. White, and M. J. Coon, (1978), Biochem. Biophys. Res. Commun. **81**, 154–160.
Groves, J. T. and R. S. Myers, (1983), J. Am. Chem. Soc. **105**, 5191–5196.
Groves, J. T., R. Quinn, T. J. McMurry, M. Nakamura, G. Lang, and B. Boso, (1985), J. Am. Chem. Soc. **107**, 354–360.
Groves, J. T. and D. V. Subramanian, (1984), J. Am. Chem. Soc. **106**, 2177–2181.
Groves, J. T. and Y. Watanabe, (1986), J. Am. Chem. Soc. **108**, 507–508.
Guengerich, F. P., (1977), J. Biol. Chem. **252**, 3970–3979.
Guengerich, F. P., (1978), J. Biol. Chem. **253**, 7931–7939.
Guengerich, F. P., (1983), Biochemistry **22**, 2811–2820.
Guengerich, F. P., (1984), Ann. Rev. Nutr. **4**, 207–231.
Guengerich, F. P., (1986), Biochem. Biophys. Res. Commun. **138**, 193–198.
Guengerich, F. P., (1987a), in: Mammalian Cytochromes P-450, (F. P. Guengerich, ed.), CRC Press, Boca Raton, Florida, Vol.1, 1–24.
Guengerich, F. P., (1987b), in: Progress in Drug Metabolism, Vol. 10 (J. W. Bridges, L. F. Chasseaud, and G. G. Gibson, eds.), Taylor & Francis, London, Vol. 10, 1–54.
Guengerich, F. P., G. A. Dannan, S. T. Wright, M. V. Martin, and L. S. Kaminsky, (1982a), Biochemistry **21**, 6019–6030.
Guengerich, F. P. and D. C. Liebler, (1985), CRC Crit. Rev. Toxicol. **14**, 259–307.
Guengerich, F. P. and T. L. Macdonald, (1984), Acc. Chem. Res. **17**, 9–16.
Guengerich, F. P., M. V. Martin, P. H. Beaune, P. Kremers, T. Wolff, and D. J. Waxman, (1986a), J. Biol. Chem. **261**, 5051–5060.
Guengerich, F. P., D. Müller-Enoch, and I. A. Blair, (1986b), Mol. Pharmacol. **30**, 287–295.
Guengerich, F. P., P. Wang, and N. K. Davidson, (1982b), Biochemistry **21**, 1698 to 1706.
Guengerich, F. P., R. J. Willard, J. P. Shea, L. E. Richards, and T. L. Macdonald, (1984), J. Am. Chem. Soc. **106**, 6446–6447.
Guengerich, F. P., D. R. Umbenhauer, P. F. Churchill, P. H. Beaune, R. Böcker, R. G. Knodell, M. V. Martin, and R. S. Lloyd, (1987), Xenobiotica, **17, 311–316.**
Guillouzo, A., P. Beaune, M-N. Gascoin, J-M. Begue, J.-P. Campion, F. P. Guengerich, and C. Guguen-Guillouzo, (1985), Biochem. Pharmacol. **34**, 2991–2995.
Gut, J., T. Catin, P. Dayer, T. Kronbach, U. Zanger, and U. A. Meyer, (1986), J. Biol. Chem. **261**, 11734–11743.
Gut, J., R. Gasser, P. Dayer, T. Kronbach, T. Catin, and U. A. Meyer, (1984), Fed. Eur. Biol. Soc. Lett. **173**, 287–290.
Gut, J., U. T. Meier, T. Catin, and U. A. Meyer, (1987), Biochim. Biophys. Acta, **884**, 435–447.
Halpert, J. R., C. Balfour, N. E. Miller, and L. S. Kaminsky, (1986), Mol. Pharmacol. **30**, 19–24.
Halpert, J., B. Näslund, and I. Betnér, (1983), Mol. Pharmacol. **23**, 445–452.
Hammons, G. J., F. P. Guengerich, C. C. Weis, F. A. Beland, and F. F. Kadlubar, (1985), Cancer Res. **45**, 3578–3585.

HANKINSON, O., R. D. ANDERSEN, B. W. BIRREN, F. SANDER, M. NEGISHI, and D. W. NEBERT, (1985), J. Biol. Chem. **260**, 1790—1795.
HARADA, N., G. T. MIWA, J. S. WALSH, and A. Y. H. LU, (1984), J. Biol. Chem. **259**, 3005—3010.
HARADA, N. and M. NEGISHI, (1984a), J. Biol. Chem. **259**, 1265—1271.
HARADA, N. and M. NEGISHI, (1984b), J. Biol. Chem. **259**, 12285—12290.
HARDWICK, J. P., P. LINKO, and J. A. GOLDSTEIN, (1985), Mol. Pharmacol. **27**, 676 to 682.
HARDWICK, J. P., B-J. SONG, E. HUBERMAN, and F. J. GONZALEZ, (1987), J. Biol. Chem., **262**, 801—810.
HAUGEN, D. A., T. A. VAN DER HOEVEN, and M. J. COON, (1975), J. Biol. Chem. **250**, 3567—3570.
HEINEMANN, F. S. and J. OZOLS, (1982), J. Biol. Chem. **257**, 14988—14999.
HEINEMANN, F. S. and J. OZOLS, (1983), J. Biol. Chem. **258**, 4195—4201.
HEINRICH, B., (1986), Diplomarbeit, University of Munich.
HIGASHI, Y., H. YOSHIOKA, M. YAMANE, O. GOTOH, and Y. FUJII-KURIYAMA, (1986), Proc. Natl. Acad. Sci. U.S.A. **83**, 2841—2845.
HUANG, M.-T., S. B. WEST, and A. Y. H. LU, (1976), J. Biol. Chem. **251**, 4659—4665.
HUANG, Y. Y., T. HARA, S. G. SLIGAR, M. J. COON, and T. KIMURA, (1986), Biochemistry **25**, 1390—1394.
ICHIHARA, K., K. ISHIHARA, M. KAKU, K. OGITA, S. YAMAMOTO, and M. KUSENOSE, (1983), Biochem. Int. **7**, 179—186.
ICHIHARA, K., K. ISHIHARA, E. KUSENOSE, and M. KUSENOSE, (1981), J. Biochem. (Tokyo) **89**, 1821—1827.
INGELMAN-SUNDBERG, M., I. JOHANSSON, and A. HANSON, (1979), Acta Biol. Med. Germ. **38**, 379—388.
JAISWAL, A. K., F. J. GONZALEZ, and D. W. NEBERT, (1985a), Nucleic Acids Res. **13**, 4503—4520.
JAISWAL, A. K., F. J. GONZALEZ, and D. W. NEBERT, (1985b), Science **228**, 80—83.
JAISWAL, A. K., D. W. NEBERT, and F. J. GONZALEZ, (1986), Nucleic Acids Res. **14**, 6773—6774.
JANSSON, I., J. MOLE, and J. B. SCHENKMAN, (1985a), J. Biol. Chem. **260**, 7084—7093.
JANSSON, I., P. P. TAMBURINI, L. V. FAVREAU, and J. B. SCHENKMAN, (1985b). Drug Metab. Disp. **13**, 453—458.
JERINA, D. M., D. P. MICHAND, R. J. FELDMANN, R. N. ARMSTRONG, K. P. VYAS. D. R. THAKKER, D. YAGI, P. E. THOMAS, D. E. RYAN, and W. LEVIN, (1982), in: Microsomes, Drug Oxidations, and Drug Toxicity, (R. SATO and R. KATO, eds.), Wiley-Interscience, New York, 195—201.
JI, S., J. J. LEMASTERS, V. CHRISTENSON, and R. G. THURMAN, (1982), Proc. Natl. Acad. Sci. U.S.A. **79**, 5415—5419.
JOHN, M. E., T. OKAMURA, A. DEE, B. ADLER, M. C. JOHN, P. C. WHITE, E. R. SIMPSON, and M. R. WATERMAN, (1986), Biochemistry **25**, 2846—2853.
JOHNSON, E. F., (1980), J. Biol. Chem. **255**, 304—309.
JOHNSON, E. F. and A. MULLER-EBERHARD, (1977), Biochem. Biophys. Res. Commun. **76**, 644—651.
JONES, D. P. and H. S. MASON, (1978), J. Biol. Chem. **253**, 4874—4880.
JONES, P. B. C., A. G. MILLER, D. I. ISRAEL, D. R. GALEAZZI, and J. P. WHITLOCK. Jr., (1984), J. Biol. Chem. **259**, 12357—12363.
JONES, P. B. C., D. R. GALEAZZI, J. M. FISHER, and J. P. WHITLOCK, Jr., (1985), Science **227**, 1499—1502.
KAKU, M., K. ICHIHARA, E. KUSENOSE, K. OGITA, S. YAMAMOTO, I. YANO, and M. KUSENOSE, (1984), J. Biochem. (Tokyo) **96**, 1883—1891.
KAKU, M., E. KUSENOSE, S. YAMAMOTO, S. ICHIHARA, K. ICHIHARA, and M. KUSENOSE, (1985), J. Biochem. (Tokyo) **97**, 663—670.

KAMATAKI, T., K. MAEDA, M. SHIMADA, K. KITANI, T. NAGAI, and R. KATO, (1985), J. Pharmacol. Exp. Therap. **233**, 222—228.
KAMATAKI, T., K. MAEDA, Y. YAMAZOE, N. MATSUDA, K. ISHII, and R. KATO, (1983a), Mol. Pharmacol. **24**, 146—155.
KAMATAKI, T., K. MAEDA, Y. YAMAZOE, T. NAGAI, and R. KATO, (1983b), Arch. Biochem. Biophys. **225**, 758—770.
KAMINSKY, L. S., G. A. DANNAN, and F. P. GUENGERICH, (1984), Eur. J. Biochem. **141**, 141—148.
KAMINSKY, L. S., M. J. FASCO, and F. P. GUENGERICH, (1980), J. Biol. Chem. **255**, 85—91.
KAWAJIRI, K., O. GOTOH, K. SOGAWA, Y. TAGASHIRA, M. MURAMATSU, and Y. FUJII-KURIYAMA, (1984), Proc. Natl. Acad. Sci. U.S.A. **81**, 1649—1653.
KAWAJIRI, K., J. WATANABE, O. GOTOH, Y. TAGASHIRA, K. SOGAWA, and Y. FUJII-KURIYAMA, (1986), Eur. J. Biochem. **159**, 219—225.
KELLERMAN, G., C. R. SHAW, and M. LUYTEN-KELLERMAN, (1973), New Engl. J. Med. **289**, 934—937.
KIMURA, S., F. J. GONZALEZ, and D. W. NEBERT, (1984), J. Biol. Chem. **259**, 10705 to 10713.
KIMURA, S. and D. W. NEBERT, (1986), Nucleic Acids Res. **14**, 6765—6766.
KITADA, M., T. KAMATAKI, K. ITAHASHI, T. RIKIHISA, R. KATO, and Y. KANAKUBO, (1985), Arch. Biochem. Biophys. **241**, 275—280.
KLEINBLOESEM, C. H., P. VAN BRUMMELEN, H. FABER, M. DANHOF, N. P. E. VERMEULEN, and D. D. BREIMER, (1984), Biochem. Pharmacol. **33**, 3721—3724.
KOMORI, M., Y. IMAI, and R. SATO, (1984), J. Biochem. (Tokyo) **95**, 1379—1388.
KOOP, D. R. and M. J. COON, (1979), Biochem. Biophys. Res. Commun., **91**, 1075 to 1081.
KOOP, D. R., E. T. MORGAN, G. E. TARR, and M. J. COON, (1982), J. Biol. Chem. **257**, 8472—8480.
KRONBACH, T., P. DAYER, and U. A. MEYER, (1985), Experientia **41**, 822.
KUNZE, K., B. L. K. MANGOLD, C. WHEELER, H. S. BEILAN, and P. R. ORTIZ DE MONTELLANO, (1983), J. Biol. Chem. **258**, 4202—4207.
KÜPFER, A. and R. PREISIG, (1983), Semin. Liver Dis. **3**, 341—354.
LAMBETH, J. D., D. W. SEYBERT, J. R. LANCASTER, Jr., J. C. SALERNO, and H. KAMIN, (1982), Mol. Cell. Biochem. **45**, 13—31.
LARREY, D., L. M. DISTLERATH, G. A. DANNAN, G. R. WILKINSON, and F. P. GUENGERICH, (1984), Biochemistry **23**, 2787—2795.
LECAMWASAM, D. S., C. FRANKLIN, and P. TURNER, (1975), Br. J. Clin. Pharmacol. **2**, 275—279.
LIEBLER, D. C. and F. P. GUENGERICH, (1983), Biochemistry **22**, 5482—5489.
LIEM, H. H., U. MULLER-EBERHARD, and E. F. JOHNSON, (1980), Mol. Pharmacol. **18**, 565—570.
LUBET, R. A., R. T. MAYER, J. W. CAMERON, R. W. NIMS, M. D. BURKE, T. WOLFF, and F. P. GUENGERICH, (1985), Arch. Biochem. Biophys. **238**, 43—48.
MACGEOCH, C., E. T. MORGAN, J. HALPERT, and J-Å. GUSTAFSSON, (1984), J. Biol. Chem. **259**, 15433—15439.
MACLEOD, S. M., H. G. GILES, B. BENGERT, F. F. LIU, and E. M. SELLERS, (1979), J. Clin. Pharmacol. **19**, 15—19.
MAHGOUB, A., J. R. IDLE, L. G. DRING, R. LANCASTER, and R. L. SMITH, (1977), Lancet **2**, 584—586.
MANSUY, D., J. LECLAIRE, M. FONTECAVE, and M. MOMENTEAU, (1984), Biochem. Biophys. Res. Commun. **119**, 319—325.
MATHIS, J. M., R. A. PROUGH, R. N. HINES, E. BRESNICK, and E. R. SIMPSON, (1986), Arch. Biochem. Biophys. **246**, 439—448.
MCMARTIN, D. N., J. A. O'CONNOR, Jr., M. J. FASCO, and L. S. KAMINSKY, (1980), Toxicol. Appl. Pharmacol. **54**, 411—419.

MIKI, N., T. SUGIYAMA, T. YAMANO, and Y. MIYAKE, (1981), Biochem. Int. **3**, 217.
MIKI, N., R. MIURA, T. SUGIYAMA, T. YAMANO, and Y. MIYAKE, (1982), Biochem. Int. **5**, 511–517.
MIWA, G. T., J. S. WALSH, W. J. A. VANDENHEAVEL, B. ARISON, E. SESTOKAS, R. BUHS, A. ROSEGAY, S. AVERMITILIS, A. Y. H. LU, M. A. R. WALSH, R. W. WALKER, R. TAUB, and T. A. JACOB, (1982), Drug Metab. Disp. **10**, 268–274.
MILLER, R. E. and F. P. GUENGERICH, (1982), Biochemistry **21**, 1090–1097.
MIZUKAMI, Y., K. SOGAWA, Y. SUWA, M. MURAMATSU, and Y. FUJII-KURIYAMA, (1983), Proc. Natl. Acad. Sci. U.S.A. **80**, 3958–3962.
MOLOWA, D. T., E. G. SCHUETZ, S. A. WRIGHTON, P. B. WATKINS, P. KREMERS, G. MENDEZ-PICON, G. A. PARKER, and P. S. GUZELIAN, (1986), Proc. Natl. Acad. Sci. U.S.A. **83**, 5311–5315.
MORGAN, E. T., D. R. KOOP, and M. J. COON, (1982), J. Biol. Chem. **257**, 13951–13957.
MURAKAMI, K. and K. OKUDA, (1981), J. Biol. Chem. **256**, 8658–8662.
NEBERT, D. W., M. ADESNIK, M. J. COON, R. W. ESTABROOK, F. J. GONZALEZ, F. P. GUENGERICH, I. C. GUNSALUS, E. F. JOHNSON, B. KEMPER, W. LEVIN, I. R. PHILLIPS, R. SATO, and M. R. WATERMAN, (1987), DNA 6, 1–11.
NEGISHI, M. and D. W. NEBERT, (1979), J. Biol. Chem. **254**, 11015–11023.
OGITA, K., E. KUSENOSE, S. YAMAMOTO, K. ICHIHARA, and M. KUSENOSE, (1983), Biochem. Int. **6**, 191–198.
OHYAMA, T., D. W. NEBERT, and M. NEGISHI, (1984), J. Biol. Chem. **259**, 2675–2682.
OKINO, S. T., L. C. QUATTROCHI, H. J. BARNES, S. OSANTO, K. J. GRIFFIN, E. F. JOHNSON, and R. H. TUKEY, (1985), Proc. Natl. Acad. Sci. U.S.A. **82**, 5310–5314.
OMNIECINSKI, C. J., (1986), Nucleic Acids Res. **14**, 1525–1539.
ORTIZ DE MONTELLANO, P. R. and M. A. CORREIA, (1983), Ann. Rev. Pharmacol. Toxicol. **23**, 481–503.
OTTON, S. V., T. INABA, and W. KALOW, (1984), Life Sci. **34**, 73–80.
OZOLS, J., (1986), J. Biol. Chem. **261**, 3965–3979.
OZOLS, J., F. S. HEINEMANN, and E. F. JOHNSON, (1985), J. Biol. Chem. **260**, 5427 to 5434.
PARK, B. K., (1981), Br. J. Clin. Pharmacol. **12**, 97–102.
PHILLIPS, I. R., E. A. SHEPARD, A. ASHWORTH, and B. R. RABIN, (1985), Proc. Natl. Acad. Sci. U.S.A. **82**, 983–987.
PICKETT, C. B., R. L. JETER, J. MORIN, and A. Y. H. LU, (1981), J. Biol. Chem. **256**, 8815–8820.
POLAND, A. and J. C. KNUTSON, (1982), Ann. Rev. Pharmacol. Toxicol. **22**, 517–554.
POULOS, T. L., B. C. FINZEL, I. C. GUNSALUS, G. C. WAGNER, and J. KRAUT, (1985), J. Biol. Chem. **260**, 16122–16130.
QUATTROCHI, L. C., U. R. PENDURTHI, S. T. OKINO, C. POTENZA, and R. H. TUKEY, (1986), Proc. Natl. Acad. Sci. U.S.A. **83**, 6731–6735.
RAMPERSAUD, A., D. J. WAXMAN, D. E. RYAN, W. LEVIN, and F. J. WALZ, Jr., (1985), Arch. Biochem. Biophys. **243**, 174–183.
RICHARDSON, C. J., K. L. N. BLOCKA, R. T. ROSS, and R. K. VERBEECK, (1985), Clin. Pharmacol. Therap. **37**, 13–18.
ROBERTS, R. K., P. V. DESMOND, G. R. WILKINSON, and S. SCHENKER, (1979), Clin. Pharmacol. Therap. **25**, 826–831.
RYAN, D. E., S. IIDA, A. W. WOOD, P. E. THOMAS, C. S. LIEBER, and W. LEVIN, (1984), J. Biol. Chem. **259**, 1239–1250.
RYAN, D. E., L. RAMANATHAN, S. IIDA, P. E. THOMAS, M. HANIU, J. E. SHIVELY, C. S. LIEBER, and W. LEVIN, (1985), J. Biol. Chem. **260**, 6385–6393.
RYAN, D. E., P. E. THOMAS, D. KORZENIOWSKI, and W. LEVIN, (1979), J. Biol. Chem. **254**, 1365–1374.
RYAN, D. E., P. E. THOMAS, and W. LEVIN, (1980), J. Biol. Chem. **255**, 7941–7955.
RYAN, D. E., P. E. THOMAS, L. M. REIK, and W. LEVIN, (1982a), Xenobiotica **12**, 727–744.

RYAN, D. E., A. W. WOOD, P. E. THOMAS, F. G. WALZ, Jr., P-M. YUAN, J. E. SHIVELY, and W. LEVIN, (1982b), Biochim. Biophys. Acta **709**, 273–283.

SADANO, H. and T. OMURA, (1983), Biochem. Biophys. Res. Commun. **116**, 1013–1019.

SADANO, H. and T. OMURA, (1985), J. Biochem. (Tokyo) **98**, 1321–1331.

SAKAKI, T., K. OEDA, M. MIYOSHI, and H. OHKAWA, (1985), J. Biochem. **98**, 167–175.

SAKAKI, T., K. OEDA, Y. YABUSAKI, and H. OHKAWA, (1986), J. Biochem. (Tokyo) **99**, 741–749.

SANGHVI, A., E. GRASSI, and W. DIVEN, (1983), Proc. Natl. Acad. Sci. U.S.A. **80**, 2175–2178.

SCHWAB, G. E. and E. F. JOHNSON, (1985a), Arch. Biochem. Biophys. **237**, 17–26.

SCHWAB, G. E. and E. F. JOHNSON, (1985b), Biochemistry **24**, 7222–7226.

SCHWAB, G. E. and E. F. JOHNSON, (1987), in: Mammalian Cytochromes P-450, (F. P. GUENGERICH, ed.), CRC Press, Boca Raton, Florida, Vol. **1**, 55–105.

SCHUETZ, E. G., S. A. WRIGHTON, S. H. SAFE, and P. S. GUZELIAN, (1986), Biochemistry **25**, 1124–1133.

SERABJIT-SINGH, C. J., P. W. ALBRO, I. G. C. ROBERTSON, and R. M. PHILPOT, (1983), J. Biol. Chem. **285**, 12827–12834.

SERABJIT-SINGH, C. J., J. R. BEND, and R. M. PHILPOT, (1985), Mol. Pharmacol. **28**, 72 to 79.

SHIMADA, T. and F. P. GUENGERICH, (1985), Mol. Pharmacol. **28**, 215–219.

SHIMADA, T., K. S. MISONO, and F. P. GUENGERICH, (1986), J. Biol. Chem. **261**, 909 to 921.

SIMMONS, D. L., P. MCQUIDDY, and C. B. KASPER, (1987), J. Biol. Chem., **262**, 326–332.

SOGAWA, K., A. FUJISAWA-SEHARA, M. YAMANE, and Y. FUJII-KURIYAMA, (1986), Proc. Natl. Acad. Sci. U.S.A. **83**, 8044–8048.

SONG, B-J., H. V. GELBOIN, S-S. PARK, C. S. YANG, and F. J. GONZALEZ, (1987), J. Biol. Chem., **261**, 16679–16697.

STEARNS, R. A. and P. R. ORTIZ DE MONTELLANO, (1985), J. Am. Chem. Soc. **107**, 4081 to 4082.

STEWARD, A. R., S. A. WRIGHTON, D. S. PASCO, J. B. FAGAN, D. LI, and P. S. GUZELIAN, (1985), Arch. Biochem. Biophys. **241**, 494–508.

SUNDHEIMER, D. W., M. B. CAVENESS, and J. A. GOLDSTEIN, (1983), Arch. Biochem. Biophys. **226**, 548–557.

TANG, P. M. and J. Y. L. CHIANG, (1986), Biochem. Biophys. Res. Commun. **134**, 797–802.

TARR, G. E., S. D. BLACK, V. S. FUJITA, and M. J. COON, (1983), Proc. Natl. Acad. Sci. U.S.A. **80**, 6552–6556.

THAKKER, D. R., W. LEVIN, H. YAGI, H. J. C. YEH, D. E. RYAN, P. E. THOMAS, A. H. CONNEY, and D. M. JERINA, (1986), J. Biol. Chem. **261**, 5404–5413.

THOMAS, P. E., D. KORZENIOWSKI, D. RYAN, and W. LEVIN, (1979), Arch. Biochem. Biophys. **192**, 524–532.

THOMAS, P. E., L. M. REIK, D. E. RYAN, and W. LEVIN, (1983), J. Biol. Chem. **258**, 4590–4598.

THURMAN, R. G. and F. C. KAUFFMAN, (1980), Pharmacol. Rev. **31**, 229–251.

TRAYLOR, T. G., Y. IANAMOTO, and T. NAKANO, (1986), J. Am. Chem. Soc. **108**, 3529 to 3531.

TRZASKOS, J., S. KAWATA, and J. L. GAYLOR, (1986), J. Biol. Chem. **261**, 14651–14657.

TUKEY, R. H., S. T. OKINO, H. J. BARNES, K. J. GRIFFIN, and E. F. JOHNSON, (1985), J. Biol. Chem. **260**, 13347–13354.

UENG, T-H. and A. P. ALVARES, (1982), Mol. Pharmacol. **22**, 221–228.

UMBENHAUER, D. R., M. V. MARTIN, R. S. LLOYD, and F. P. GUENGERICH, (1987), Biochemistry **26**, 1094–1099.

VAN BLADEREN, P. J., R. N. ARMSTRONG, D. COBB, D. R. THAKKER, D. E. RYAN, P. E. THOMAS, N. D. SHARMA, D. R. BOYD, W. LEVIN, and D. M. JERINA, (1982), Biochem. Biophys. Res. Commun. **106**, 602–609.

WANG, P. P., P. BEAUNE, L. S. KAMINSKY, G. A. DANNAN, F. F. KADLUBAR, D. LARREY, and F. P. GUENGERICH, (1983), Biochemistry **22**, 5375—5383.

WANG, P., P. S. MASON, and F. P. GUENGERICH, (1980), Arch. Biochem. Biophys. **199**, 206—219.

WATKINS, P. B., S. A. WRIGHTON, P. MAUREL, E. G. SCHUETZ, G. MENDEZ-PICON, G. A. PARKER, and P. S. GUZELIAN, (1985), Proc. Natl. Acad. Sci. U.S.A. **82**, 6310—6314.

WATKINS, P. B., S. A. WRIGHTON, E. G. SCHUETZ, P. MAUREL, and P. S. GUZELIAN, (1986), J. Biol. Chem. **261**, 6264—6271.

WAXMAN, D. J., (1986), in: Cytochrome P-450, (P. R. ORTIZ DE MONTELLANO, ed.), Plenum Press, New York, 525—539.

WAXMAN, D. J., G. A. DANNAN, and F. P. GUENGERICH, (1985a), Biochemistry **24**, 4409—4417.

WAXMAN, D. J., E. J. HOLSZTYNSKA, M. KRISHNAN, and C. ATTISANO, (1985b), Pharmacologist **27**, 115.

WHITE, P. C., M. I. NEW, and B. DUPONT, (1984), Proc. Natl. Acad. Sci. U.S.A. **81**, 7505—7509.

WHITLOCK, J. P., Jr., (1986), Ann. Rev. Pharmacol. Toxicol., **26**, 333—369.

WHITLOCK, J. P., Jr. and H. V. GELBOIN, (1979), Pharmacol. Ther. **4**, 587—599.

WILLIAMS, D. E., S. E. HALE, R. T. OKITA, and B. S. S. MASTERS, (1984), J. Biol. Chem. **259**, 14600—14608.

WOLF, C. R., M. M. SZUTOWSKI, C. M. BALL, and R. M. PHILPOT, (1978), Chem.-Biol. Interact. **21**, 29—43.

WOLFF, T., L. M. DISTLERATH, M. WORTHINGTON, J. GROOPMAN, G. J. HAMMONS, F. F. KADLUBAR, R. A. PROUGH, M. V. MARTIN, and F. P. GUENGERICH, (1985), Cancer Res. **45**, 2116—2122.

WRIGHTON, S. A., P. MAUREL, E. G. SCHUETZ, P. B. WATKINS, B. YOUNG, and P. S. GUZELIAN, (1985), Biochemistry **24**, 2171—2178.

WRIGHTON, S. A., C. CAMPANILE, P. E. THOMAS, S. L. MAINES, P. B. WATKINS, G. PARKER, G., MENDEZ-PICON, M. HANIU, J. E. SHIVELY, W. LEVIN, and P. S. GUZELIAN, (1986a), Mol. Pharmacol. **29**, 405—410.

WRIGHTON, S. A., P. E. THOMAS, D. T. MOLOWA, M. HANIU, J. E. SHIVELY, S. L. MAINES, P. B. WATKINS, G. PARKER, G. MENDEZ-PICON, W. LEVIN, and P. S. GUZELIAN, (1986b), Biochemistry **25**, 6731—6735.

YABUSAKI, Y., M. SHIMIZU, H. MURAKAMI, H. NAKAMURA, K. OEDA, and H. OHKAWA, (1984), Nucleic Acids Res. **12**, 2929—2938.

YAMAMOTO, S., E. KUSENOSE, K. OGITA, M. KAKU, K. ICHIHARA, and M. KUSENOSE, (1984), J. Biochem. (Tokyo) **96**, 593—603.

YANG, C. S., D. R. KOOP, T. WANG, and M. J. COON, (1985), Biochem. Biophys. Res. Commun. **128**, 1007—1013.

YANG, S. K., M. MUSHTAQ, and P-L. CHIU, (1985), in: Polycyclic Hydrocarbons and Carcinogens, American Chemical Society Symposium Series No. 283, (R. G. HARVEY, ed.), American Chemical Society, Washington, D. C., 19—34.

YOSHIOKA, H., T. MIYATA, and T. OMURA, (1984), J. Biochem. (Tokyo) **95**, 937—947.

YOSHIOKA, H., K. I. MOROHASHI, K. SOGOWA, T. MIYATA, K. KAWAJIRI, T. HIROSE, S. INAYAMA, Y. FUJII-KURIYAMA, and T. OMURA, (1987), J. Biol. Chem., **262**, 1706 to 1711.

YUAN, P-M., D. E. RYAN, W. LEVIN, and J. F. SHIVELY, (1983), Proc. Natl. Acad. Sci. U.S.A. **80**, 1169—1173.

ZUBER, M. X., M. E. JOHN, T. OKAMURA, E. R. SIMPSON, and M. R. WATERMAN, (1986), J. Biol. Chem. **261**, 2475—2482.

Chapter 4
Multiple Activities of Cytochrome P-450

A. I. Archakov and A. A. Zhukov

1. Introduction

Table 1 lists functions performed by proteins containing heme as the prosthetic group. Oxygenases catalyze reactions involving oxygen insertion into an organic molecule and are subdivided into di- and monooxygenases depending on how many atoms from the dioxygen molecule are found in the oxygenated product. In the reactions catalyzed by oxidases, dioxygen acts only as an electron acceptor. There are three types of oxidase reactions, but heme enzymes are classically considered capable of only four-electron dioxygen reduction to give two molecules of water. In addition, hemoproteins participate in electron transport in the respiratory and other chains and can reversibly bind oxygen.

Table 1. Biological functions of hemoproteins

Enzyme type	Reaction	Example
I. Oxygenases		
1. Dioxygenases	$A + O_2 \rightarrow A(O)_2$	tryptophan-2,3-dioxygenase
2. Monooxygenases	$DH_2 + O_2 + AH \rightarrow D + AOH + H_2O$	cytochrome P-450
II. Oxidases		
1. 1-electron	$DH_2 + 2O_2 \rightarrow D + 2H^+ + 2O_2^-$	?
2. 2-electron	$DH_2 + O_2 \rightarrow D + H_2O_2$	?
3. 4-electron	$2DH_2 + O_2 \rightarrow 2D + 2H_2O$	cytochrome oxidase
III. Peroxidases	$ROOH + 2DH_2 \rightarrow ROH + 2DH^{\cdot} + H_2O$	
IV. Electron carriers	$D_{red} + Fe^{3+} \rightarrow D_{ox} + Fe^{2+}$ $A_{ox} + Fe^{2+} \rightarrow A_{red} + Fe^{3+}$	cytochrome c, cytochrome b_5
V. Oxygen carriers	$Fe^{2+} + O_2 \rightarrow (FeO_2)^{2+}$	myoglobin, hemoglobin

A is organic substrate, DH_2 and D_{red} are electron donors, A_{ox} is electron acceptor, Fe is heme iron.

Cytochromes P-450 are typical external monooxygenases and catalyze oxygenation of various organic substrates using NAD(P)H and molecular oxygen as cosubstrates. At the same time, these enzymes are known to be capable of utilizing organic hydroperoxides and hydrogen peroxide as cosubstrates in hydroxylation reactions; that is, to act as peroxidases. Cytochrome P-450 can also reduce certain compounds, thus acting as an electron carrier. P-450-dependent monooxygenations are accompanied by the release of superoxide radicals and hydrogen peroxide, i.e. oxidase reaction. Water formation is characteristic of P-450 functioning as a monooxygenase. That is why the enzyme is also termed a mixed function oxidase. However, P-450 has recently been shown capable of acting as a true four-electron oxidase, water being the only product derived from molecular oxygen. Finally, indirect evidence is available indicating that P-450 can participate in dioxygenase reactions and even in cellular oxygen transport.

The present article reviews the data available on multiple activities of cytochrome P-450. Since questions concerned with the monooxygenase reactions of the enzyme have been repeatedly and extensively reviewed (see for example White and Coon, 1980; Sligar et al., 1984; Guengerich and McDonald, 1984; Metelitza, 1984; Ortiz de Montellano, 1986) cytochrome P-450 activities other than monooxygenase will be given primary consideration in this chapter.

2. Oxidase activity

2.1. Stoichiometry

After $^{18}O_2$ was shown to be incorporated into substrate by a non-specific NADPH-dependent liver microsomal hydroxylase (Udenfriend et al., 1956; Posner et al., 1961; Baker and Chaykin, 1962; McMahon et al., 1969; Parli et al., 1971) it became clear that the enzyme represents a mixed function oxidase and the reaction must proceed according to the following equation,

$$NADPH + H^+ + O_2 + AH = NADP^+ + H_2O + AOH \quad (1)$$

In other words, there must be a 1:1:1 ratio between NADPH oxidized, O_2 consumed, and product formed. Such stoichiometry is easy to show for the reactions catalyzed by P-450 CAM from *Pseudomonas putida*. Camphor oxidation in this system accounts for 95% of NADH utilized (Sligar, 1976). With mammalian P-450-dependent systems, the situation is much more complex. In this case, the ratio of substrate oxidized to NADPH or O_2 consumed is usually smaller than unity (see Table 2 for several examples). In these systems, in contrast to bacterial ones, NADPH oxidation occurs also in the absence of added substrate (so called NADPH-oxidase activity, or NADPH free oxidation) (Gillette et al., 1957; Archakov et al., 1975; Estabrook and Werringloer, 1974; Zhukov and Archakov, 1982) at a rate that is often higher than the oxidation rate of certain substrates. In early studies, free NADPH oxidation and that associated with hydroxylation were regarded as proceeding independently. Stoichiometry was calculated based on the assumption that substrate addition does not affect reactions in which NADPH was oxidized prior to the addition, the difference between NADPH oxidation rates in the presence and in the absence of substrate being taken into account. Sometimes this approach gave stoichiometry close to that expected. Thus, nearly 1:1

Table 2. Stoichiometry of P-450-dependent monooxygenations

System	Substrate	ΔAOH/ΔNADPH	Reference
Rat liver microsomes	ethylmorphine	0.68	(HOLTZMAN, 1970)
	ethylmorphine	0.84	(STRIPP et al., 1972)
	benzphetamine	0.55	(STRIPP et al., 1972)
Rabbit liver microsomes	benzphetamine	0.54	(ZHUKOV and ARCHAKOV, 1985a)
	cyclohexane	0.85	(STAUDT et al., 1974)
	n-hexane	0.46	(STAUDT et al., 1974)
	perfluorohexane	0	(STAUDT et al., 1974)
Microsomes from adrenal cortex	17-hydroxyprogesterone	0.3	(COOPER et al., 1965)
	androst-4-ene-3,17-dione	0	(NARASIMHULU, 1971)
Reconstituted Systems:			
P-450 LM2	benzphetamine	0.43	(NORDBLOM and COON, 1977)
	cyclohexane	0.76	(NORDBLOM and COON, 1977)
P-450 CAM	camphor	0.95	(GELB et al., 1982a)
	dehydrocamphor	0.75	(GELB et al., 1982a)
	ketopericyclocamphanone	0	(SLIGAR, 1976)
P-450 11β	deoxycorticosterone	0.25	(MARTSEV et al., 1985)

ratios of substrate oxidized to cosubstrates (NADPH or O_2) consumed were obtained for trimethylamine *N*-oxidation in pig liver microsomes (BAKER and CHAYKIN, 1962), C_{21}-hydroxylation of 17-hydroxyprogesterone in adrenal microsomes (COOPER et al., 1965), ethylmorphine and aminopyrine *N*-demethylation in liver microsomes of untreated (HOLTZMAN, 1970; HOLTZMAN et al., 1977) and phenobarbital-induced (Orrenius, 1965) rats, respectively, as well as benzphetamine oxidation in the system reconstituted from partially purified P-450, NADPH-dependent reductase, and microsomal lipid (LU et al., 1970). Such a stoichiometry, however, is by no means always observed. As demonstrated by STRIPP et al. (1972) the oxidation rate of a number of substrates is substantially higher than the increase in NADPH oxidation produced by their addition. The ratio is 2.0 with ethylmorphine and imipramine, 4.3 with aminopyrine, 2.6 with *p*-chloromonomethylaniline. Moreover, added substrate may produce no increase in NADPH uptake at all but still undergo oxidation (GILLETTE et al., 1957). These results indicate that substrate can compete for NADPH with the routes of its free oxidation and they disprove the assumption about the independence of substrate-free and substrate-stimulated oxidation of the cosubstrate. Of greater importance in the light of the problem under consideration here is, however, the fact that substrate-stimulated NADPH uptake may greatly exceed the rate of substrate oxidation (ESTABROOK and COHEN, 1969; GILLETTE, 1966; COHEN and ESTABROOK, 1971) that is, substrate itself can stimulate NADPH oxidation not associated with the monooxygenase reaction. While the above data could be accounted for by, for instance, heterogeneity of microsomal P-450s, with the substrate-bound and substrate-free fractions independently catalyzing monooxygenase and oxidase reactions, respectively, the latter results clearly indicate that even substrate-bound P-450 can combine the properties of a monooxygenase and an oxidase. JEFFERY and MANNERING (1974, 1979) as well as BUENING and FRANKLIN (1974) have shown, however, that the NADPH uptake increase may be overestimated due to the pyrophosphatase reaction proceeding in microsomes (especially with uninduced preparations). But NORDBLOM and COON in their study of a reconstituted system composed of P-450 LM2, NADPH-dependent reductase, and dilauroylphosphatidylcholine have demonstrated the stoichiometry observed as being a combination of two parallel P-450-catalyzed reactions, of which the first represents substrate monooxygenation (eq. 1) and the second is hydrogen peroxide formation by the oxidase mechanism:

$$NADPH + O_2 + H^+ \rightarrow NADP^+ + H_2O_2 \qquad (2)$$

Substrate addition to the system increased the rate of H_2O_2 generation (NORDBLOM and COON, 1977).

Thus, the data available are more consistent with the concept which allows the uncoupling of P-450-catalyzed reactions. This concept implies that all P-450 substrates can be subdivided into true ones, partial uncouplers, and complete uncouplers (pseudosubstrates) depending on the ratio of NADPH fractions consumed for substrate oxidation (monooxygenase reaction) and for the formation of reduced oxygen species (oxidase reactions).

The uncoupling phenomenon, when a fraction of activated oxygen is released from the complex with an enzyme without reacting with substrate, is well known in monooxygenase biochemistry. Thus, for the FAD-containing, NADH-dependent bacterial enzyme salicylate hydroxylase, benzoate and its derivatives serve as complete uncouplers, while many substituted salicylates are partial uncouplers, NADH in their presence being partly consumed for the oxygenase reaction and partly for H_2O_2 formation (WHITE-STEVENS and KAMIN, 1970, 1972; KATAGIRI et al., 1973). 6-*H*ydroxynicotinate (STEENNIS et al., 1973; HOWELL and MASSEY, 1970) and 3,4-dihydroxybenzoate (SPECTOR and MASSEY, 1972) are complete uncouplers of *p*-hydroxybenzoate hydroxylase from *Pseudomonas*. With orcinol hydroxylase from *Pseudomonas putida*, resorcinol is a partial and *m*-cresol a complete uncoupler (OTHA and RIBBONS, 1970). 2,5-, 2,6-, 3,4-dichloro, and 2,4,5-trichlorophenol are pseudosubstrates for 2,4-dichlorophenol hydroxylase from *Acinetobacter* (BEADLE and SMITH, 1982). Many of the substrates of the pteridine-dependent rat liver enzyme phenylalanine hydroxylase fail to provide tight coupling between substrate and cosubstrate oxidation. The extent of coupling depends also on tetrahydropterin structure (STORM and KAUFMAN, 1968) and the presence of a specific "stimulating" factor of the protein nature (FISHER and KAUFMAN, 1973a, b). BERNHARDT et al., (1973) have studied in detail the conversion of a number of benzoate derivatives catalyzed by 4-methoxybenzoate hydroxylase from *Pseudomonas putida* (putidamonooxin). In addition to 4-methoxybenzoate, 3,4-dimethoxy- and 4-ethoxybenzoate are also true substrates. On the other hand, benzoate and 4-trifluoromethylbenzoate serve as complete uncouplers, H_2O_2 being the only product formed. The other compounds tested, 4-methylamino-, 3-methoxy-, 4-hydroxy-3-methoxy-, 3- and 4-hydroxybenzoate, combined the properties of both the substrate and uncoupler, monooxygenase reaction accounting for 21–83% of the oxygen consumed, the rest being directly reduced to H_2O_2 without intermediate release of free superoxide radicals (BERNHARDT and KUTHAN, 1981).

The same situation appears to be typical of cytochromes P-450. NARASIMHULU (1971) has found that though not hydroxylated by C_{21}-hydroxylase from bovine adrenal microsomes, androst-4-ene-3,17-dione produces spectral changes characteristic of type I substrates and stimulates NADPH and oxygen consumption with a stoichiometry of 2:1 that is, it serves as a complete uncoupler. The role of complete uncouplers for the liver microsomal monooxygenase system can be played by substrate analogues containing the chemically inert C–F bonds instead of the C–H ones. These include, for example, perfluoro-n-hexane and perfluorocyclohexane (ULLRICH and DIEHL, 1971). Other halogenated hydrocarbons, such as 1,1,1-trichloroethane, can also act as uncouplers (TAKANO et al., 1985). Apparently, P-450 substrates can be arranged in a continuous series from true substrates to complete uncouplers. Thus, liver microsomes of phenobarbital-treated rabbits oxidize cyclohexane with the stoichiometry NADPH:O_2:cyclohexanol of 1:1:0.85. Perfluorohexane acts as a complete uncoupler, while in the presence of n-hexane 46% of NADH consumed is accounted for by product formation (STAUDT et al., 1974). Analogously, substrate oxidation by P-450 CAM accounts for 95% of NADH consumed with camphor and for 75% with dehydrocamphor as a substrate, while in the presence of 6-ketopericyclocamphanone nearly all electron equivalents from NADH reduce oxygen to form H_2O_2 (SLIGAR, 1976; GELB et al., 1982a).

While the uncoupling of microsomal monooxygenase reactions was first demonstrated in the early 70ies, its mechanism remained unclear for a long time. Any of the intermediate complexes between oxygen and ferrous heme may be a point of uncoupling occurring in the case of electron transfer to oxygen followed by dissociation. If P-450-dependent monooxygenase reactions are

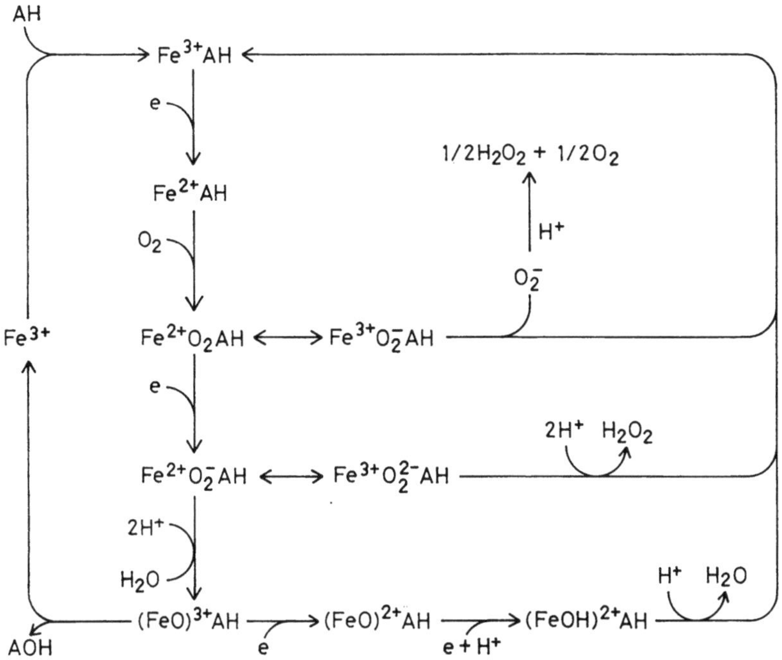

Fig. 1. Monooxygenase and oxidase catalytic cycles of cytochrome P-450. Fe and AH denote heme iron and organic substrate, respectively.

assumed to proceed via the oxenoid mechanism, according to which oxygen insertion into substrate is preceded by the formation of the $[FeO]^{3+}$ species, at least three ways for uncoupling are feasible (Fig. 1). The first is P-450 oxy-complex $[FeO_2]^{2+}$ autoxidation to give superoxide anion. The second is protonation and decay of the hypothetical peroxycomplex $[FeO_2]^{1+}$ to yield hydrogen peroxide, and the third involves two-electron reduction of the oxenoid species to give water. It should be taken into account that, in addition to the above direct routes of H_2O_2 and water formation, there are indirect ones. Thus, hydrogen peroxide can result from spontaneous or superoxide dismutase catalyzed dismutation of O_2^-

$$NADPH + 2O_2 \rightarrow NADP^+ + 2O_2^- + H^+ \quad (3)$$

$$2O_2^- + 2H^+ \rightarrow H_2O_2 + O_2 \quad (4)$$

$$NADPH + O_2 + H^+ \rightarrow NADP^+ + H_2O_2 \quad (5)$$

and water can arise from H_2O_2 breakdown under the action of catalase

$$2NADPH + 2O_2 + 2H^+ \rightarrow 2NADP^+ + 2H_2O_2 \quad (6)$$

$$2H_2O_2 \rightarrow O_2 + 2H_2O$$

$$2NADPH + O_2 + 2H^+ \rightarrow 2NADP^+ + 2H_2O \quad (7)$$

These reactions hamper measurement of the stoichiometry of oxidase reactions. Thus, to determine the amount of H_2O_2 formed via direct two-electron oxygen reduction and that of O_2^-, one must exclude reactions 4 and 6. While this can be easily done with reaction 6 by adding specific catalase inhibitors, such as azide, hydroxylamine, etc., no means of inhibiting reaction 4 is available. Therefore, straightforward determination of O_2 and directly formed H_2O_2 is impossible. The O_2^- formation rate can be calculated by extrapolating the experimentally obtained values to infinite concentration of the indicator, and the amount of direct H_2O_2 formation can then be determined by subtracting half the O_2^- from the net H_2O_2 formation rate.

2.2. Superoxide anion formation

O_2^- formation has been demonstrated repeatedly using various approaches, both in microsomes (AUST et al., 1972; DEBEY and BALNY, 1973; SASAME et al., 1975; NELSON et al., 1976; BARTOLI et al., 1977; UEMURA et al., 1977; AUCLAIR et al., 1978; POWIS, 1979; KUTHAN and ULLRICH, 1982; SOODAEVA et al., 1982; ZHUKOV and ARCHAKOV, 1982, 1985a) and reconstituted P-450-containing systems (KUTHAN et al., 1978; INGELMAN-SUNDBERG and JOHANSSON, 1980; SOODAEVA et al., 1982). The question of the site of O_2^- formation in microsomes has been discussed for a long time, an important role in O_2^- generation being ascribed to such components of the oxygenase system as NADPH-dependent reductase and cytochrome b_5. More recently, however, it has been shown that such commonly used indicators as epinephrine and nitroblue tetrazolium can either catalyze autoxidation of electron-transport proteins or themselves undergo oxidation to yield O_2^-, leading to considerable overestimation of the reductase and cytochrome b_5 ability to generate O_2^- (BERMAN et al., 1976; BORS et al., 1978; SCHENKMAN et al., 1979). The development of a procedure utilizing the non-autoxidizable cytochrome c subjected to succinylation in order to reduce its affinity to the reductase (KUTHAN et al., 1982) has solved this problem. Using this procedure KUTHAN and ULLRICH have shown microsomal O_2^- formation to be carbon monoxide sensitive with maximal reversibility by visible light observed at 450 nm. The reactions of O_2^- formation and 7-ethoxycoumarin deethylation show equal sensitivity to CO indicating that O_2^- formation is completely due to P-450, the contribution of other microsomal electron carriers being small (KUTHAN and ULLRICH, 1982). The same is indicated by sharp enhancement of O_2^- formation on the addition of P-450 to a reconstituted system containing NADPH-dependent reductase and NADPH (SOODAEVA et al., 1982). In microsomes, antibodies to cytochrome b_5 increase the O_2^- generation rate (DYBING et al., 1976), while NADH inhibits the reaction (SASAME et al., 1975). O_2^- generation is also retarded by the addition of cytochrome b_5 to a reconstituted system containing the NADPH-dependent reductase, P-450, NADPH, and *p*-nitroanisole (INGELMAN-SUNDBERG and JOHANSSON, 1980). These data indicate that O_2^- results from the decay of oxycytochrome P-450 whose concentration and, hence, decomposition rate is decreased by cytochrome b_5-mediated reduction. The fact that the addition of cytochrome b_5 to the reconstituted system improves coupling between NADPH and substrate oxidation without affecting NADPH oxidation rate (IMAI and SATO, 1977; INGELMAN-SUNDBERG and JOHANSSON, 1980; IMAI, 1981; GORSKY and COON, 1986) can also be accounted for by a decreased fraction of NADPH consumed for O_2^- formation via oxycytochrome P-450 decay. Isolated cytochrome P-450 CAM can also undergo autoxidation to give O_2^-. The oxygenated complex of this hemoprotein is relatively stable and decomposes with a first-order rate constant of 0.008 s^{-1} (SLIGAR et al., 1977). However, no O_2^- formation can be detected during the autoxidation of the rat liver microsomal P-450 isozyme LM_4 and direct formation of H_2O_2 and water takes place (OPRIAN et al., 1983). Substrates stabilize oxycytochrome P-450 thus saving it for reduction by a second electron. Camphor binding increases oxycytochrome P-450 CAM half-life 12-fold (OPRIAN et al., 1983), and cholesterol binding to cytochrome P-450 CAM increases the oxycomplex half-life 15-fold (TUCKEY and KAMIN, 1982).

2.3. Hydrogen peroxide formation

H_2O_2 generation during NADPH oxidation in microsomes was first detected in 1957 (GILLETTE et al., 1957) and repeatedly confirmed later both in the presence and in the absence of P-450 substrates (THURMAN et al., 1972; BOVERIS et al., 1972; HILDEBRANDT et al., 1973, 1982; STAUDT et al., 1974; NORDBLOM and COON, 1977; ZHUKOV and ARCHAKOV 1982, 1985a; KUTHAN and ULLRICH, 1982; HILDEBRANDT and ROOTS, 1975; ROESSING et al., 1985; RITTER et al., 1985). As with O_2^- generation, the majority of investigators agree that H_2O_2 is produced mainly by cytochrome P-450. This is indicated by equal activation energies, affinities for NADPH, and sensitivities to carbon monoxide for the reactions of ethylmorphine deethylation and H_2O_2 production (ESTABROOK and WERRINGLOER, 1974). Maximum reversibility of CO inhibition by visible light is observed at 450 nm (KUTHAN and ULLRICH, 1982). H_2O_2 generation is inhibited by the typical P-450 inhibitor metyrapone (ROOTS et al., 1980; RITTER et al., 1985). The formation of a complex containing an NADPH and a substrate molecule bound, apparently, to the reductase and P-450, respectively, is required for H_2O_2 generation to proceed (HILDEBRANDT et al., 1982).

P-450 substrates influence H_2O_2 formation differently. The effect depends on both the substrate nature and P-450 isozyme involved. Aminopyrine, for instance, does not affect the initial rate of H_2O_2 formation in the microsomes of phenobarbital-induced and untreated rats, but decreases it in the case of induction by 3-methylcholanthrene and increases it with pregnenolone-16α-carbonitrile induction (HILDEBRANDT et al., 1973). Hexobarbital stimulates H_2O_2 production in the microsomes of phenobarbital- and pregnenolone-16α-carbonitrile-treated animals, while ethylmorphine shows this effect only in the latter case and does not affect the rate in the former (ESTABROOK and WERRINGLOER, 1974). Benzphetamine enhances H_2O_2 production in the microsomes of phenobarbital- but not 3-methylcholanthrene-induced and untreated rats (MANNERING, 1981). H_2O_2 production can be stimulated by the product of substrate oxidation on cytochrome P-450 rather than by substrate itself as shown, for instance, for propranolol (HILDEBRANDT et al., 1982). The reaction in this case is characterized by a distinct lag-period. The same action is exhibited by the benzphetamine demethylation product, benzylamphetamine (NORDBLOM and COON, 1977). Stimulation of H_2O_2 formation may underlie the inhibitory action of certain compounds on oxygenase reactions. Thus, low concentrations of the naturally occurring flavonoid quercetin inhibit the oxidation of ethoxyresorufin, *p*-nitroanisole and benzopyrene in liver microsomes of rats treated with β-naphthoflavone. The substance does not affect the rates of NADPH and oxygen consumption but doubles the rate of H_2O_2 formation (SOUSA and MARLETTA, 1985).

Some authors considered the observed uncoupling in general and H_2O_2 production in particular as an experimental artefact arising from disintegration of endoplasmic reticulum during isolation of the microsomal fraction. It has been recently shown, however, that aminopyrine administration to rats stimulates H_2O_2 production *in vivo*, especially against the background of phenobarbital pretreatment. With guinea pigs, phenobarbital induction alone stimulated H_2O_2 production without administration of substrates (PREMEREUR et al., 1985).

As pointed out above, two routes of P-450-dependent H_2O_2 formation are theoretically feasible, namely protonation and dissociation of the peroxy-complex (direct route) and dismutation of preformed superoxide anions (indirect route). The choice between these two alternatives may be made by measuring the stoichiometry of O_2^- and H_2O_2 generation. In the case of indirect H_2O_2 formation, the ratio of the O_2^- to H_2O_2 formation rate should equal 2 according to the stoichiometry of reaction 4, while if H_2O_2 is in part formed by the direct route the ratio will be less than 2 and the closer to zero the greater the fraction of directly formed H_2O_2.

The stoichiometry of O_2^- and H_2O_2 formation was first studied by KUTHAN et al., (1978), who used succinylated cytochrome c as an O_2^- trap. The ΔO_2^-: ΔH_2O_2 ratios measured in reconstituted systems containing NADPH-dependent reductase, P-450, and 7-ethoxycoumarin in different combinations were considerably greater than two, which apparently points to the incorrectness of the procedure used for calculating the O_2^- generation rate. Later, such a procedure was developed in detail in the same laboratory and ratios close to unity were obtained both in the absence and in the presence of substrates (KUTHAN et al., 1982). The authors, failed however, to demonstrate a linear dependence between the O_2^- formation rate and concentration of microsomal protein, and the results were extrapolated to zero concentration, where the ratio of 2 indicative of the indirect reaction was found. Since the validity of such extrapolation is not obvious, a somewhat different calculation procedure was used in our laboratory, and the dependence obtained was linear over the whole range of protein concentrations tested (Zhukov and Archakov, 1985a). Using this procedure and liver microsomes of phenobarbital-induced rabbits, we found that the $\Delta O_2^-:\Delta H_2O_2$ ratio was essentially smaller than 2, which points to direct H_2O_2 formation as proceeding in parallel with the indirect reaction. The contribution of the latter increases with pH (along with an increase in the oxy-P-450 steady-state level), ionic strength and on the addition of Mg^{2+} ions (Zhukov and Archakov, 1985a, b).

One of the objections to the possibility of direct H_2O_2 formation via peroxy-complex decomposition was the lack of synergism of NADPH and NADH (Estabrook and Werringloer, 1974; Estabrook et al., 1979; Kuthan and Ullrich, 1982) observed normally in the oxidation of the majority of P-450 substrates proceeding through the intermediate formation of the peroxy-complex. In our opinion, however, the lack of synergism does not point to oxy-P-450 autoxidation as the only route of H_2O_2 formation. If this were the case, NADH, by accelerating oxy-P-450 reduction and thus decreasing its concentration, would decrease the rate of indirect H_2O_2 formation, and, hence, antagonism rather than an additive effect would be observed. It has been shown in our laboratory that antagonism is indeed observed at high pH when, as judged from the stoichiometry, H_2O_2 results mainly from O_2^- dismutation. But at low pH, when the direct route of H_2O_2 formation is the major one, NADPH and NADH act as antagonists (Zhukov and Archakov, 1985a, b). The additive effect observed at neutral pH (Estabrook and Werringloer, 1974; Estabrook et al., 1979; Kuthan and Ullrich, 1982), when the fractions of directly and indirectly formed H_2O_2 are approximately equal, is apparently due to the decrease in the oxy-P-450 level being balanced by the enhancement of direct reaction via peroxy-complex decomposition.

When considering the direct route of H_2O_2 formation, the second electron is normally assumed to be supplied by cytochrome b_5 or NADPH-dependent reductase. However, an alternative viewpoint exists ascribing the role of an electron donor to substrate. Heinemeyer et al. (1980), in their study of the stoichiometry of hexobarbital oxidation, have demonstrated that the sum of H_2O_2 formed and hexobarbital consumed was not accounted for by NADPH oxidized. They therefore assumed that the substrate was consumed for the reduction of oxy-P-450, and the peroxy-complex thus formed decomposed to give H_2O_2. Estabrook believes that this mechanism, which he terms "carbon activation", is operating in H_2O_2 formation during the oxidation of arachidonic acid by P-450 (Estabrook, 1982; Estabrook et al., 1982). Generally speaking, the relative probability of oxy-P-450 being reduced by substrate or the reductase (cytochrome b_5) is considered to depend on the substrate's ability to act as an electron donor (Estabrook, 1982).

2.4. Water formation

It has been demonstrated in a number of investigations dealing with the stoichiometry of P-450-catalyzed reactions that a fraction of NADPH oxidized and oxygen consumed may remain unaccounted for by substrate oxidation and H_2O_2 formation (ESTABROOK and WERRINGLOER, 1974; ESTABROOK et al., 1979; HEINEMEYER et al., 1980; MORGAN et al., 1982; ZHUKOV and ARCHAKOV, 1982). Only half of the NADPH oxidized in microsomes in the absence of substrate corresponds to H_2O_2 formation (ESTABROOK and WERRINGLOER, 1974; ZHUKOV and ARCHAKOV, 1982; HEINEMEYER et al., 1980), and in the presence of benzphetamine formaldehyde and H_2O_2 formed account for about 40% of NADPH consumed (ESTABROOK et al., 1979). Since both substrate oxidation and H_2O_2 formation require equimolar consumption of both NADPH and oxygen, one can conclude that the cosubstrates participate in a third reaction. This may be either the monooxygenation of a certain endogenous substrate or the direct four-electron reduction of the dioxygen molecule to yield two molecules of water. The alternatives can be distinguished by calculating the stoichiometry of NADPH and oxygen consumption unaccounted for by substrate oxidation and H_2O_2 formation. The oxidase mechanism (equation 7) requires the NADPH to O_2 ratio to equal 2, while it should be unity for the monooxygenase reaction (equation 1). The ratio of 2 has been obtained by STAUDT et al. in the presence of the pseudosubstrate perfluorohexane (STAUDT et al., 1974). The conclusion about direct water formation was, however, not valid since no measures were taken to block microsomal catalase, in the presence of which the obtained stoichiometry could also be observed with indirect reaction involving the intermediate formation of free hydrogen peroxide (see above). With sodium azide added to block catalase, the present authors have demonstrated that direct water formation through complete four-electron reduction of the dioxygen molecule is indeed catalyzed by liver microsomes of phenobarbital-induced rabbits both under conditions of NADPH-free oxidation and in the presence of various cytochrome P-450 substrates and pseudosubstrates (ZHUKOV and ARCHAKOV, 1982, 1985a, b). Similar results have been obtained by GORSKY et al. with reconstituted systems containing NADPH-dependent reductase, phospholipid, and different isozymes of rabbit liver cytochrome P-450. Water formation was not demonstrated to involve reversible dissociation of peroxy-P-450 to yield H_2O_2 (GORSKY et al., 1984). Perfluorinated hydrocarbons are distinguished from true substrates by their ability to sharply stimulate the oxidase reaction of water formation both in microsomes and the reconstituted system. The reaction accounts for more than 80% of NADPH, consumed compared to 50% under the conditions of free oxidation (GORSKY et al., 1984; ZHUKOV and ARCHAKOV, 1985a).

Interestingly, with some substrates, water formation is observed only in microsomes, while in the reconstituted system NADPH consumption is entirely accounted for by substrate oxidation and H_2O_2 formation. This was shown, for instance, for benzphetamine and dimethylaniline in the case of their oxidation in rabbit liver microsomes (ZHUKOV and ARCHAKOV, 1985b) and in a reconstituted system containing P-450 LM2 (NORDBLOM and COON, 1977; GORSKY et al., 1984). Water formation in the presence of these substrates is probably catalyzed by P-450 isozymes other than P-450 LM2.

As to the precise mechanism of water formation, the first molecule is apparently released during peroxy-P-450 decay to yield the oxenoid species $[FeO]^{3+}$. The further fate of the oxenoid oxygen, which can be either inserted into substrate or reduced to yield the second water molecule, appears to depend both on the substrate reactivity and on the specificity of the P-450 isozyme to this substrate. With such inert compounds as perfluorinated hydrocarbons, insertion is impossible and further reduction takes place, while with more reactive substrates, both reactions may run in parallel. Good experimental support for this interpretation is provided by recent findings of Miwa et al. (1985). They studied 7-ethoxycoumarin oxidation in liver microsomes of phenobarbital-induced rabbits and showed that deuteration of C_1 in the ethyl group decreases 5-fold the rate of 7-deethylation, but does not affect those of 6-hydroxylation, NADPH oxidation, oxygen consumption, and H_2O_2 formation. The authors conclude that a pathway for $[FeO]^{3+}$ decay exists in addition to oxygen insertion into substrate, the reaction taking this way when the insertion is hampered (Miwa et al., 1985). This pathway is likely to represent oxygen reduction to water.

The model proposed allows one to account, without rejecting the feasibility of direct H_2O_2 formation, for the fact that perfluorinated hydrocarbons fail to increase the rate of H_2O_2 formation over that observed in the presence of their non-fluorinated analogues (Kuthan and Ullrich, 1982). Indeed, if peroxy-P-450 decay is assumed to begin from the release of a water molecule and the formation of oxenoid species $[FeO]^{3+}$, then in the case of the coupled reaction the second oxygen atom will be inserted into substrate, while uncoupling will result in the formation of the second water molecule rather than hydrogen peroxide.

The sequence and number of steps leading to the release of the second water molecule cannot be unequivocally elucidated on the basis of the data available, and Fig. 1 therefore shows only one of the alternatives. The third and fourth electrons can be supplied, for instance, by cytochrome b_5. This is consistent with the fact that the greater the uncoupling, the lower the level of reduced b_5 in microsomes and the higher the rate of its oxidation (Staudt et al., 1974).

3. Peroxidase activity

Kadlubar et al. (1973) have first found that liver microsomes can catalyze oxidative dealkylation of amines in the presence of organic hydroperoxides. In later studies summarized in several reviews (O'Brien, 1978; White and Coon, 1980; Metelitza, 1984; Sligar et al., 1984), it has been shown that the role of cosubstrate may be played by peracids (Blake and Coon, 1980) and hydrogen peroxide (Nordblom et al., 1976; Werringloer and Kawano, 1980; Karuzina et al., 1983) as well as by compounds containing a single oxygen atom, such as sodium periodate (Hrycay et al., 1975a), sodium chlorite (Nordblom et al., 1976), iodosobenzene (Lichtenberger et al.,

1976), iodosobenzene diacetate (GUSTAFSSON et al., 1979), pyridine *N*-oxide (SLIGAR et al., 1980). All the reactions catalyzed by P-450 in the presence of NADPH and oxygen may also proceed in the presence of the above oxidants. These are aromatic hydroxylation (RAHIMTULA and O'BRIEN, 1974; METELITZA et al., 1979), *O*- and *N*-dealkylation (KADLUBAR et al., 1973; RAHIMTULA and O'BRIEN, 1975; NORDBLOM et al., 1976; SCHELLER et al., 1976), epoxidation (AKHREM et al., 1977), oxidation of saturated and heterocyclic compounds (ELLIN and ORRENIUS, 1975; GUSTAFSSON and BERGMAN, 1976) and alcohols (RAHIMTULA and O'BRIEN, 1977a).

The main question concerning the hydroperoxide-dependent reactions of P-450 as compared to classical peroxidases is that of the mechanism of the O—O bond scission after the formation of cytochrome-hydroperoxide complex and of the further fate of the active iron-oxo species so formed. Cytochromes P-450 are well known to differ from peroxidases in that they contain cysteine thiolate as the fifth axial ligand of heme iron instead of histidine (BRUICE, 1986). The mechanism of peroxidase reactions has been studied in detail. At the first step of the reaction the iron-protoporphyrin moiety Fe(III)PrIX undergoes two-electron oxidation by a peroxide to form the oxo-liganded π-cation radical $OFe(IV)PrIX^{+}$, termed "Compound I", the second oxygen atom from the peroxide being released as water (or an alcohol). Compound I then oxidizes substrate by electron or hydrogen abstraction to give substrate radical and ferro-oxo species OFe(IV)PrIX termed "Compound II". Compound II abstracts another electron from substrate to give water, oxidized substrate, and the enzyme in the initial state. Thus the classical peroxidase reaction necessarily involves the step of one-electron substrate oxidation to give a free radical.

With P-450, the formation of a compound I-like species is also beyond doubt, but due to the presence of a thiolate-anion in the P-450 active site instead of a histidine, the enzyme is capable of oxygen insertion into the substrate C—H bond in addition to electron abstraction. The reaction can also involve intermediate formation of a substrate radical which then reacts with the iron-bound hydroxyl radical ("oxygen rebound" mechanism) (GROVES and MCCLUSKY, 1976; GROVES, 1986)

$$OFe(IV)S^{\cdot} + AH \rightarrow HOFe(IV)S^{-} + A^{\cdot} \rightarrow Fe(III)S^{-} + AOH.$$

The difference between peroxidases and cytochrome P-450 is that with the former, hydrogen is abstracted from substrate by porphyrin cation radical, while with the latter it is the oxenoid species that attacks substrate.

However, the mechanism of O—O bond cleavage by P-450 has not yet been definitely established. A considerable body of evidence favours a heterolytic mechanism with the formation of a compound I-like species. Thus, when added to microsomes or isolated P-450 LM2, hydroperoxides produce difference spectra identical to those of compound I of certain peroxidases (such as cytochrome c peroxidase) (YONETANI and RAY, 1965). Mixing microsomes with cumene hydroperoxide (CHP) gives rise to a signal in the ESR spectrum with $g = 2$ that is, a radical species similar to compound I is formed (RAHIMTULA et al., 1974). [4-^{3}H]Acetanilide oxidation in the presence of hydroperoxides and in the complete system is characterized by equal values of NIH-shift, which points to a common mechanism of oxygen insertion into substrate (RAHIMTULA et al., 1978). Oxygenation of a variety of substrates in the above two systems shows close values of the kinetic and thermodynamic parameters (AKHREM et al., 1977). The formation of a compound I-like species is favoured by the fact that oxygenation can be carried out by single-oxygen donors such as iodosobenzene in which homolytic I—O bond cleavage is difficult to believe. That heterolytic iodosobenzene breakdown does indeed occur was shown in ESR studies with chromium- and manganese-substituted porphyrins (GROVES et al., 1980; GROVES and HAUSHALTER, 1981). A complex with similar spectral properties is formed on iodosobenzene interaction with P-450 CAM containing manganese-substituted porphyrin (GELB et al., 1982b). Iodosobenzene-dependent camphor oxygenation by P-450 CAM being inhibited by excess substrate (HEIMBROOK and WHITCOMBE, 1981) is indicative of the ping-pong kinetic mechanism, which means that the iodobenzene skeleton cannot directly participate in substrate oxygenation since it cannot be present in the active site together with substrate. EPR and optical spectroscopy suggest a common intermediate for iodosobenzene-, hydroperoxide-, and peracid-driven reactions (ULLRICH, 1977; LICHTENBERGER and ULLRICH, 1977; GROVES et al., 1981).

The possibility of a homolytic O—O bond cleavage is inferred from experiments in which the rate of CHP-dependent oxygenation has been shown to depend on the nature of substituents in the molecule of both substrate and cosubstrate (BLAKE and COON, 1981). This can be accounted for by the participation in the rate-limiting step of cumyloxy-radical formed through the CHP O—O bond homolysis. Benzo[a]pyrene conversion in the NADPH- and CHP-supported systems proceeds by different ways, phenols being the major products in the former and quinones in the latter case (CAPDEVILA et al., 1980), which may also be due to different mechanisms.

In our laboratory, CHP-dependent reactions were demonstrated to substantially differ from NADPH-dependent ones in their sensitivity to inhibitors, much greater similarity being observed between the NADPH- and H_2O_2-supported systems (KARUZINA et al., 1983).

Peracid homolysis will give a carboxy-radical which shows a tendency to decarboxylation. With peroxyphenylacetic acid used by WHITE et al., decarboxylation will give a benzyl radical which, upon recombination with the iron-oxo complex, will give benzyl alcohol. These authors demonstrated that benzyl alcohol is indeed formed during the oxygenation of various substrates by peroxyphenylacetic acid in both hepatic and bacterial P-450-dependent systems (WHITE et al., 1980). Interestingly, the ability to catalyze peracid decarboxylation is a unique property of P-450 not found in other hemoproteins (horseradish peroxidase, chloroperoxidase, catalase, myoglobin) (McCARTHY and WHITE, 1983a). Later, however, the same authors showed by kinetic and inhibitory analysis that peroxyacetic acid decarboxylation and substrate oxygenation by cytochrome P-450 proceed via entirely independent routes without formation of any common intermediate (McCarthy and White, 1983b). Therefore, the mechanism of O—O bond cleavage during peracid-supported hydroxylation remains unclear.

Thus, no definite conclusion can at present be drawn concerning the mechanism of O—O bond cleavage in the peroxidase reactions of P-450. The majority of results favours heterolysis. At the same time, the data of Blake and Coon conclusively indicate that homolytic mechanism is operating with CHP (Blake and Coon, 1981). The mechanism could depend on the oxidant properties. Thus, it has been shown in the studies of interaction between hydroperoxides and iron(III)porphyrins that the second-order rate constant for oxygen transfer to the porphyrin depends on the p*K* of the leaving group (ROH). The dependence shows break at p*K* 11, which might result from a heterolytic mechanism being operational at p$K < 11$ (peracids) and a homolytic one at p$K > 11$ (hydroperoxides) (Lee and Bruice, 1985).

In view of the above data, one could expect the H_2O_2-dependent system to model the NADPH/O_2-dependent one best. However, different product patterns and responses to methylcholanthrene treatment were observed with NADPH- and H_2O_2-dependent oxygenation of prostaglandins PGE_1 and PGE_2 in rat liver microsomes (Holm et al., 1985). Care should, however, be taken in treating results obtained with microsomes since differences in regioselectivity observed may well arise from different P-450 isozymes being involved in NADPH- and hydroperoxide-dependent reactions. Thus, the fact that induction of rats by phenobarbital, methylcholanthrene, and Arochlor 1254 affects NADPH- but not CHP-dependent benzo[*a*]pyrene oxidation might be due to the involvement of constitutive isozymes in the latter case (Wong et al., 1986).

As pointed out above, cytochrome P-450 differs from peroxidases in its ability to insert oxygen into the substrate molecule, while with peroxidases the pathway involving two subsequent one-electron reductions of the oxenoid species is more common. With P-450, however, the reaction can also take the latter pathway. Thus, P-450 catalyzes cytochrome b_5 oxidation in the presence of CHP (Rahimtula and O'Brien, 1977b). NADH is also oxidized by hydroperoxides in the presence of microsomes (Hrycay and O'Brien, 1974). In the experiments with the reconstituted system, this reaction has been shown to require cytochrome b_5 and NADH-dependent reductase for maximal activity (Hrycay et al., 1975b). NADPH is also oxidized under these conditions (Hrycay et al., 1975b). These results indicate that oxenoid species can be reduced by electrons supplied from cytochrome b_5 or NADPH-dependent reductase, which appears to occur not only in the presence of hydroperoxides, but also during the P-450-catalyzed four-electron oxidase reaction (see Fig. 1).

The role of hydrogen donor in P-450-dependent peroxidase reactions may be played by such compounds as tetra- and dimethylphenylenediamine (Hrycay and O'Brien, 1971), pyrogallol (McCarthy and White, 1983a), the 1,4-dihydropyridine derivative felodipine (BAARNHIELM et al., 1984). In the latter two cases the reaction may also proceed in the presence of NADPH and oxygen and, as shown for felodipine, is inhibited by carbon monoxide (Baarnhielm et al., 1984). While the mechanisms of substrate oxidation in hydroperoxide- and NADPH/O_2-dependent systems may be identical, in the latter case the reaction should be treated as a four-electron oxidase rather than peroxidase one since the dioxygen molecule appears to undergo four-electron reduction to yield two molecules of water.

Though P-450 ability to utilize hydrogen peroxide and organic hydroperoxides as cosubstrates undoubtedly allows the enzyme to be formally considered a peroxidase, the question still remains as to the degree of similarity between P-450 and "classical peroxidases" such as horseradish peroxidase, catalase, and cytochrome c peroxidase. This point has been recently discussed in a review by MARNETT et al., (1986) where the authors point to considerable differences between the above classical peroxidases on the one hand and such hemoproteins as cytochrome P-450 and metmyoglobin on the other. Thus, turnover numbers of P-450 and metmyoglobin are close and several orders of magnitude lower than those of peroxidases. This appears to result from the fact that, contrary to peroxidases, P-450 reacts slowly with hydroperoxides and the formation of the higher oxidation state rather than its reduction is rate-limiting. The reason for such different reactivities might be the availability in peroxidases of distal histidine participating in general base-general acid catalysis as well as of an arginine capable of stabilizing the negative charge formed in the transition state for hydroperoxide heterolysis (POULOS and KRAUT, 1980), while such residues are lacking in P-450 and its heme environment is nonpolar. Certain other peculiarities, such as different products formed from some hydroperoxides on P-450 and peroxidases also distinguishes the former from classical peroxidases (MARNETT et al., 1986).

4. Dioxygenase activity

CHEN and LIN (1969) have demonstrated that fluorene hydroxylation in rat liver microsomes proceeds in two steps. First, 9-hydroperoxyfluorene is formed, which is then reduced to the 9-hydroxy derivative. Tetralin oxidation proceeds by the same way (CHEN and LIN, 1968). With dimethylbenzanthracene, endoperoxide is formed at the first step which is then reduced to a diol (CHEN and TU, 1977). Later, the same reaction sequence was demonstrated for 9-methylfluorene whose hydroperoxide intermediate is more stable. Both steps required P-450 and NADPH (CHEN et al., 1985). 9-Hydroperoxymethylfluorene was also formed in the presence of CHP (CHEN and GURKA, 1985). ESTABROOK et al. (1982) believe that one of the pathways for P-450-dependent oxidation of arachidonic acid also involves intermediate hydroperoxide formation. It is quite possible that P-450 also participates in the biosynthesis of prostaglandins at the step of arachidonate cyclization to yield PGG_2 (GRAF and ULLRICH, 1982).

Thus, in the above cases P-450 can activate substrate for dioxygen addition, exhibiting dioxygenase activity. The mechanism of such activation remains, however, unclear. Estabrook et al., (1982) believe that activation involves hydrogen abstraction from arachidonic acid by oxy-P-450 to yield substrate radical and hydrogen peroxide. On the other hand, the possibility of NADPH and O_2 substitution for CHP in the formation of 9-hydroperoxy-9-methylfluorene (Chen and Gurka, 1985) indicates that another species, such as the oxenoid one, may activate substrate.

The further fate of the incorporated dioxygen depends on the chemical nature of the substrate. Hydroperoxides may be stable and exist in the cell as such. Otherwise, they can undergo further catalytic or spontaneous conversion with fixation of both oxygen atoms (formation of diols and carboxyls, cyclization) or decompose to release one or both oxygen atoms as water or singlet oxygen (Porter et al., 1984; Ullrich, 1984).

5. Reductase activity

Not only oxygen can serve as an electron acceptor for P-450. Rat liver microsomes catalyze the reduction of azo dyes, the reaction being partly inhibited by carbon monoxide and stimulated by phenobarbital and 3-methylcholanthrene treatment (Hernandez et al., 1967). Similar sensitivity to CO and induction of P-450 is observed with the reduction of nitro compounds to amines in liver microsomes of mice, rats, and rabbits (Gillette et al., 1968). *N*-Oxides can also serve as substrates for P-450-catalyzed reduction (Sugiura et al., 1974, 1976; Iwasaki et al., 1977). Unlike azo- and nitroreduction, only P-450 but not NADPH-dependent reductase or cytochrome b_5 can catalyze this reaction (Sugiura et al., 1976; Iwasaki et al., 1977). *N*-Oxides form complexes with P-450 apparently through oxygen coordination with the heme iron and undergo two-electron reduction to give water and an amine. A stoichiometry of 1:1 exists between NADPH oxidation and *N*-oxide reduction (Sugiura et al., 1976). Finally, P-450 catalyzes reductive dehalogenation of a variety of haloalkanes, such as halothane (2-bromo-2-chloro-1,1,1-trifluoroethane) (Fujii et al., 1981, 1982; Ahr et al., 1982), carbon tetrachloride (Ahr et al., 1980; Nastainczyk et al., 1982), hexa- and pentachloroethane (Nastainczyk et al., 1982), dibromoethane (Tomasi et al., 1983).

Ullrich et al., have proposed a scheme for P-450-dependent metabolism of haloalkanes (Ahr et al., 1982; Nastainczyk et al., 1982). The scheme implies that upon the formation of the enzyme-substrate complex it is reduced by an electron supplied by the NADPH-dependent reductase to yield a substrate free radical and release a halogenide anion. Either the radical may abstract hydrogen from other compounds or the enzyme molecule, which will result in lipid peroxidation, mutagenesis and enzyme inactivation or undergo a second reduction, preferably via cytochrome b_5. In the latter case carbanion is formed tending to β-elimination. The ratio of two- to one-electron reduction products depends mainly on substrate. Carbon tetrachloride is metabolized entirely via the one-electron pathway (AHR et al., 1980), with halothane the ratio is close to two (AHR et al., 1982), while with hexa- and pentachloroethane the contribution of two-electron reduction is 99.5 and 96%, respectively (NASTAINCZYK et al., 1982). A stoichiometry of 1:1 is observed in the latter case between NADPH consumed and product formed. NADPH and NADH exhibit synergism as is the case with monooxygenase reactions (NASTAINCZYK et al., 1982).

The property shared by the majority of P-450-dependent reductase reactions *in vitro* is that they do not proceed under aerobic conditions since oxygen appears to compete efficiently for electrons supplied to the hemoprotein. With haloalkanes, metabolism in the presence of oxygen proceeds via the monooxygenase pathway, and halothane, for instance, is converted to trifluoroacetic acid (KARASHIMA et al., 1977), though the formation of trichloromethyl radical from CCl_4 is still possible under aerobic conditions. However, the results obtained *in vitro* cannot be extended to physiological conditions since oxygen tension in tissues is considerably lower (about 4%) (KRATZ and STAUDINGER, 1965). P-450-dependent hexachloroethane reduction at this tension is inhibited by less than 50% (NASTAINCZYK et al., 1982). Other data also indicate that monooxygenation and reduction of haloalkanes can proceed in parallel. Thus, the products of halothane covalent binding retained both labelled chlorine and ^3H, which points to reductive elimination of bromine (VAN DYKE and GANDOLFI, 1976; GANDOLFI et al., 1980). Deuteration of this compound did not affect toxicity, but decreased the rate of oxygenation (Sipes et al., 1980).

6. Oxygen transport

LONGMUIR et al. (1971) in their studies on respiration of tissue slices have found that with liver slices, unlike brain ones, the dependence of respiration rate on the partial pressure of oxygen is inconsistent with the model suggesting passive oxygen diffusion in tissues. The data obtained could be accounted for if an oxygen carrier providing facilitated diffusion was assumed to exist in hepatocytes. Since the liver is distinguished from the brain by much higher P-450 content and since oxygen diffusion in tissues proceeds mainly along the membranes of endoplasmic reticulum (LONGMUIR, 1981), the authors have suggested that the role of reversible oxygen binder might be played by ferrous P-450. In the experiment when oxygen was added to a dithionite-reduced microsomal suspension, subsequent addition of CO resulted in oxygen evolution, the amount of oxygen released being closely correlated with that of P-450 in the suspension. This was considered a result of oxygen displacement by CO from the complex with ferrous P-450 (LONGMUIR et al., 1973). These results cannot be treated as the definite proof for P-450 participation in facilitated oxygen diffusion in the cell. However, taking into account that the rate constant of oxy-P-450 dissociation without charge transfer is sufficiently high (about 1 s^{-1} (PEDERSON and

Gunsalus, 1977)) and that about 30% of total P-450 amount is found as oxy-complex in the steady state (Werringloer, 1982), such a possibility cannot be rejected.

7. Oxene transferase activity and singlet oxygen formation

Ullrich et al. (Ullrich et al., 1982; Graf et al., 1983; Ullrich, 1984) have demonstrated that, when catalyzing the formation of hydroperoxides from PGH_2, cytochromes P-450 termed thromboxane A_2 synthase and prostacyclin synthase act as the carriers of active oxygen atoms, using endoperoxides as substrates. Thromboxane synthase isolated from pig aorta exhibited all properties of P-450 except that it could not bind typical substrates or be reduced by NADPH in the microsomal fraction (Ullrich et al., 1981; Graf and Ullrich, 1982). The reaction is likely to involve intermediate formation of an oxenoid species $[FeO]^{3+}$ and similar to the iodosobenzene supported peroxidase reaction. The heme iron does not undergo reduction in this reaction, and CO shows no inhibition. As a matter of fact, the reaction represents PGH_2 isomerisation to prostacyclin or thromboxane A_2 (Ullrich et al., 1985). The formation of an oxenoid species in the course of PGG_2 oxidation is also suggested by the possibility of cooxygenation of such a typical P-450 substrate as cyclohexane in this reaction. When added to the mixture containing prostaglandin endoperoxide synthase and PGG_2, cyclohexane was converted to cyclohexanol, the isotope effect being the same as that observed with iodosobenzene (Gross et al., 1984). A characteristic feature of both reactions is the formation of singlet oxygen as follows (Cadenas et al., 1983; Ullrich, 1984; Nastainczyk and Ullrich, 1984),

$$Fe(III) \xrightarrow[ROH,\ PhI]{ROOH,\ PhIO} [FeO]^{3+} \xrightarrow[ROH,\ PhI]{ROOH,\ PhIO} {}^1O_2 + Fe(III)$$

This appears to be the only enzyme reaction conclusively demonstrated as accompanied by the formation of singlet oxygen in the enzyme active center. Thus, in prostaglandin biosynthesis P-450 acts not only as a dioxygenase at the step of PGG_2 formation from arachidonic acid (cyclooxygenase) and an oxene transferase in the formation of thromboxane A_2 and prostacyclin from PGH_2 (thromboxane A_2 synthase and prostacyclin synthase), but also as an enzyme capable of generating singlet oxygen as, apparently, a side product.

8. References

Ahr, H. J., L. J. King, W. Nastainczyk, and V. Ullrich, (1980), Biochem. Pharmacol. **29**, 2855–2861.

Ahr, H. J., L. J. King, W. Nastainczyk, and V. Ullrich, (1982), Biochem. Pharmacol. **31**, 383–390.

Akhrem, A. A., D. I. Metelitza, S. H. Bielski, P. A. Kiselev, M. E. Skurko, and S. A. Usanov, (1977), Croatica chem. acta **49**, 223–235.
Archakov, A. I., G. F. Zhirnov, and I. I. Karuzina, (1975), Vopr. med. khim. **21**, 281–285.
Auclair, C., D. de Prost, and D. Hakim, (1978), Biochem. Pharmacol. **27**, 355–358.
Aust, S. D., D. L. Roerig, and T. C. Pederson, (1972), Biochem. Biophys. Res. Commun. **47**, 1133–1137.
Baarnhielm, C., I. Scanberg, and K. O. Borg, (1984), Xenobiotica **14**, 719–726.
Baker, J. R. and S. Chaykin, (1962), J. Biol. Chem. **237**, 1309–1313.
Bartoli, G. M., T. Galeotti, G. Palombini, G. Parisi, and A. Azzi, (1977), Arch. Biochem. Biophys. **184**, 276–281.
Beadle, C. A. and A. R. W. Smith, (1982), Eur. J. Biochem. **123**, 323–332.
Berman, M. C., C. M. Adams, K. M. Ivanetich, and J. M. Kench, (1976), Biochem. J. **157**, 237–246.
Bernhardt, F.-H., N. Erdin, H. Staudinger, and V. Ullrich, (1973), Eur. J. Biochem. **35**, 126–134.
Bernhardt, F.-H. and H. Kuthan, (1981), Eur. J. Biochem. **120**, 547–555.
Blake, R. C. and M. J. Coon, (1980), J. Biol. Chem. **255**, 4100–4111.
Blake, R. C. and M. J. Coon, (1981), J. Biol. Chem. **256**, 12127–12133.
Bors, W., M. Saran, E. Lengfelder, C. Michell, C. Fuchs, and C. Frenzel, (1978), Photochem. Photobiol. **28**, 629–638.
Boveris, A., N. Oshino, and B. Chance, (1972), Biochem. J. **128**, 617–630.
Bruice, T. C., (1986), Ann. N. Y. Acad. Sci. **471**, 83–98.
Buening, M. K. and M. R. Franklin, (1974), Mol. Pharmacol. **10**, 999–1003.
Cadenas, E., H. Sies, and W. Nastainczyk, (1983), Hoppe Seyler's Z. Physiol. Chem. **364**, 519–528.
Capdevila, J., R. W. Estabrook, and R. A. Prough, (1980), Arch. Biochem. Biophys. **200**, 186–195.
Chen, C. and C. C. Lin, (1968), Biochim. biophys. acta **170**, 366–374.
Chen, C. and C. C. Lin, (1969), Biochim. biophys. acta **184**, 634–640.
Chen, C. and M. Tu, (1977), J. Biochem. **160**, 805–808.
Chen, C. and D. P. Gurka, (1985), Arch. Biochem. Biophys. **238**, 187–194.
Chen, C., T. I. Ruo, and D. P. Gurka, (1985), Arch. Biochem. Biophys. **238**, 195–205.
Cohen, B. S. and R. W. Estabrook, (1971), Arch. Biochem. Biophys. **143**, 46–53.
Cooper, D. Y., R. W. Estabrook, and O. Rosenthal, (1965), J. Biol. Chem. **238**, 1320–1323.
Debey, P. and C. Balny, (1973), Biochimie **55**, 329–332.
Dybing, E., S. D. Nelson, J. R. Mitchell, H. A. Sasame, and J. R. Gillette, (1976), Mol. Pharmacol. **12**, 911–920.
Einstein, L., P. Debey, and P. Douzou, (1977), Biochem. Biophys. Res. Commun. **77**, 1377–1383.
Ellin, A. and S. Orrenius, (1975), FEBS Lett. **50**, 378–381.
Estabrook, R. W. and B. Cohen, (1969), in: Microsomes and Drug Oxidations, (Gillette, J. R. et al., eds.), Acad. Press, New York, 95–109.
Estabrook, R. W. and J. Werringloer, (1974), in: Microsomes and Drug Oxidations, (Ullrich, V. et al., eds.), Pergamon Press, Oxford, 748–757.
Estabrook, R. W., S. Kawano, J. Werringloer, H. Kuthan, H. Tsuji, H. Graf, and V. Ullrich, (1979), Acta biol. med. germ. **38**, 423–434.
Estabrook, R. W., (1982), in: Microsomes, Drug Oxidations, and Drug Toxicity, (Sato, R. and R. Kato, eds.), Japan Scientific Societes Press, Tokyo, 133–138.
Estabrook, R. W., N. Chacos, C. Martin-Wixtrom, and J. Capdevila, (1982), in: Oxygenases and Oxygen Metabolism, (Nozaki, M. et al., eds.), Academic Press, New York, 371–384.
Fisher, D. B. and S. Kaufman, (1973a), J. Biol. Chem. **248**, 4300–4304.
Fisher, D. B. and S. Kaufman, (1973b), J. Biol. Chem. **248**, 4345–4353.

FUJII, K., N. MIKI, T. SUGIYAMA, M. MORIO, T. YAMANO, and Y. MIYAKE, (1981), Biochem. Biophys. Res. Commun. **102**, 507–512.
FUJII, K., N. MIKI, M. KANASHIRO, R. MIURA, T. SUGIYAMA, M. MORIO, T. YAMANO, and Y. MIYAKE, (1982), J. Biochem. **91**, 415–418.
GANDOLFI, A. J., R. D. WHITE, I. G. SIPES, and L. R. POHL, (1980), J. Pharmacol. Exp. Ther. **214**, 721–725.
GELB, M. H., P. J. MALKONEN, and S. G. SLIGAR, (1982a), Biochem. Biophys. Res. Commun. **104**, 853–858.
GELB, M. H., W. A. TOSCANO, Jr., and S. G. SLIGAR, (1982b), Proc. Nat. Acad. Sci. USA. **79**, 5788–5792.
GILLETTE, J. R., B. B. BRODIE, and B. N. LA DU, (1957), J. Pharmacol. Exp. Ther. **119**, 532–540.
GILLETTE, J. R., (1966), Adv. Pharmacol. **4**, 219–261.
GILLETTE, J. R., H. KAMIN, and H. A. SASAME, (1968), Mol. Pharmacol. **4**, 541–548.
GORSKY, L. D., D. R. KOOP, and M. J. COON, (1984), J. Biol. Chem. **259**, 6812–6817.
GORSKY, L. D. and M. J. COON, (1986), Drug Metab. Dispos. **14**, 89–96.
GRAF, H. and V. ULLRICH, (1982), in: Cytochrome P-450. Biochemistry, Biophysics, and Environmental Implications, (HIETANEN, E. et al., eds.), Elsevier, Amsterdam. 103 to 106.
GRAF, H., H. H. RUF, and V. ULLRICH, (1983), Angew. Chemie Int. Ed. **22**, 487.
GROSS, H., W. NASTAINCZYK, and V. ULLRICH, (1984), in: Oxygen Radicals in Chemistry and Biology, (BORS, W. et al., eds.), Gruyter & Co., Berlin, 435–439.
GROVES, J. T. and G. A. MCCLUSKY, (1976), J. Am. Chem. Soc. **98**, 859–861.
GROVES, J. T., W. J. KRUPER, Jr., and R. C. HAUSHALTER, (1980), J. Am. Chem. Soc. **102**, 6375–6377.
GROVES, J. T. and R. C. HAUSHALTER, (1981), J. Chem. Soc. Chem. Commun. 1165–1166.
GROVES, J. T., R. C. HAUSHALTER, M. NAKAMURA, T. E. NEMO, and B. J. EVANS, (1981), J. Am. Chem. Soc. **103**, 2884–2886.
GROVES, J. T., (1986), Ann. N. Y. Acad. Sci. **471**, 99–107.
GUENGERICH, F. P. and J. KRAUT, (1980), Acc. Chem. Res. **17**, 9–16.
GUSTAFSSON, J.-A. and J. BERGMAN, (1976), FEBS Lett. **70**, 276–279.
GUSTAFSSON, J.-A., L. RONDAHL, and J. BERGMAN, (1979), Biochemistry **18**, 865–870.
HEIMBROOK, D. C. and T. WHITCOMBE, (1981), Fed. Proc. **40**, 1963.
HEINEMEYER, G., A. G. HILDEBRANDT, I. ROOTS, S. NIGAM, (1980), in: Microsomes, Drug Oxidations, and Chemical Carcinogenesis, (COON, M. J. et al., eds.), Acad. Press, New York, 331–334.
HERNANDEZ, P. H., P. MAZEL, and J. R. GILLETTE, (1967), Biochem. Pharmacol. **16**, 1877–1888.
HILDEBRANDT, A. G., M. SPECK, and I. ROOTS, (1973), Biochem. Biophys. Res. Commun. **54**, 968–975.
HILDEBRANDT, A. G. and I. ROOTS, (1975), Arch. Biochem. Biophys. **171**, 385–397.
HILDEBRANDT, A. G., C. BERGS, G. HEINEMEYER, E. SCHLEDE, I. ROOTS, B. ABBAS-ALI, and A. SCHMOLDT, (1982), Adv. Exp. Med. Biol. **136A**, 179–198.
HOLM, K. A., R. J. ENGELL, and D. KUPFER, (1985), Arch. Biochem. Biophys. **237**, 477–489.
HOLTZMAN, J. L., (1970), Biochemistry **9**, 995–1000.
HOLTZMAN, J. L., R. P. MASON, and R. P. ERICKSON, (1977), in: Microsomes and Drug. Oxidations, (ULLRICH, V. et al., eds.), Pergamon Press, Oxford, 331–338.
HOWELL, L. G. and V. MASSEY, (1970), Biochem. Biophys. Res. Commun. **40**, 887–893.
HRYCAY, E. G. and P. J. O'BRIEN, (1971), Arch. Biochem. Biophys. **147**, 14–27.
HRYCAY, E. G. and P. J. O'BRIEN, (1974), Arch. Biochem. Biophys. **160**, 230–245.
HRYCAY, E. G., J.-A. GUSTAFSSON, M. INGELMAN-SUNDBERG, and L. ERNSTER, (1975a), Biochem. Biophys. Res. Commun. **66**, 209–216.
HRYCAY, E. G., H. G. JONEN, A. Y. H. LU, and W. LEVIN, (1975b), Arch. Biochem. Biophys. **166**, 145–151.

IMAI, Y. and R. SATO, (1977), Biochem. Biophys. Res. Commun. **75**, 420–426.
IMAI, Y., (1981), J. Biochem. **89**, 351–362.
INGELMAN-SUNDBERG, M. and I. JOHANSSON, (1980), Biochem. Biophys. Res. Commun. **97**, 582–589.
IWASAKI, K., H. NOGUCHI, R. KATO, Y. IMAI, and R. SATO, (1977), Biochem. Biophys. Res. Commun. **77**, 1143–1149.
JEFFERY, E. and G. J. MANNERING, (1974), Mol. Pharmacol. **10**, 1004–1008.
JEFFERY, E. and G. J. MANNERING, (1979), Mol. Pharmacol. **15**, 396–409.
KADLUBAR, F. F., K. C. MORTON, and D. M. ZIEGLER, (1973), Biochem. Biophys. Res. Commun. **54**, 1255–1261.
KATAGIRI, M., S. TAKEMORI, and M. NAKAMURA, (1973), in: Oxidases and Related Redox Systems, vol. 1, (KING, T. E. et al., eds.), Univ. Park Press, Baltimore, 164–189.
KARASHIMA, D., Y. HIROKATA, A. SHIGEMATSU, and T. FURUKAWA, (1977), J. Pharmacol. Exp. Ther. **203**, 409–416.
KARUZINA, I. I., A. I. VARENITSA, and A. I. ARCHAKOV, (1983), Biokhimiya **48**, 1788 to 1793.
KRATZ, F. and H. STAUDINGER, (1965), Hoppe Seyler's Z. Physiol. Chem. **343**, 27–34.
KUTHAN, H., H. TSUJI, H. GRAF, V. ULLRICH, J. WERRINGLOER, and R. W. ESTABROOK, (1978), FEBS Lett. **91**, 343–345.
KUTHAN, H. and V. ULLRICH, (1982), Eur. J. Biochem. **126**, 583–588.
KUTHAN, H., V. ULLRICH, and R. W. ESTABROOK, (1982), Biochem. J. **203**, 551–558.
LEE, W. A. and T. C. BRUICE, (1985), J. Am. Chem. Soc. **107**, 513–514.
LICHTENBERGER, F., W. NASTAINCZYK, and V. ULLRICH, (1976), Biochem. Biophys. Res. Commun. **70**, 939–945.
LICHTENBERGER, F. and V. ULLRICH, (1977), in: Microsomes and Drug Oxidations, (ULLRICH et al., eds.), Pergamon Press, Oxford, 218–231.
LONGMUIR, I. S., D. C. MARTIN, H. J. GOLD, and S. SUN, (1971), Microvascular Res. **3**, 125–134.
LONGMUIR, I. S., S. SUN, and W. SOUCIE, (1973), in: Oxidases and Related Redox Systems, vol. 2, (KING, T. E. et al., eds.), Univ. Park Press, Baltimore, 451–461.
LONGMUIR, I. S., (1981), in: Adv. Physiol. Sci., vol. 25, (KOVACH, A. G. B. et al., eds.), Akademiai Kiado, Budapest, 19–22.
LU, A. Y. H., H. W. STROBEL, and M. J. COON, (1970), Mol. Pharmacol. **15**, 396–409.
MANNERING, G. J., (1981), in: Concepts in Drug Metabolism. Part B. (JENNER, P. and TESTA, B., eds.), Marcel Dekker, New York, 53–166.
MARNETT, J. M., P. WELLER, and J. R. BATTISTA, (1986), in: Cytochrome P-450. Structure, Mechanism, Biochemistry, (ORTIZ DE MONTELLANO, P. R., ed.), Plenum Press, New York, 29–76.
MARTSEV, S. P., V. L. CHASHCHIN, and A. A. AKHREM, (1985), Biokhimiya **50**, 243–257.
MCCARTHY, M. and R. E. WHITE, (1983a), J. Biol. Chem. **258**, 9153–9158.
MCCARTHY, M. and R. E. WHITE, (1983b), J. Biol. Chem. **258**, 11610–11616.
MCMAHON, R. E., H. R. SULLIVAN, J. C. CRAIG, and W. E. PEREIRA, (1969), Arch. Biochem. Biophys. **132**, 575–577.
METELITZA, D. I., A. A. AKHREM, A. N. ERJOMIN, M. A. KISSEL, and S. A. USANOV, (1979), Acta biol. med. germ. **38**, 511–518.
METELITZA, D. I., (1984), in: Modelirovaniye okislitel'no-vosstanovitelnykh fermentov (Modelling of oxido-reductases), Nauka i Tekhnika, Minsk, 122–147.
MIWA, G. T., N. HARADA, and A. Y. H. LU, (1985), Arch. Biochem. Biophys. **239**, 155–162.
MORGAN, E. T., D. R. KOOP, and M. J. COON, (1982), J. Biol. Chem. **257**, 13951–13957.
NARASIMHULU, S., (1971), Arch. Biochem. Biophys. **147**, 384–390.
NASTAINCZYK, W., H. J. AHR, and V. ULLRICH, (1982), Biochem. Pharmacol. **31**, 391–396.
NASTAINCZYK, W. and V. ULLRICH, (1984), in: Oxygen Radicals in Chemistry and Biology, (BORS, W. et al., eds.), Gruyter & Co., Berlin, 441–445.

NELSON, S. D., J. R. MITCHELL, E. DYBING, and H. A. SASAME, (1976), Biochem. Biophys. Res. Commun. **70**, 1157–1165.
NORDBLOM, G. D., R. E. WHITE, and M. J. COON, (1976), Arch. Biochem. Biophys. **175**, 524–533.
NORDBLOM, G. D. and M. J. COON, (1977), Arch. Biochem. Biophys. **180**, 343–347.
O'BRIEN, P. J., (1978), Pharmacol. Ther. A2, 517–536.
OPRIAN, D. D., L. D. GORSKY, and M. J. COON, (1983), J. Biol. Chem. **258**, 8684–8691.
ORRENIUS, S., (1965), J. Cell. Biol. **26**, 713–723.
ORTIZ DE MONTELLANO, P. R., (1986), in: Cytochrome P-450. Structure, Mechanism, Biochemistry, (ORTIZ DE MONTELLANO, ed.), Plenum Press, New York, 217–271.
OTHA, Y. and D. W. RIBBONS, (1970), FEBS Lett. **11**, 189–192.
PARLI, C. J., N. WANG, and R. E. MCMAHON, (1971), Biochem. Biophys. Res. Commun. **43**, 1204–1209.
PEDERSON, T. C. and I. C. GUNSALUS, (1977), Fed. Proc. **36**, 663.
PORTER, N. A., L. S. LEHMAN, D. G. WUZEK, and P. M. GROSS, (1984), in: Oxygen Radicals in Chemistry and Biology, (BORS, W. et al., eds.), Gruyter & Co., Berlin, 235–247.
POSNER, H. S., C. MITOMA, S. ROTHBERG, and S. UDENFRIEND, (1961), Arch. Biochem. Biophys. **94**, 280–290.
POULOS, T. L. and J. KRAUT, (1980), J. Biol. Chem. **255**, 8199–8205.
POWIS, G., (1979), Biochem. Pharmacol. **28**, 83–89.
PREMEREUR, N., C. Van den BRANDEN, and F. ROELS, (1985), Arch. Int. Physiol. Biochim. **93**, 241–248.
RAHIMTULA, A. D. and P. J. O'BRIEN, (1974), Biochem. Biophys. Res. Commun. **60**, 440–447.
RAHIMTULA, A. D., P. J. O'BRIEN, E. G. HRYCAY, J. A. PETERSON, and R. W. ESTABROOK, (1974), Biochem. Biophys. Res. Commun. **60**, 695–702.
RAHIMTULA, A. D. and P. J. O'BRIEN, (1975), Biochem. Biophys. Res. Commun. **62**, 268–275.
RAHIMTULA, A. D. and P. J. O'BRIEN, (1977a), Eur. J. Biochem. **77**, 201–209.
RAHIMTULA, A. D. and P. J. O'BRIEN, (1977b), in: Microsomes and Drug Oxidations, (ULLRICH, V. et al., eds.), Pergamon Press, Oxford, 210–217.
RAHIMTULA, A. D., P. J. O'BRIEN, H. E. SEIFRIED, and D. M. JERINA, (1978), Eur. J. Biochem. **89**, 133–139.
RITTER, J., R. KAHL, and A. G. HILDEBRANDT, (1985), Res. Commun. Chem. Pathol. Pharmacol. **47**, 48–58.
ROESSING, D., R. KAHL, and A. G. HILDEBRANDT, (1985), Toxicology **34**, 67–77.
ROOTS, I., G. LASCHINSKY, A. G. HILDEBRANDT, G. HEINEMEYER, and S. NIGAM, (1980), in: Microsomes, Drug Oxidation, and Chemical Carcinogenesis, (COON, M. J. et al., eds.), Acad. Press, New York, 375–378.
SASAME, H. A., J. R. MITCHELL, and J. R. GILLETTE, (1975), Fed. Proc. **34**, 729.
SCHELLER, F., R. RENNEBERG, P. MOHR, G.-R. JÄNIG, and K. RUCKPAUL, (1976), FEBS Lett. **71**, 309–314.
SCHENKMAN, J. B., I. JANSSON, G. POWIS, and H. KAPPUS, (1979), Mol. Pharmacol. **15**, 428–432.
SIPES, I. G., A. J. GANDOLFI, L. R. POHL, G. KRISHNA, and B. R. BROWN, (1980), J. Pharmacol. Exp. Ther. **214**, 716–720.
SLIGAR, S. G., J. D. LIPSCOMB, P. G. DEBRUNNER, and I. C. GUNSALUS, (1974), Biochem. Biophys. Res. Commun. **61**, 290–296.
SLIGAR, S. G., (1976), Biochemistry **15**, 5399–5406.
SLIGAR, S. G., B. S. SHASTRY, and I. C. GUNSALUS, (1977), in: Microsomes and Drug Oxidations, (ULLRICH, V. et al., eds.), Pergamon Press, Oxford, 202–209.
SLIGAR, S. G., K. A. KENNEDY, and D. C. PEARSON, (1980), in: Oxidases and Related Redox Systems, (KING, T. E. et al., eds.), Pergamon Press, New York, 837–856.
SLIGAR, S. G., M. H. GELB, and D. C. HEIMBROOK, (1984), Xenobiotica, **14**, 63–86.

SOODAEVA, S. K., E. D. SKOTZELYAS, A. A. ZHUKOV, and A. I. ARCHAKOV, (1982), in: Cytochrome P-450. Biochemistry, Biophysics, and Environmental Implications, (HIETANEN, E. et al., eds.), Elsevier, Amsterdam, 615–618.
SOUSA, R. I. and M. A. MARLETTA, (1985), Arch. Biochem. Biophys. **240**, 345–357.
SPECTOR, T. and V. MASSEY, (1972), J. Biol. Chem. **247**, 4679–4687.
STAUDT, H., F. LICHTENBERGER, and V. ULLRICH, (1974), Eur. J. Biochem. **46**, 99–106.
STEENNIS, P. J., M. M. CORDES, J. G. H. HILKENS, and F. MUELLER, (1973), FEBS Lett. **36**, 177–180.
STORM, C. B. and S. KAUFMAN, (1968), Biochem. Biophys. Res. Commun. **32**, 788–793.
STRIPP, B., N. ZAMPAGLIONE, M. HAMRICK, and J. R. GILLETTE, (1972), Mol. Pharmacol. **8**, 189–196.
SUGIURA, M., K. IWASAKI, and H. NOGUCHI, (1974), Life Sci. **15**, 1433–1442.
SUGIURA, M., K. IWASAKI, and R. KATO, (1976), Mol. Pharmacol. **12**, 322–334.
TAKANO, T., Y. MIYAZAKI, and Y. MOTOHASHI, (1985), J. Toxicol. Sci. **10**, 249.
THURMAN, R. G., H. G. LEY, and R. SCHOLZ, (1972), Eur. J. Biochem. **25**, 420–430.
TOMASI, A., E. ALBANO, M. U. DIANZANI, T. F. SLATER, and V. VANNINI, (1983), FEBS Lett. **160**, 191–194.
TUCKEY, R. C. and H. KAMIN, (1982), J. Biol. Chem. **257**, 9309–9314.
UDENFRIEND, S., C. MITOMA, and H. S. POSNER, (1956), Am. Chem. Soc. Abstr. 130th Meeting, 54c–55c.
UEMURA, T., E. CHIESARA, and D. COVA, (1977), Mol. Pharmacol. **13**, 196–215.
ULLRICH, V. and H. DIEHL, (1971), Eur. J. Biochem. **20**, 509–512.
ULLRICH, V., (1977), in: Microsomes and Drug Oxidations, (ULLRICH, V. et al., eds.), Pergamon Press, Oxford, 192–201.
ULLRICH, V., L. CASTLE, and P. WEBER, (1981), Biochem. Pharmacol. **7**, 159–167.
ULLRICH, V., L. CASTLE, and M. HAURAND, (1982), in: Oxygenases and Oxygen Metabolism, (NOZAKI, M. et al., eds.), Acad. Press, New York, 497–509.
ULLRICH, V., (1984), in: Oxygen Radicals in Chemistry and Biology, (BORS, W. et al., eds.), Gruyter & Co., Berlin, 391–404.
ULLRICH, V., H. GRAF, and M. HAURAND, (1985), in: Microsomes and Drug Oxidations, (BOOBIS, A. R. et al., eds.), Taylor & Francis, London, 95–104.
VAN DYKE, K. A. and A. J. GANDOLFI, (1976), Drug Metab. Dispos. **4**, 40.
WERRINGLOER, J. and S. KAWANO, (1980), in: Microsomes, Drug Oxidations, and Chemical Carcinogenesis, (COON, M. J. et al., eds.), Acad. Press, New York, 403–407.
WERRINGLOER, J., (1982), in: Microsomes, Drug Oxidations, and Drug Toxicity, (SATO, R. and KATO, R., eds.), Japan Scientific Societes Press, Tokyo, 171–178.
WHITE, R. E. and M. J. COON, (1980), Ann. Rev. Biochem. **49**, 315–356.
WHITE, R. E., S. G. SLIGAR, and M. J. COON, (1980), J. Biol. Chem. **255**, 11108–11111.
WHITE-STEVENS, R. H. and H. KAMIN, (1970), Biochem. Biophys. Res. Commun. **38**, 882–889.
WHITE-STEVENS, R. H. and H. KAMIN, (1972), J. Biol. Chem. **247**, 2358–2370.
WONG, A. K. L., E. CAVALIERI, and E. ROGAN, (1986), Biochem. Pharmacol. **35**, 1583–1588.
YONETANI, T. and G. S. RAY, (1965), J. Biol. Chem. **240**, 4509–4516.
ZHUKOV, A. A. and A. I. ARCHAKOV, (1982), Biochem. Biophys. Res. Commun. **109**, 813–818.
ZHUKOV, A. A. and A. I. ARCHAKOV, (1985a), Biochemistry U. S. S. R. **50**, 1659–1673.
ZHUKOV, A. A. and A. I. ARCHAKOV, (1985b), in: Cytochrome P-450, Biochemistry, Biophysics and Induction, (VERECZKEY, L. and MAGYAR, K., eds.), Akademiai Kiado, Budapest, 75–78.

Chapter 5

The Hormonal and Molecular Basis of Sexually Differentiated Hepatic Drug and Steroid Metabolism in the Rat

E. T. Morgan and J.-Å. Gustafsson

1. Introduction.

Sex-dependent expression of cytochrome P-450 isozymes

In the late 1970s, the concept of sexually differentiated isozymes of hepatic cytochrome P-450 catalyzing sex-specific drug and steroid hydroxylations in the rat received increasing attention due to the following observations: (a) cytochrome P-450 is a family of structurally and catalytically distinguishable isozymes which display different responses to different xenobiotic inducers (Lu and West, 1980); (b) most drug and many steroid hydroxylations in the liver are P-450[1] catalyzed (Kuntzman et al., 1964; Conney, 1967; Lu and West, 1980) and (c) many rat hepatic drug- and steroid-metabolizing activities show significant sex differences. It should be noted that the terms "sex-specific" and "sexually differentiated" used here are usually not intended to mean a total lack of an isozyme in one sex, but rather a significant quantitative difference in the level of expression.

In 1980, we reported the partial purification of a P-450 fraction from female rat liver which possessed the female-specific 5α-androstane-$3\alpha,17\beta$-diol disulfate 15β-hydroxylase activity (Gustafsson et al., 1980; Lefevre et al., 1980). The major protein-staining band in the electrophoretic analysis of this fraction was absent from the equivalent fraction purified from male rats (Gustafsson et al., 1980; Lefevre et al., 1980). At the same time, Chung and Chao (1980) reported that the male-specific testosterone 16α-hydroxylase activity was present in a partially purified P-450 fraction from male rat liver, but not in the same fraction from females. These early studies strongly indicated that at least two sexually differentiated hepatic microsomal enzyme activities were attributable to distinct, sex-specific isozymes of cytochrome P-450.

At least four isozymes of P-450 are now known to be expressed in a sexually differentiated manner in the rat, and as discussed below there is evidence for a fifth, namely P-450 DEa. These isozymes and their nomenclatures in various laboratories are shown in Table 1, together with some preferred substrates of each isozyme. Only those isozymes showing well-characterized, several-fold sex differences are included, although smaller sex differences in the expression of other isozymes may occur.

2. Properties of sex-specific isozymes

2.1. P-450 16α

Cheng and Schenkman (1982) were the first to purify a sex-specific P-450 isozyme to electrophoretic homogeneity, but were unaware at that time that

[1]The abbreviations used are: P-450, cytochrome P-450; GH, growth hormone; GHSP, growth hormone secretory pattern; Hx, hypophysectomized

Table 1. Sex specific isozymes of cytochrome P-450 in rat liver

Isozyme	Alternative nomenclature	Dominant gender	Preferred substrates
P-450 g[1]	RLM_3[3]	male	benzphetamine
P-450 16α[2]	h[1], RLM_5[3], UT-A[4], 2c[5], male[6]	male	steroids (2-, 16α-), 7-propoxycoumarin, aminopyrine, benzphetamine, ethylmorphine, *p*-tolethylsulfide
P-450 15β[7]	i[1], 2d[5], UT-I[8], female[6]	female	steroid sulfates (15β-), aminopyrine, ethylmorphine
P-450 PCN[9]	PCN-E[4], PB-2a[8], p[10]	male	ethylmorphine, estradiol (2-, 4-), androstenedione (6β-)
P-450 DEa[7]	—	female	ethylmorphine

[1] Ryan et al. (1984); [2] Morgan et al. (1985b); [3] Cheng and Schenkman (1982); [4] Guengerich et al. (1982); [5] Waxman (1984); [6] Kamataki et al. (1983); [7] MacGeoch et al. (1984); [8] Waxman et al. (1985); [9] Elshourbagy and Guzelian (1980); [10] Wrighton et al. (1985). It should be noted that the isozymes equated with P-450 PCN here may be different homologs and all are not necessarily sexually differentiated.

P-450 RLM_5 and RLM_3 were male-specific proteins. Using a modification of their protocol, we purified P-450 RLM_5 from male rat liver and showed it to be male-specific using an immunoadsorbed polyclonal antibody preparation which specifically recognizes RLM_5 on Western blots (Morgan et al., 1985a). These antibodies also inhibited 60% of the testosterone 16α- and 2α-hydroxylase activity of male rat liver microsomes, but had no effect on the much lower activities in female microsomes or on testosterone 6β-hydroxylase activity of microsomes from either sex (Morgan et al., 1985a). Our results were in agreement with those of Waxman (1984) for P-450 2c. We concluded that P-450 RLM_5/2c was the male-specific testosterone and 4-androstene-3,17-dione 16α-hydroxylase, and renamed it P-450 16α (Morgan et al., 1985a, b). As indicated in Table 1, P-450 16α has also been named UT-A, h and male in various laboratories.

P-450 16α catalyzes the 16α-hydroxylation of testosterone, androstenedione and progesterone (Cheng and Schenkman, 1982; Waxman, 1984; Morgan et al., 1985a), as well as the 2α-hydroxylation of testosterone and progesterone (Waxman, 1984), and appears to be responsible for the well-characterized sex-differences in these microsomal activities (Waxman, 1984; Morgan et al., 1985a). Cheng and Schenkman (1982) also reported that P-450 16α effectively catalyzes the 2-hydroxylation of 17β-estradiol, a fact confirmed by RYAN et al. (1984) and DANNAN et al. (1986b). The latter authors also showed that P-450 16α contributes significantly to the sex-specific microsomal estrogen 2-hydroxylase activity of rat liver. This conclusion is supported by the observation that the microsomal estrogen 2-hydroxylase activity is subject to growth hormone regulation (JELLINCK et al., 1985), as is P-450 16α expression (MORGAN et al., 1985b).

Recently, ANDERSSON and JÖRNVALL (1986) reported that P-450 16α catalyzes the 25-hydroxylations of vitamin D_3 and 5β-cholestane-3α,7α,12α-triol, and showed that this isozyme is responsible for these microsomal activities in male rat liver. The lower but significant activities in female liver microsomes were attributed to other isozymes (ANDERSSON and JÖRNVALL, 1986). P-450 16α contributes significantly to the masculine microsomal metabolism of 7-propoxycoumarin, benzo[*a*]pyrene and aminopyrine, but other isozymes are also significant in these reactions (KATO et al., 1986). In addition, purified P-450 16α metabolizes *p*-tolethylsulfide, toluene and dimethylaniline (WAXMAN, 1984), benzphetamine, aminopyrine and ethylmorphine (MORGAN et al., 1985a), *R*- and *S*-warfarin (GUENGERICH et al., 1982), hexobarbital, zoxazolamine and *p*-nitroanisole (RYAN et al., 1984) with moderate to high efficiency. Such a versatile catalyst could conceivably account for many of the known male-specific drug-metabolizing activities in rat liver, and for others not yet documented. However, in order to unequivocally ascribe any given sex-dependent activity to P-450 16α or any other isozyme, detailed antibody inhibition studies will be required.

In Sprague-Dawley rats, P-450 16α is present as two allelic variants, distinguishable by two-dimensional gel electrophoresis (RAMPERSAUD et al., 1985). Some populations of this strain express both allozymes and some express only one. Several inbred strains of rat were each found to express only a single variant, and in some strains there was a third form not found in Sprague-Dawley rats (RAMPERSAUD et al., 1985). As the authors noted, it will be interesting to learn whether there are any differences in the function of these allozymes, or in the developmental and hormonal regulation of their expression.

NEGISHI and co-workers have purified a male-specific testosterone 16α-hydroxylase from 129/J mouse liver (HARADA and NEGISHI, 1985). This isozyme is 10 times more active in testosterone 16α-hydroxylation than is rat P-450 16α, and is responsible for $>90\%$ of the microsomal activity in mice of either sex. Its structural and genetic relationship to rat P-450 16α has not yet been reported.

2.2. P-450 15β

Although we had obtained a partially purified female P-450 fraction which catalyzed the 15β-hydroxylation of 5α-androstane-3α,17β-diol 3,17-disulfate as early as 1980 (GUSTAFSSON et al., 1980; LEFEVRE et al., 1980), we were unable to further purify the enzyme to homogeneity from this fraction, due to the persistence of a high-Mr contaminant.

During the purification of P-450 16α, we noticed that the contaminant seen in our 15β-hydroxylase preparations was removed at an early stage in the purification procedure. Accordingly, we tried to use the same protocol for the purification of the 15β-hydroxylase (MacGeoch et al., 1984). Using a minor modification of this method, we obtained an electrophoretically homogeneous P-450 isozyme which catalyzed the 15β-hydroxylation of 5α-androstane-3α,17β-diol 3,17-disulfate at a rate which was five times faster than that of female rat liver microsomes (MacGeoch et al., 1984). Immunoabsorbed antibodies to this isozyme were used to demonstrate its very low abundance in male rat liver and its regulation by growth hormone (MacGeoch et al., 1984; 1985). We concluded that this enzyme is the female specific steroid sulfate 15β-hydroxylase of rat liver, and named it P-450 15β. Although polyclonal antibodies to P-450 15β inhibited neither the 15β-hydroxylase activities of microsomes nor the purified enzyme (MacGeoch et al., 1984), our conclusion was supported by our later observation that monoclonal antibodies to P-450 15β inhibit the 15β-hydroxylase activity of female rat liver microsomes by 70% (Morgan et al., 1987). P-450 15β has also been purified in other laboratories and named P-450 female, 2d or i (Table 1).

P-450 15β also catalyzes the metabolism of some drug substrates in a reconstituted system containing the purified isozyme. These include aminopyrine, ethylmorphine and benzphetamine (MacGeoch et al., 1984; Kamataki et al., 1983; Waxman, 1984). However, none of these activities is female-specific. Indeed, their microsomal metabolism is more efficient in males. Thus, 15β-hydroxylation of steroid sulfates is the only female-specific activity which has been attributed to P-450 15β.

2.3. P-450 DEa

During purification of the female-specific P-450 15β, another P-450 isozyme was isolated from female rat liver, namely P-450 DEa (MacGeoch et al., 1984). Figure 1 shows the relative mobilities of purified P-450 isozymes 16α, 15β and

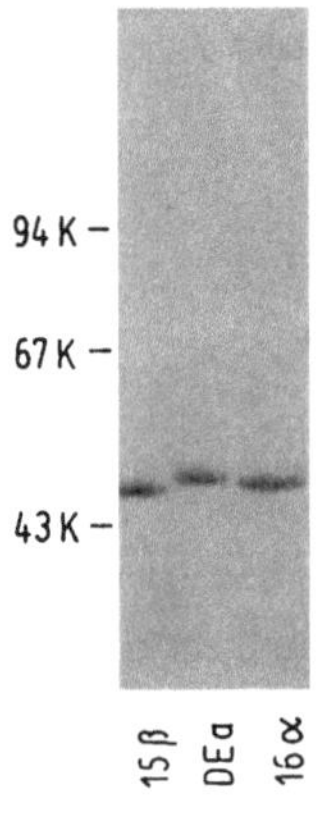

Fig. 1. SDS-polyacrylamide gel electrophoresis of purified P-450 isozymes.
1.0 μg of each isozyme was electrophoresed on a 7.5% polyacrylamide gel containing 0.1% SDS.

DEa on SDS-polyacrylamide gel electrophoresis. P-450 15β has an Mr of 50,000, while purified DEa and 16α both have Mr values of 52,000 in this system. P-450 DEa has been named arbitrarily by us on the basis of its elution profile relative to P-450 15β on a DEAE-Sepharose column. Although we have purified it from female rat liver, we have not yet prepared specific antibodies to P-450 DEa. Consequently, we have been unable to determine whether it is sex-specific or not.

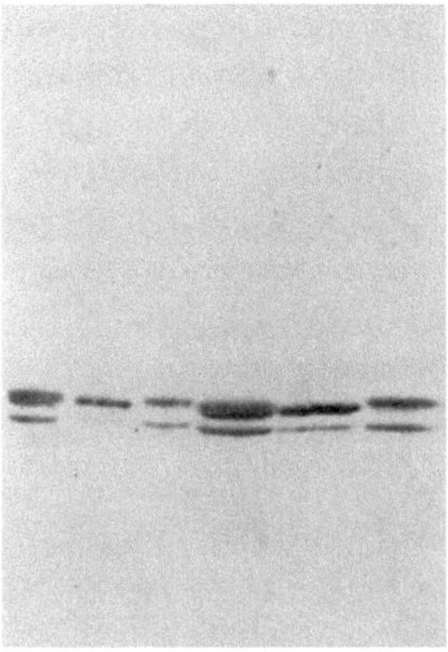

Fig. 2. Western blot showing apparent sex-specificity of P-450 DEa.
Microsomes from the indicated sources were electrophoresed on a 7% polyacrylamide gel containing SDS. The proteins were blotted onto a nitrocellulose filter and probed with a polyclonal antibody to P-450 16α. The antibody-antigen complex was visualized by incubation with goat anti-rabbit IgG followed by a horseradish peroxidase-rabbit antiperoxidase complex. The microsomal samples in each *lane* were as follows: 1) 2 μg male + 2 μg female, 2) 2 μg male, 3) 2 μg female, 4) 5 μg male + 5 μg female, 5) 5 μg male, 6) 5 μg female.

Purified P-450 DEa is electrophoretically indistinguishable from P-450 16α and is structurally related to this isozyme as observed by immunological cross-reactivity and peptide mapping (Morgan et al., 1985a). However, P-450 16α and DEa can also be discriminated by the latter two criteria. Ryan et al., (1984) have purified a P-450 isozyme, P-450 f, which is also electrophoretically, immunologically and structurally similar to P-450 16α, and it is possible that P-450 f and P-450 DEa are the same enzyme. A cDNA for P-450 f has been cloned, and used to show that P-450 f expression is about twice as high in female rat liver as in males (Gonzalez et al., 1986a). A similar sex difference in the microsomal P-450 f apoprotein levels has been reported (Bandiera et al., 1986). However, P-450 DEa purified in our laboratory is a low-spin hemoprotein (MacGeoch et al., 1984), whereas P-450 f isolated by Ryan et al. (1984) is in the high-spin state. This suggests that the two isozymes may be different, although the difference in spin state could be explained by the presence or absence of a ligand bound *in vivo*, possibly originating from the diet or environment.

Figure 2 shows the result of a recent experiment designed to answer the question of whether P-450 DEa is a sex-specific isozyme. We have consistently noted that although purified P-450 16α and DEa are difficult to resolve by SDS-polyacrylamide gel electrophoresis, the two proteins can be resolved in partially purified or microsomal preparations containing either isozyme. Figure 2 is a Western blot of male and female rat liver microsomes run separately or together on a 7% SDS-polyacrylamide gel, blotted on nitrocellulose and probed with a polyclonal antibody to P-450 16α. This antibody cross-reacts with P-450 DEa, P-450 15β, and a male-specific protein presumed to be P-450 g. The **upper** immunologically-stained bands in each lane correspond to P-450 16α and P-450 DEa in liver microsomes from male and female rats, respectively (Lanes 2, 3, 4 and 6). A distinctly higher mobility of P-450 16α compared to P-450 DEa is noted. Lane 1 shows that in a mixture of male and female rat liver microsomes, the upper band is a doublet. Thus, microsomal P-450 16α and DEa can be resolved in this system. The presence of only a single upper band in male rat liver microsomes, corresponding to P-450 16α, indicates that P-450 DEa is absent from, or expressed at low levels in male rat liver (compare lanes 1 and 2). This experiment does not conclusively demonstrate that P-450 DEa is female-specific, but strongly indicates that further investigation with a specific P-450 DEa antibody is needed.

2.4. P-450 g

P-450 g is a male-specific rat liver isozyme isolated by RYAN et al. (1984) and is apparently equivalent to the RLM_3 of CHENG and SCHENKMAN (1982). It catalyzes the 6β-hydroxylation of testosterone, as does the male-specific isozyme P-450 PCN (below). P-450 g differs from the latter isozyme, however, in both catalytic and structural properties (RYAN et al., 1984; HANIU et al., 1984). While the sex-specificity of P-450 g, and its developmental induction in postpubertal males have been clearly demonstrated (BANDIERA et al., 1986), its contributions to any sexually differentiated microsomal drug- or steroid-metabolizing activities have yet to be shown.

Interestingly, BANDIERA et al. (1986) have found that adult male outbred rats can be divided into two populations expressing P-450 g at different levels. The levels of P-450 g in the livers of these animals are not dependent on their serum testosterone concentrations. Genetic studies using inbred strains of rats expressing high and low levels of P-450 g in the males have shown that the gene regulating this trait is cis acting, autosomal, and inherited additively (RAMPERSAUD et al., 1987). In addition, McLELLAN-GREEN et al. (1987) showed that while female Sprague-Dawley and male Fischer rats both have negligible levels of translatable hepatic P-450 g mRNA, male Sprague-Dawley rats expressing high or low levels of the isozyme have similar levels of the mRNA. This indicates a translational or postranslational regulation of P-450 g expression in Sprague-Dawley rats, that differs between the two populations.

2.5. P-450 PCN

P-450 PCN is an isozyme first isolated from the livers of rats treated with pregnenolone-16α-carbonitrile, and is induced by that compound (ELSHOURBAGY and GUZELIAN, 1980). Isozymes with highly similar physical and immuno-

Table 2. Monoclonal antibodies to P-450 16α, 15β and DEa[a]

Antigen	Number of growing clones	Number of positive wells	Number of specific clones[b]
16α	1224	161	1
15β + DEa	456	60	5[c]

[a] Female BALB/c mice were immunized with either pure P-450 16α or a partially purified female P-450 fraction containing P-450 15β and DEa as the major components. After fusion, growing hybridomas were screened by ELISA for antibodies to the injected antigen. Positives were rescreened immediately for cross-reactivity to the antigen from the other sex.

[b] Number of clones producing antibodies which did not cross-react in the second ELISA screen.

[c] All of these clones produced antibodies specific for P-450 15β.

logical properties to P-450 PCN have been isolated in various laboratories, and have been designated P-450 PB-2a (WAXMAN et al., 1985), PB/PCN-E or PCN-E (GUENGERICH et al., 1982; WAXMAN et al., 1985), or P-450 p (WRIGHTON et al., 1985). P-450 PB-2a/PCN is inducible by phenobarbital as well as by pregnenolone-16α-carbonitrile, while P-450 p has been shown to be induced by glucocorticoids and macrolide antibiotics (SCHUETZ et al., 1984; WRIGHTON et al., 1985).

P-450 PB-2a/PCN is found at much higher levels in hepatic microsomes from male rats than from female rats (WAXMAN et al., 1985), and is neonatally imprinted and androgen-regulated (DANNAN et al., 1986b). Based on these observations and the fact that antibodies to this isozyme effect an 87% inhibition of male rat liver microsomal androstenedione 6β-hydroxylase activity (WAXMAN et al., 1985), these authors have proposed that P-450 PB-2a/PCN is the isozyme responsible for this sexually differentiated activity in male rat liver. The purified isozyme is not an effective catalyst of 6β-hydroxylation, but this is explained by the isozyme's propensity to lose catalytic activity during purification (ELSHOURBAGY and GUZELIAN, 1980). Similarly, P-450 PCN/PB-2a/PCN-E can be shown by antibody inhibition to contribute significantly to the male-specific microsomal ethylmorphine demethylase (ELSHOURBAGY and GUZELIAN, 1980) and 17β-estradiol 2- and 4-hydroxylase activities (DANNAN et al., 1986a) despite relatively low activities of the purified enzyme towards these substrates.

Recently, GONZALEZ et al. (1986b) have found two P-450 PCN-related mRNAs to be expressed in rat liver, one of which is PCN-inducible and one of which is constitutive. The constitutive mRNA is sexually differentiated and follows the same developmental pattern as P-450 PB-2a/PCN-E. However, the NH_2-termini of the cloned mRNAs are distinct from those of P-450 PCN-E and P-450 p (GONZALEZ et al., 1986b). Thus, it is possible that P-450 PCN, p, PB-2a and PCN-E isozymes are not identical, and/or contain two or more highly homologous forms of the isozyme which might be under different hormonal and xenobiotic regulation. It is still apparent, however, that one or more members of the P-450 PCN gene family are responsible for the male-specific 6β-hydroxylase activity.

3. Some sex-specific P-450 isozymes are structurally related

Cytochrome P-450 isozymes are the products of a multigene superfamily (WHITLOCK, 1986), which can be classified into different families on the basis of protein and DNA sequence homology. Members of a gene family have greater than 50% amino acid sequence homology, and members of a subfamily are at least 70% homologous (DAYHOFF, 1976). Peptide mapping, NH_2-terminal amino acid sequencing and immunological cross-reactivities indicate that some sexually differentiated P-450 isozymes have greater structural similarities to each other than to other known P-450 isozymes, indicating that they may belong to a new gene subfamily. cDNA cloning and sequencing have demonstrated that P-450 16α belongs to the same subfamily as P-450 PB-i and f, part of the P-450 b/e gene family.

3.1. Immunochemical evidence

Four independent laboratories have reported cross-reactivity of polyclonal antibodies to isozymes from the group P-450 16α, P-450 15β, P-450 f, P-450 g and P-450 DEa with one or more other members of the group. Using a Western blot technique, we showed that rabbit antibodies to P-450 16α cross-react with P-450 15β and DEa (MORGAN et al., 1985a), and that antibodies raised to P-450 15β cross-react with P-450 16α and DEa (MACGEOCH et al., 1984). This followed a report by KAMATAKI et al. (1983) that antibodies to P-450 16α cross-react with P-450 15β in Ouchterlony immunodiffusion analysis, although these authors saw no cross-reactivity of anti-P-450 15β serum with P-450 16α. Both laboratories reported that cross-reactivity could be removed from the antibody preparations by immunoadsorption with a heterologous crude antigen i. e. partially purified P-450 (MACGEOCH et al., 1984; MORGAN et al., 1985a) or microsomes (KAMATAKI et al., 1983) from rats of the opposite sex. In contrast, WAXMAN (1984) observed residual cross-reactivity of anti-P-450 16α with P-450 15β, measured by an ELISA, even after immunoadsorption. He also reported that antibodies to isozyme PB-1, a protein moderately induced by phenobarbital, cross-reacted significantly with P-450 15β (WAXMAN, 1984).

BANDIERA et al. (1985) also found significant immunochemical similarities among sex-specific P-450 isozymes of rat liver. They raised antibodies to P-450 f and the male-specific P-450 g. Not only did antibodies to these two isozymes cross-react with each other, but they also recognized P-450 16α and P-450 15β in both Ouchterlony immunodiffusion, ELISA and Western blot assays (Bandiera et al., 1985). Anti-P-450 f also showed weak cross-reactivity with the phenobarbital-inducible isozymes P-450 b and e, complementing an earlier observation that anti-P-450 b recognizes P-450 f (Ryan et al., 1984). The same authors were able to render the anti-P-450 f and g antibodies specific for their respective antigens by immunoadsorbing them with the cross-reactive antigens in partially purified form (Bandiera et al., 1986).

The above observations indicate that there may be extensive amino acid sequence homology among various members of the P-450 isozyme group 16α, 15β, f, g and DEa, and also indicate homology of at least P-450f and P-450 15β with P-450b and PB-1, the phenobarbital-inducible isozymes. Bandiera et al. (1985) have noted that other examples of immunological cross-reactivity among P-450 isozymes or other globular proteins are accompanied by a >60% amino acid sequence homology.

It is conceivable that some or all of the cross-reactivities discussed above could be due to the presence of the cross-reacting antigens in the purified proteins used for immunization. Bandiera et al. (1985) showed that the anti-P-450 f and g antibodies did not recognize any contaminating proteins in the purified antigen preparations, arguing against this possibility. However, the level of contamination required to produce the cross-reacting antibodies would not necessarily be the same as that required for detection of the contaminants by Western blotting. We have produced monoclonal antibodies to P-450 16α and 15β which significantly cross-react with the heterologous antigen and/or P-450 DEa, proving that the observed immunological cross-reactivities are indeed due to common antigenic determinants among the antigens. These studies are described below (Morgan et al., 1987).

Purified P-450 16α was used to immunize female BALB/c mice. Spleen cells were prepared from mice exhibiting a high serum titer to P-450 16α and fused to SP^2/O myeloma cells. Cells were plated in microtiter wells, and hybridomas selected in medium containing hypoxanthine, aminopterin and thymidine. Growing hybridomas were screened by ELISA for production of antibodies to P-450 16α. Clones giving a positive signal in this assay were then screened by ELISA for cross-reactivity with a mixture of female rat liver proteins containing P-450 15β and DEa as the major components. Table 2 shows that one in eight growing hybridomas produced a P-450 16α antibody, but only one of these antibodies (< 1%) did not also recognize the P-450 15β/DEa mixture. Similarly, when mice were immunized with the P-450 15β/DEa mixture, one in eight clones produced an antibody to this antigen. Only five of these (8%) did not recognize P-450 16α, and all five were specific for P-450 15β (Table 2). Andersson and Jörnvall (1986) have also reported a low frequency of specific monoclonal antibodies to P-450 16α. Only two of 27 positive hybridomas produced antibodies that were monospecific for the isozyme.

We found that even among those monoclonal antibodies which specifically recognized P-450 15β or P-450 16α in the initial ELISA screens, significant cross-reactivity could be observed depending on the assay system used. Anti-P-450 15β monoclonal antibodies F22 and F23 were specific for P-450 15β in ELISA, Western blotting and Sepharose-coupled immunoadsorption assays, whereas antibodies F3, F1 and F20 each showed significant cross-reactivity with P-450 16α and/or P-450 DEa in at least one of these assays. It is of interest to note that all five of the P-450 15β-specific monoclonal antibodies were able to compete for each other's binding to P-450 15β, and all inhibited the catalytic activity of microsomal P-450 15β by up to 70%. These observations indicate (*a*) that those antigenic sites on P-450 15β which are not shared by P-450 16α and P-450 DEa are confined to one part of the molecule and (*b*) that this part of the molecule is near the P-450 15β active site. The observed inhibition of microsomal 15β-hydroxylase provides the proof that P-450 15β is indeed the isozyme responsible for this activity.

The significant cross-reactivities observed with F1, F3 and F20 antibodies in some systems, but not in others, argues that even those epitopes which distinguish P-450 15β from P-450 16α and DEa bear some structural resemblance to those on their counterparts. Even the apparently specific F22 and F23 antibodies showed some affinity for P-450 16α, since they effectively inhibited the testosterone 16α- and 2α-hydroxylase activities of male rat liver microsomes. We propose that F22 and F23 recognize epitopes at the active sites of both P-450 15β and P-450 16α, but with a significant difference in affinity. Thus, under conditions favoring high-affinity binding, i. e. procedures with many washing steps such as Western blotting, ELISA and immunoaffinity adsorption, P-450 15β binds the antibodies with much greater efficiency. However, under conditions involving constant incubation of the antibodies with the enzymes, lower affinity interactions are detected, and thus inhibition of P-450 16α catalytic activity is observed.

3.2. Protein sequencing and peptide mapping

Several laboratories have studied the structural relationship among two or more of the sex-specific P-450 isozymes by electrophoretic analysis of the peptides formed by limited proteolysis of the isozymes by different peptidases ("peptide mapping").

We compared the peptide maps of P-450 16α and DEa formed by α-chymotrypsin and *S. aureus* V_8 protease (Morgan et al., 1985a). Although significant differences were observed in the sizes of some of the peptides generated, many similarities could be seen in the peptide patterns, indicating structural similarities between these isozymes. Waxman (1984) also found similarities in the peptide maps of P-450 16α and 15β generated by α-chymotrypsin. In a comparison by Ryan et al. (1984) of the peptide maps of P-450 f, g and 16α with those of five other isozymes, similarities between the patterns for P-450 g and 16α can be discerned, while those of P-450 f appear to be different.

The NH_2-terminal sequences of P-450 f, g, 16α and 15β have been elucidated

(Haniu et al., 1984). Of the first 15 NH_2-terminal amino acid residues, P-450 16α shares 10—11 with P-450 f, g and 15β. P-450 g also shares 12/15 residues with P-450 15β. However, homology in this region is less marked between P-450 f and P-450 g (7/15) and P-450 f and P-450 15β (8/15), suggesting that P-450 f may be more distantly related to the sex-specific isozymes. The NH_2-terminal sequence of P-450 16α also showed high homology to P-450 PB-1 (13/20 residues), the phenobarbital-inducible isozyme.

3.3. cDNA sequencing

Gonzalez et al. (1986a) have cloned full length cDNAs for P-450 f and P-450 PB-1. The deduced amino acid sequences show 75% similarity, placing these two isozymes in the same gene subfamily. Both isozymes are 50% homologous to P-450 b, indicating that the new PB-1/f subfamily belongs in the P-450 b family.

We recently cloned and sequenced the cDNA for P-450 16α (Ström et al., unpublished observations), using polyclonal antibodies to screen a library created in λgt 11. Identification of the cDNA as P-450 16α was based on (*a*) comparison of the nucleotide sequence with the NH_2-terminal protein sequence (*b*) recognition of the expressed cDNA-β-galactosidase fusion protein by specific polyclonal and monoclonal antibodies and (*c*) hormonal regulation of the mRNA (see below). The sequence of our cDNA is identical to that recently published for P-450 (M-1) (Yoshioka et al., 1987). P-450 16α is 69 and 73% homologous to P-450 f and PB-1, respectively, placing it in the PB-1/f subfamily. Interestingly, Rampersaud and Walz (1987) mapped P-450 b and e to a different chromosome from P-450 g and h, suggesting that b/e and g/h represent different P-450 gene families. This conclusion is obviously at odds with the deduced sequences for P-450 b and h, which are 52% similar.

There is some question as to the size of the PB-1/f subfamily. Gonzalez et al. (1986a) favored a maximum of three members, based on genomic blot data. Friedberg et al. (1986) using the same technique, found evidence for a fairly large PB-1/f subfamily. Yoshioka et al. (1987) have reported that a cDNA isolated from female rat liver shows high sequence similarity with P-450 16α, but the information given does not preclude its identity with P-450 f or P-450 PB-1 cDNAs. The immunochemical and physical evidence available indicate that P-450 g, 16α and 15β share some features of both primary and secondary structure. Immunochemical evidence for structural similarities between P-450 15β and P-450 PB-1 has also been presented (Waxman, 1984). The familial relationships among these isozymes will be revealed when the cDNA sequences for all of them become available. To that end, we have cloned a cDNA for P-450 15β (MacGeoch et al., 1987) and are currently determining its nucleotide sequence.

4. Hormonal regulation of sex-specific P-450 isozymes

4.1. P-450-dependent enzyme activities

During the 1970s and early 80s, the Gustafsson laboratory established that pituitary growth hormone is the primary hormone involved in maintenance of sexually differentiated steroid and drug metabolism in rat liver (for a review, see Gustafsson et al., 1983). Growth hormone secretion is in turn regulated by the action of gonadal steroids, establishing a gonadal-hypothalamo-pituitary-liver axis.

Blood GH levels in the rat display a sexually dimorphic, temporal variation (Edén, 1979). In the male, this is characterized by regular peaks every 3—4 h, with troughs of almost zero levels between the peaks. In the female blood GH levels are more constant, with smaller peaks occurring at irregular intervals and higher baseline levels than in the male. These fluctuations in blood GH have been attributed to episodic pituitary GH secretion, and we speak of the sexually differentiated GH secretory pattern (GHSP). It should be noted, however, that a direct connection between the blood GH pattern and episodic pituitary GH secretion has not been demonstrated formally.

The hepatic drug- and steroid-metabolizing activities of a male or hypophysectomized rat liver can be feminized by continuous infusion of GH, mimicking the female GHSP (Mode et al., 1981). However, intermittent injection of the same daily dose of GH to Hx rats has no such effect. There are no sex differences in Hx rats. These observations clearly demonstrate the major role of the sexually dimorphic GHSP in regulation of these enzyme activities.

Sex steroids can also modulate hepatic drug- and steroid-metabolizing activities, but this action is achieved via the hypothalamo-pituitary axis, since gonadectomy or gonadal hormone treatments have no effect in Hx rats (Gustafsson and Stenberg, 1976). In the male rat, testicular androgens have a dual role. Androgens secreted in the neonatal period imprint the hypothalamus to direct a masculine GHSP (Jansson et al., 1985), and thus a masculine pattern of hepatic drug and steroid metabolism, from the advent of puberty (Einarsson et al., 1973; Gustafsson et al., 1983). Thus, males castrated after puberty still exhibit an essentially masculine profile of hepatic activities. The second role of androgens is in maintaining the masculine pattern of hepatic drug and steroid metabolism from puberty onwards (Einarsson et al., 1973; Gustafsson et al., 1983). It has been clearly demonstrated that androgen treatment of unimprinted, mature rats can produce a normal male GHSP (Jansson and Frohman, 1987) and a fully masculine hepatic steroid-metabolizing profile (Jansson et al., 1985), disproving the previous concept that only neonatally imprinted rats can fully respond to androgen administered in the adult phase.

The female GHSP, and hence the female pattern of hepatic drug and steroid metabolism, appears to be more dependent on the absence of androgen than on the presence of estrogen, since gonadectomy of female rats has relatively little effect compared to that seen in males (GUSTAFSSON et al., 1983). However, estrogen administration to intact or castrated male rats does feminize hepatic steroid-metabolizing activities (BERG and GUSTAFSSON, 1973; GUSTAFSSON and INGELMAN-SUNDBERG, 1974).

4.2. P-450 isozyme levels

The male-specific microsomal steroid 16α-hydroxylase and female-specific steroid sulfate 15β-hydroxylase activities of rat liver are two activities regulated by GH and testosterone as described above. With the purification of the P-450 isozymes catalyzing these reactions in untreated adult rats, and development of specific immunoassays for them, we and others have been able to study the hormonal regulation of expression of these isozymes in rat liver. These independent studies have clearly established that developmental and hormonal regulation of hepatic microsomal 16α- and 15β-hydroxylase activities is achieved by modulation of the levels of P-450 16α and P-450 15β in the endoplasmic reticulum, and that the GHSP is the major determining factor of the hepatic P-450 16α/15β phenotype. Thus, expression of P-450 16α and P-450 15β is low until puberty, when the surge of sex hormones is responsible for the development of sex-specific expression around age 4—5 weeks (MACGEOCH et al., 1985; MORGAN et al., 1985b; MAEDA et al., 1984; WAXMAN et al., 1985). The male phenotype is neonatally imprinted by androgen, inducible by androgen in the adult (MORGAN et al., 1985a; MACGEOCH et al., 1985; WAXMAN et al., 1985), and inducible by discontinuous GH administration to Hx rats (MORGAN et al., 1985b; KATO et al., 1986). Similarly, the feminine phenotype is favored by a lack of neonatal androgen imprinting and is inducible by estrogen in intact male rats or gonadectomized rats of either sex (MACGEOCH et al., 1985; MORGAN et al., 1985b; DANNAN et al., 1986a). Continuous GH infusion to intact male or Hx rats, mimicking the female GHSP, also feminizes the hepatic P-450 phenotype (MACGEOCH et al., 1985; MORGAN et al., 1985b; KATO et al., 1986). Hx rats exhibit no sex difference, and an intermediate phenotype (KAMATAKI et al., 1985; MACGEOCH et al., 1985; MORGAN et al., 1985b). Sex steroids have no effect in Hx rats, confirming that they act via the hypothalamo-pituitary axis (KAMATAKI et al., 1985). The roles of the pituitary gland and of constant blood levels of GH in maintaining the female P-450 16α/15β phenotype are shown in Figure 3, where it can be seen that hypophysectomy of male and female rats abolishes the sex differences in P-450 16α and 15β expression. Infusion of human GH induces the female phenotype in intact male rats or Hx rats of either sex.

Male-specific expression of P-450 PCN is also neonatally imprinted by androgen (WAXMAN et al., 1985) but the development of sex-specific expression of P-450 PCN is different from that of P-450 16α and P-450 15β. P-450 PCN expression is near adult male levels in immature rats of both sexes, then begins to fall steadily in the female from between weeks 2 and 4 to week 8 (WAXMAN et al., 1985). P-450 PCN is GH-regulated, but is not dependent on the GHSP since it is induced in Hx males regardless of whether administration is intermittent or continuous (YAMAZOE et al., 1986).

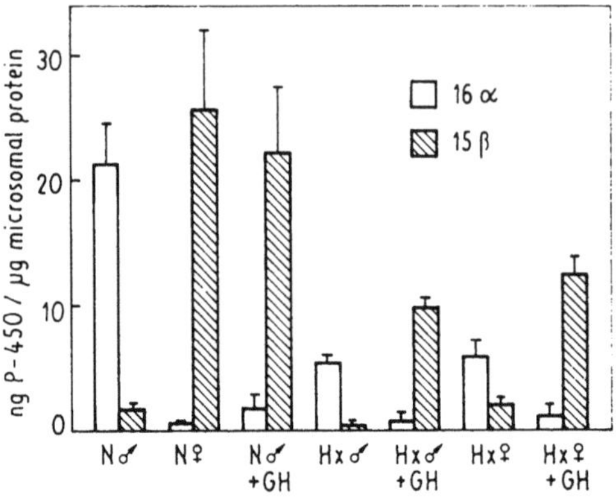

Fig. 3. Effects of hypophysectomy and GH infusion on hepatic expression of P-450 16α and P-450 15β.
Human GH was infused from osmotic minipumps at 5 µg/h for 7 days to sexually mature, intact (N) or Hx male and female rats. The P-450 isozymes were quantified in hepatic microsomes using a Western blot immunoassay. Values represent the mean ± S. E. M. of values from three animals.

The hormonal regulation of P-450 g expression has not been characterized, but RAMPERSAUD et al. (1987) have described the development of its male-specific expression in inbred rats expressing either high or low levels of P-450 g. The development of adult levels of P-450 g expression was found to coincide with puberty in both strains, as observed for P-450 16α.

Interestingly, our recent unpublished experiments, in collaboration with Dr. SKETT of Glasgow University, have indicated a role for insulin in regulation of P-450 16α expression. P-450 16α levels are reduced 20-fold in streptozotocin diabetic male rats within three days of drug administration. This is consistent with a similar observation by FAVREAU and SCHENKMAN (1987). It is possible that this effect is merely due to the known decrease in GH secretion that occurs in diabetes (TANNENBAUM, 1981), but we have preliminary evidence that the effects of diabetes are not the same as those of hypophysectomy, indicating a possible direct role of insulin in maintaining P-450 16α expression.

5. Growth hormone regulates P-450 16α and P-450 15β at a pretranslational level

Using specific polyclonal antibodies to P-450 15β to screen a female rat liver cDNA library in pBR322, we isolated a partial P-450 15β cDNA clone (MACGEOCH et al., 1987). This cDNA recognized a single size class of female-specific mRNA on Northern blots, and was used to characterize the hormonal regulation of expression of the P-450 15β mRNA. The results showed a good correlation between P-450 15β mRNA levels and apoprotein levels or catalytic activities in microsomes of intact male and female rats, Hx females, females injected with androgen, and males infused with human GH (MACGEOCH et al.,

1987). In subsequent experiments (STRÖM et al., 1987), using a cloned P-450 16α cDNA probe, we also determined that P-450 16α mRNA correlates with the microsomal levels of P-450 16α under the same hormonal manipulations. Sex- and tissue-dependent expression of P-450 16α mRNA in mature male rat livers has been reported by YOSHIOKA et al. (1987). Noshiro and Negishi (1986) have demonstrated the pretranslational regulation by GH of the murine P-450 16α and P-450 15α mRNAs. These authors showed that in GH-deficient mice, GH is a masculinizing factor when given intermittently, but that continuous GH has no effect. They have suggested that continuous GH is not a feminizing factor as such in the rat, but rather acts to suppress the male peaks. However, this theory is not supported by the fact that GH infusion of Hx rats, which have no peaks to suppress, also feminizes P-450 16α and P-450 15β. There may be a species difference in the mechanisms of GH regulation of sex-dependent P-450 isozymes.

6. Unanswered questions

The GHSP regulates hepatic P-450 16α and P-450 15β expression pretranslationally in the rat, but this could reflect either an effect on rates of transcription of the relevant genes or on RNA processing or stability. Only by measuring these parameters will the question be solved. Both transcription and post-transcriptional mechanisms are known for regulation of other P-450 isozymes by xenobiotics (WHITLOCK, 1986; SIMMONS et al., 1987). In the only other known instance of peptide hormone regulation of P-450 isozymes, adrenocorticotropin regulates transcription of adrenal steroidogenic P-450 isozymes in a cyclic AMP-dependent process involving de novo synthesis of a "steroid hydroxylase inducing protein" (JOHN et al., 1985; ZUBER et al., 1986). Also favoring the transcriptional model is the fact that α2u globulin, a hepatic secretory protein whose male-specific secretion is GHSP-dependent (HUSMAN et al., 1985), is transcriptionally regulated by GH (KULKARNI et al., 1985).

Perhaps the most intriguing question is how the liver recognizes the different patterns of blood GH. At present there is no prevailing theory, and little indication of possible mechanisms. It is worth noting, however, that a large number of hepatic proteins are under GHSP control in the rat in addition to P-450 isozymes. These include α2u globulin, steroid 5α-reductase (JANSSON et al., 1985), isozymes II and III of carbonic anhydrase (JEFFERY et al., 1986) and GH and prolactin receptors (BAXTER et al., 1982; ROY and CHATTERJEE, 1983; NORSTEDT and PALMITER, 1984). The cloning of the genomic genes and 5'-flanking regions for GHSP-regulated P-450 isozymes, and identification of *cis* and *trans* acting regulatory factors, may yield the necessary clues to the identities of the cellular mediators of the GHSP response.

Acknowledgements

We thank our colleagues Drs. A. Mode, P. Zaphiropoulos and A. Ström for supplying unpublished data on the molecular cloning of cDNA for P-450 16α and its hormonal regulation. This work was supported by a grant from the Swedish Medical Research Council (Nr. 13X-2819).

7. References

Andersson, S. and H. Jörnvall, (1986), J. Biol. Chem. **261**, 16932–16936.

Bandiera, S., D. E. Ryan, W. Levin, and P. E. Thomas, (1985), Arch. Biochem. Biophys. **240**, 478–482.

Bandiera, S., D. E. Ryan, W. Levin, and P. E. Thomas, (1986), ibid **248**, 658–676.

Baxter, R. C., Z. Zaltsman, and J. R., Turtle, (1982), Endocrinology **111**, 1020–1022.

Berg, A. and J-Å. Gustafsson, (1973), J. Biol. Chem. **248**, 6559–6567.

Cheng, K-C. and J. B. Schenkman, (1982), J. Biol. Chem. **257**, 2378–2385.

Cheng, K-C. and J. B. Schenkman, (1984), Drug Metab. Disp. **12**, 222–234.

Chung, L. W. K. and H. Chao, (1980), Mol. Pharmacol. **18**, 543–549.

Conney, A. H., (1967), Pharmacol. Rev. **19**, 317–366.

Dannan, G. A., F. P. Guengerich, and D. J. Waxman, (1986a), J. Biol. Chem. **261**, 10728–10735.

Dannan, G. A., D. J. Porubek, S. D. Nelson, D. J. Waxman, and F. P. Guengerich, (1986b), Endocrinology **118**, 1952–1960.

Dayhoff, M. O., (1976), in: Atlas of Protein Sequence and Structure. Vol. 5, Suppl. 2. National Biomedical Research Foundation, Washington D. C., **9**.

Edén, S., (1979), Endocrinology **105**, 555–560.

Einarsson, K., J-Å. Gustafsson, and Å. Stenberg, (1973), J. Biol. Chem. **248**, 4987 to 4997.

Elshourbagy, N. and P. S. Guzelian, (1980), J. Biol. Chem. **255**, 1279–1285.

Favreau, L. V. and J. B. Schenkman, (1987), Biochem. Biophys. Res. Commun. **142**, 623–630.

Friedberg, T., D. J. Waxman, M. Atchison, A. Kumar, T. Haaparanta, C. Raphael, and M. Adesnik, (1986), Biochemistry **25**, 7975–7983.

Gonzalez, F. J., S. Kimura, B. J. Song, J. Pastewka, H. V. Gelboin, and J. P. Hardwick, (1986a), J. Biol. Chem. **261**, 10667–10672.

Gonzalez, F. J., B. J. Song, and J. P. Hardwick, (1986b), Mol. Cell Biol. **6**, 2969–2976.

Guengerich, F. P., G. A. Dannan, S. T. Wright, M. V. Martin, and L. S. Kaminsky, (1982), Biochemistry **21**, 6019–6030.

Gustafsson, J-Å. and M. Ingelman-Sundberg, (1974), J. Biol. Chem. **249**, 1940–1945.

Gustafsson, J-Å. and Å. Stenberg, (1976), Proc. Natl. Acad. Sci. U. S. A. **73**, 1462 to 1465.

Gustafsson, J-Å., P. Eneroth, A. Hansson, T. Hökfelt, A. Lefevre, C. MacGeoch, A. Mode, G. Norstedt, and P. Skett, (1980), in: Biochemistry, Biophysics and Regulation of Cytochrome P-450 (J-Å. Gustafsson et al., eds.), Elsevier/North Holland, 171–178.

Gustafsson, J-Å., A. Mode, G. Norstedt, and P. Skett, (1983), Ann. Rev. Physiol. **45**, 51–60.

Haniu, M., D. E. Ryan, S. Iida, C. S. Lieber, W. Levin, and J. Shively, (1984), Arch. Biochem. Biophys. **235**, 304–311.

Harada, N. and M. Negishi, (1985), Proc. Natl. Acad. Sci. USA **82**, 2024–2028.

Husman, B., G. Norstedt, A. Mode, and J-Å. Gustafsson, (1985), Mol. Cell. Endocrinol. **40**, 205–210.

JANSSON, J-O., S. EKBERG, O. ISAKSSON, A. MODE, and J-Å. GUSTAFSSON, (1985), Endocrinology **117**, 1881–1887.
JANSSON, J-O. and L. A. FROHMAN, (1987), Endocrinology **120**, 1551–1557.
JEFFERY, S., C. A. WILSON, A. MODE, J-Å. GUSTAFSSON, and N. D. CARTER, (1986), J. Endocr. **110**, 123–126.
JELLINCK, P. H., J. A. QUAIL, and C. A. CRAWLEY, (1985), Endocrinology **117**, 2274 to 2278.
JOHN, M. E., M. C. JOHN, E. R. SIMPSON, and M. R. WATERMAN, (1985), J. Biol. Chem. **261**, 5760–5767.
KAMATAKI, T., K. MAEDA, Y. YAMAZOE, T. NAGAI, and R. SATO, (1983), Arch. Biochem. Biophys. **225**, 758–770.
KAMATAKI, T., M. SHIMADA, K. MAEDA, and R. KATO, (1985), Biochem. Biophys. Res. Commun. **130**, 1247–1253.
KATO, R., Y. YAMAZOE, M. SHIMADA, N. MURAYAMA, and T. KAMATAKI, (1986), J. Biochem. **100**, 895–902.
KULKARNI, A. B., R. M. GUBITS, and P. FEIGELSON, (1985), Proc. Natl. Acad. Sci. U.S.A. **82**, 2579–2582.
KUNTZMAN, R., M. JACOBSEN, K. SCHNEIDMAN, and A. H. CONNEY, (1964), J. Pharmacol. Exp. Ther. **146**, 280–285.
LARREY, D., L. M. DISTLERATH, G. A. DANNAN, G. R. WILKINSON, and F. P. GUENGERICH, (1984), Biochemistry **23**, 2787–2795.
LEFEVRE, A., P. F. HALL, A. HANSSON, and J-Å. GUSTAFSSON, (1980), in: Biochemistry, Biophysics and Regulation of Cytochrome P-450 (J-Å. GUSTAFSSON et al., eds.), Elsevier/North Holland, 533–536.
LU, A. Y. H. and S. B. WEST, (1980), Pharmacol. Rev. **31**, 277–295.
MACGEOCH, C., E. T. MORGAN, J. HALPERT, and J-Å. GUSTAFSSON, (1984), J. Biol. Chem. **259**, 15433–15439.
MACGEOCH, C., E. T. MORGAN, and J-Å. GUSTAFSSON, (1985), Endocrinology **117**, 2085–2092.
MACGEOCH, C., E. T. MORGAN, B. CORDELL, and J-Å. GUSTAFSSON, (1987), Biochem. Biophys. Res. Commun. **143**, 782–788.
MAEDA, K., T. KAMATAKI, T. NAGAI, and R. KATO, (1984), Biochem. Pharmacol. **33**, 509–512.
MCLELLAN-GREEN, P., D. J. WAXMAN, M. CAVENESS, and J. A. GOLDSTEIN, (1987), Arch. Biochem. Biophys. **253**, 13–25.
MODE, A., G. NORSTEDT, B. SIMIC, P. ENEROTH, and J-Å. GUSTAFSSON, (1981), Endocrinology **108**, 2103–2108.
MORGAN, E. T., C. MACGEOCH, and J-Å. GUSTAFSSON, (1985a), Mol. Pharmacol. **27**, 471–479.
MORGAN, E. T., C. MACGEOCH, and J-Å. GUSTAFSSON, (1985b), J. Biol. Chem. **260**, 11895–11898.
MORGAN, E. T., M. RÖNNHOLM, and J-Å. GUSTAFSSON, (1987), Biochemistry **26, 4193** to **4200.**
NORSTEDT, G., and R. PALMITER, (1984), Cell **36**, 805–812.
NOSHIRO, N. and M. NEGISHI, (1986), J. Biol. Chem. **261**, 15923–15927.
RAMPERSAUD, A., D. J. WAXMAN, D. E. RYAN, W. LEVIN, and F. G. WALZ, (1985), Arch. Biochem. Biophys. **244**, 857–864.
RAMPERSAUD, A. and F. G. WALZ, (1987), J. Biol. Chem. **262**, 5449–5663.
RAMPERSAUG, A., S. BANDIERA, D. E. RYAN, W. LEVIN, P. E. THOMAS, and F. G. WALZ, (1987), Arch. Biochem. Biophys. **252**, 145–151.
ROY, A. K. and B. CHATTERJEE, (1983), Ann. Rev. Physiol. **45**, 37–50.
RYAN, D. E., S. IIDA, A. W. WOOD, P. E. THOMAS, C. S. LIEBER, and W. LEVIN, (1984), J. Biol. Chem. **259**, 1239–1250.
SCHUETZ, E. G., S. A. WRIGHTON, J. L. BARWICK, and P. S. GUZELIAN, (1984), J. Biol. Chem. **259**, 1999–2006.

SIMMONS, D. L., P. MCQUIDDY, and C. B. KASPER, (1987), J. Biol. Chem. **262**, 326–332.
STRÖM, A., A. MODE, E. MORGAN, and J.-Å. GUSTAFSSON (1987), Biochem. Soc. Trans. **15**, 575–576.
TANNENBAUM, G. S., (1981), Endocrinology **108**, 76–82.
WAXMAN, D. J., (1984), J. Biol. Chem. **259**, 15481–15490.
WAXMAN, D. J., G. A. DANNAN, and F. P. GUENGERICH, (1985), Biochemistry **24**, 4409–4417.
WHITLOCK, J. P., Jr., (1986), Ann. Rev. Pharmacol. Toxicol. **26**, 333–369.
WRIGHTON, S. A., P. MAUREL, E. G. SCHUETZ, P. B. WATKINS, B. YOUNG, and P. S. GUZELIAN, (1985), Biochemistry **24**, 2171–2178.
YAMAZOE, Y., M. SHIMADA, N. MURAYAMA, S. KAWANO, and R. KATO, (1986), J. Biochem. **100**, 1095–1097.
YOSHIOKA, H., K. MOROHASHI, K. SOGAWA, T. MIYATA, K. KAWAJIRI, T. HIROSE, S. INAYAMA, Y. FUJII-KURIYAMA, and T. OMURA, (1987), J. Biol. Chem. **262**, 1706 to 1711.
ZUBER, M. X., M. E. JOHN, T. OKAMURA, E. R. SIMPSON, and M. R. WATERMAN, (1986), J. Biol. Chem. **261**, 2475–2482.

Chapter 6

Evolution, Structure, and Gene Regulation of Cytochrome P-450

O. Gotoh and Y. Fujii-Kuriyama

1. Introduction

Almost three decades have elapsed since the first paper on the characterization and denomination of cytochrome P-450 was published by OMURA and SATO (1962). At that time, functional roles of this hemoprotein were not known. Elaborate work by ESTABROOK and his coworkers soon revealed the involvement of this hemoprotein in the C-21 hydroxylation of 17α-hydroxyprogesterone as a terminal oxidase of the microsomal electron pathway in adrenal cortex (ESTABROOK et al., 1963) and in the oxidative metabolism of xenobiotics in liver (COOPER et al., 1965). Since then, many biological oxidative processes have been reported to be catalyzed by the cytochrome P-450 system. These include the metabolism of endogenous substrates such as many kinds of steroids, fatty acids and prostaglandins as well as exogenous substrates such as an almost infinite number of drugs and other lipophilic chemicals including environmental pollutants and carcinogens. This prodigious metabolic versatility of the monooxygenase system is now known to result from the participation of multiple forms of cytochrome P-450, the total number of which still remains to be established. Generally speaking, P-450 species which are involved in the metabolism of endogenous substrates exhibit rigid substrate specificities, like most of other enzymes, whereas the so-called drug-metabolizing P-450s possess distinct but broad substrate specificities which in some cases overlap with one another, thus forming a sort of comprehensive protective system against the enormous range of xenobiotics, probably including even compounds never previously encountered by biological organisms.

This molecular multiplicity has become apparent only in recent years as a result of the identification, purification and subsequent sequence analysis of the N-terminal regions of various forms of P-450 and, therefore, due to products of different genes. Accordingly, extensive discussions about structural and evolutionary relations among these various forms of P-450 have had to await the very recent application of gene technology to the elucidation of complete primary structures of various forms of P-450.

Another very important feature of the cytochrome P-450 superfamily is that many of them are inducible in tissue-specific and inducer-specific fashions. In response to inducing agents, either exogenous or endogenous, the synthesis of distinct forms of P-450 is known to be enhanced in specific tissues. This inductive response plays an important role in increasing the rate of metabolism of foreign compounds to detoxified forms or in some cases to harmful reactive intermediates, and in establishing sex-dependent metabolic states of xenobiotics and intrinsic substrates such as steroids and prostaglandins. The tools of recent molecular biology are also beginning to be applied productively to the elucidation of molecular mechanisms of P-450 gene regulation.

Since many review articles (LU and WEST, 1980; GUENGERICH, 1979; NEBERT and NEGISHI, 1982; CONNEY, 1982) have been published concerning multiplicity, functions and induction mechanisms of cytochrome P-450 as revealed by biochemical and immunological techniques, we will limit our discussion, for the most part, to molecular biological studies of cytochrome P-450 using cloned cDNA and genomic DNA, and to analysis of coded protein sequence information. It is not our intention, therefore, to provide comprehensive documentation concerning the characterization of all known P-450s from different organs and different species. In addition to this article, several other reviews on cytochrome P-450 from various view points have been published (ADESNIK and ATCHISON, 1986; BLACK and COON, 1986; WHITLOCK, 1986). This review article is written by surveying the papers published until early 1987.

2. cDNA cloning of P-450s

It has been proposed that hundreds or even thousands of forms of cytochrome P-450 are responsible for the broad substrate specificities of the detoxifying system induced by xenobiotics in livers and other organs (NEBERT, 1979). Taking advantage of the molecular cloning technology, several research groups, including ours, started to address the question as to how many forms of P-450 there are in a species of organisms, how they are structurally related to one another, and how they are regulated for their expression in response to external or internal stimuli.

An early attempt of cDNA cloning of P-450 led to the first successful report of elucidating the complete primary structure of mammalian cytochrome P-450, P-450b (FUJII-KURIYAMA et al., 1982b). At the same time, HANIU et al. (1982a, 1982b) determined the primary structure of bacterial P-450 cam consisting of 412 amino acids by sequence analysis of the purified protein.

These two pieces of work on the structure of P-450 paved the way for the investigation of the structural and evolutionary aspects of P-450 gene family. Since then, cDNA cloning technology has been yielding an increasing number of primary structures of P-450 year by year.

In the early phase of cDNA cloning, a plasmid vector pBR322 was most frequently used as a cloning vector. mRNA preparations used for synthesis of double-stranded cDNA were enriched in P-450 mRNA as much as possible by size-fractionation in sucrose density gradient (FUJII-KURIYAMA et al., 1981; KAWAJIRI et al., 1983; YABUSAKI et al., 1984; OKINO et al., 1985; YOSHIOKA et al., 1986), or by immunoprecipitation of polysomes synthesizing P-450 peptide with the corresponding monoclonal or polyclonal antibody (HARDWICK et al., 1983; KIMURA et al., 1984a, 1984b; GONZALEZ et al., 1985b; HARADA et al., 1985), because the transformation of bacteria and the identification procedure of the cDNA clone were not efficient and were laborious. Differential colony hybridization, hybridization-selected translation assay, and subsequently immunological identification of the translated products were mostly used for identification of the cDNA clones. For the isolation of full length copies of cDNA, cloning procedures developed by OKAYAMA and BERG (1982) were followed by some groups (YABUSAKI et al., 1984; KIMURA et al., 1984a, 1984b). Recently, λgt11 expression vector with *lac* promoter was introduced for P-450 cDNA cloning by JAISWAL et al. (1985a), and has become a widely used vector in this field (MOLOWA et al., 1986; GONZALEZ et al., 1986a, 1986b; EVANS et al., 1986; BEAUNE et al., 1986; CHUNG et al., 1986b; SONG et al., 1986; HARDWICK et al., 1987). This phage vector has greatly improved the efficiency of transfection of cDNA and facilitated the screening procedure by identifying directly the cDNA clone which was synthesizing P-450 peptides with the corresponding antibody. The oligonucleotide probes which were synthesized on the basis of information of partial amino acid sequences of the relevant P-450 have also been introduced to expedite the identification process (MOROHASHI et al., 1984; YOSHIOKA et al., 1987). These improvements in the identification process of cDNA mean that the enrichment of P-450 mRNA is no longer a prerequisite for the cDNA cloning and have accelerated the gene cloning of many forms of cytochrome P-450.

Once a cDNA clone for a certain form of cytochrome P-450 is isolated, this cloned cDNA is easily applied for the isolation of other cDNAs or genomic DNAs with related sequences by cross-hybridization. In general, cDNAs or exonic regions for equivalent genes in different mammalian species cross-hybridize to each other. This facilitates isolation of orthologous information in other species with a cloned cDNA. Several human cDNAs and genomic DNAs have been cloned in this way (JAISWAL et al., 1985a, 1986; PHILLIPS et al., 1985; KAWAJIRI et al., 1986; HIGASHI et al., 1986; WHITE et al., 1986; QUATTROCHI et al., 1986; MOROHASHI et al., 1987a).

Cross-hybridization is also effective for isolating different members of a gene family or subfamily such as methylcholanthrene (MC)-inducible P-450s and phenobarbital (PB)-inducible P-450s; two rat MC-inducible P-450s, P-450c and P-450d (KAWAJIRI et al., 1984a, 1984b; SOGAWA et al., 1984), or mouse equivalent P-450s, P_1-450 and P_3-450 (KIMURA et al., 1984a, 1984b), cross-hybridize with each other. A similar cross-hybridization was observed in the PB-inducible P-450 subfamily, rat P-450b and P-450e, and in other families such as pregnenolone 16α carbonitrile (PCN)-inducible (GONZALEZ et al., 1985b, 1986b) and those involved in ω-hydroxylation of fatty acids (FA) (HARDWICK et al., 1987; MATSUBARA et al., 1987). IMAI (1987) has recently shown that cDNAs with as large as 50% difference in overall nucleotide sequences are detectable using a well-conserved segment as the probe under mild hybridization conditions. Thus, the cross-hybridization technique has considerably accelerated the accumulation of sequence data of P-450, although genuine correspondence between isolated message and purified protein or specific enzymatic activity is sometimes difficult due to the presence of multiple genes with similar sequences.

3. Classification of P-450s based on protein sequences

Because P-450 genes belong to a multi-gene family with many members, the interrelationships among individual genes or gene products are fairly complicated. To date nearly forty P-450 protein sequences have been determined to

Table 1. Relatedness of sequences in P-450 families and other proteins

P-450	Family	MC	PB	C21	17α	PCN	FA	SCC	11β	CAM	CO1	EH
Rat	c	**524**	28.7	25.7	27.1	21.7	17.4	18.8	18.6	12.2	13.4	11.3
Rat	b	28.2	**491**	27.2	26.0	24.8	20.6	19.0	19.6	15.4	13.0	11.3
Human	c21	23.2	23.4	**464**	31.4	21.8	24.8	21.1	18.9	18.4	13.2	12.1
Bovine	17α	26.5	23.0	29.3	**509**	24.1	18.7	18.6	17.8	13.5	12.7	13.4
Rat	pcn	15.8	16.6	16.8	15.4	**504**	23.0	20.5	18.4	15.0	10.9	13.1
Rabbit	pgω	9.7	13.2	12.8	10.6	20.3	**506**	20.0	18.1	14.1	11.2	13.6
Bovine	scc	8.9	9.6	10.4	8.9	15.4	11.9	**520**	34.9	13.0	13.0	13.4
Bovine	11β	10.0	9.8	9.7	7.4	12.1	9.8	41.8	**503**	13.5	13.8	13.5
P. putida	cam	3.1	4.3	4.8	3.1	4.1	3.4	4.9	3.0	**415**	11.2	8.7
Yeast	col	3.8	0.8	0.1	−0.2	0.1	−0.9	1.9	3.0	−0.2	**512**	12.8
Rat	eh	−1.1	1.4	0.7	1.3	0.4	1.3	0.6	0.3	−1.4	0.7	**455**

Values below the diagonal indicate significance of sequence relatedness measured in standard deviation unit. The mean and standard deviation for each comparison were estimated with 250 pairs of randomized sequences (Gotoh et al., 1983). Values above the diagonal are percent identity. Cytochrome *c* oxidase subunit I (col), epoxide hydrolase (eh), and P-450s have similar sizes as shown on the diagonal.

Table 2. P-450 sequences

Family	Gene[1]	Species	Form	Source	aa's	Mat[2]	Weight[3]	Reference
MC	IA1	Rat	c	gene	524	523	0.710	SOGAWA et al. (1984)
	IA1	Mouse	p_1	gene	524		0.710	GONZALEZ et al. (1985a)
	IA1	Human	c	gene	512		1.219	KAWAJIRI et al. (1986)
	IA1	Rabbit	6	cDNA	(519)		1.601	OKINO et al. (1985)
	IA2	Rat	d	cDNA	513	512	0.881	KAWAJIRI et al. (1984a)
	IA2	Mouse	p_3	cDNA	513		0.881	KIMURA et al. (1984b)
	IA2	Human	4	cDNA	516		1.379	QUATTROCHI et al. (1986)
	IA2	Rabbit	4	aa	515	514	1.379	OZOLS (1986)
				cDNA				OKINO et al. (1985)
PB	IIB1	Rat	b	cDNA	491		0.713	FUJII-KURIYAMA et al. (1982a)
	IIB2	Rat	e	gene	491		0.713	MIZUKAMI et al. (1983b)
	IIB1	Rabbit	2	aa	491		1.377	TARR et al. (1983)
	IIA1	Rat	a	cDNA	492		2.322	NAGATA et al. (1987)
	IIC3	Rabbit	3b	aa	490		0.922	OZOLS et al. (1985)
	IIC7	Rat	f	cDNA	490		0.415	GONZALEZ et al. (1986a)
	IIC6	Rat	pb1	cDNA	490		0.415	GONZALEZ et al. (1986a)
	IIC1	Rabbit	pc1	cDNA	(490)		0.285	LEIGHTON et al. (1984)
	IIC2	Rabbit	pc2	cDNA	490		0.285	LEIGHTON et al. (1984)
		Rat	m1	cDNA	500		0.454	YOSHIOKA et al. (1987)
	IIC5	Rabbit	1	cDNA	487		0.419	TUKEY et al. (1985)
	IIE1	Rat	j	cDNA	493		0.652	SONG et al. (1986)
	IIE1	Human	j	cDNA	493		0.727	SONG et al. (1986)
	IIE1	Rabbit	3a	cDNA	(493)	492	0.652	KHANI et al. (1987)
		Chicken	pb1	cDNA	491		1.923	HOBBS et al. (1986)
C21	XXIA1	Mouse	c21A	gene	487		3.323	CHAPLIN et al. (1986)
	XXIA2	Human	c21B	gene	494		2.608	HIGASHI et al. (1986)
	XXIA	Bovine	c21	cDNA	496		2.608	YOSHIOKA et al. (1986)

17α	XVIIA	Human	17α	cDNA	508		4.125	Chung et al. (1987)
	XVIIA	Bovine	17α	cDNA	509		4.125	Zuber et al. (1986)
PCN	IIIA1	Rat	pcn1	cDNA	504		4.248	Gonzalez et al. (1985b)
	IIIA2	Rat	pcn2	cDNA	504		4.248	Gonzalez et al. (1986b)
		Human	lp	cDNA	504		3.564	Molowa et al. (1986)
		Human	nf	cDNA	503		3.564	Beaune et al. (1986)
FA	IVA1	Rat	laω	cDNA	509		7.688	Hardwick et al. (1987)
		Rabbit	pgω	cDNA		506	7.688	Matsubara et al. (1987)
SCC	XIIA1	Human	scc	gene	521	(482)	8.322	Morohashi et al. (1987a)
	XIIA1	Bovine	scc	cDNA	520	481	8.322	Morohashi et al. (1984)
11β	XIA1	Bovine	11β	cDNA	503	479	14.531	Morohashi et al. (1987b)
CAM	CIA1	*P. putida*	cam	gene	415	414	—	Unger et al. (1986)

[1] Recommended by Nebert et al. (1987).
[2] Number of amino acids in the mature form.
[3] Used for averaging sequence information.
Numbers in parentheses are expected but not confirmed.

completion, with a dozen or more partial sequences of other P-450s. The primary purpose of this section is to clarify the evolutionary relationships among various forms of P-450 based on their amino acid sequences, though most of them were deduced from nucleotide sequences of cDNAs or genomic DNAs. We preferred amino acid sequences to nucleotide sequences for comparison, since nucleotide change rates are too rapid to measure remote evolutionary distances.

3.1. All sequence data are homologous

When only one bacterial and one mammalian sequence was known, GOTOH et al. (1983) showed that the two sequences were significantly homologous and that P-450s constitute a unique (super)family of hemoproteins. Table 1 shows results of a more extensive examination of P-450 sequences for homology. The nine sequences were selected one by one from the nine known families of P-450s (see below). The degrees of global sequence similarity measured in standard deviation units always exceed 3.0, the generally accepted threshold of significance (NEEDLEMAN and WUNSCH, 1970). This result confirms the previous conclusion that all P-450s are homologous, in other words, derived from a common ancestor.

Mouse cytochrome b_5 (BIBB et al., 1981) and three subunits of cytochrome *c* oxidases from yeast (BONITZ et al., 1980), fruit fly (DE BRUIJN, 1983) and human (ANDERSON et al., 1981) were also subjected to the homology test, since these sequences often showed significant similarity to some forms of P-450 in searches for local sequence similarity against the NBRF protein sequence database (GOTOH and TAGASHIRA, 1986). Rat epoxide hydrolase (PORTER et al., 1986) was included as a control. Only yeast cytochrome *c* oxidase subunit I showed a significant level of sequence similarity to rat P-450c, bovine P-450 11β (Table 1) and rat P-450d (data not shown). All other global sequence similarities were below the significant level (Table 1 and data not shown). The segments in P-450c or P-450 11β elicited by a new method (GOTOH, 1987) to be locally best matched to the oxidase subunit are mapped around the regions responsible for heme-binding and possible electron transfer (see next section), which might imply structural resemblance between the heme-supporting environments in these different classes of hemoprotein. Further investigations are necessary to support or reject this supposition.

3.2. Sequence data

To get a closer view of mutual relationships among various forms of P-450s, we selected 38 complete or nearly complete (>95%) sequences (Table 2). Several published sequences not subjected here to extensive analysis are listed in Table 3 in connection with their closest relatives in the 38 sequences in Table 2.

Basically we avoided mixing two or more sequences of different origin for the

Table 3a. Variant Sequences

Present Sequence			Variant Sequence			
Species	**Form**	**Residue**	**Residue**	**Form**	**Source**	**Reference**
Rat	c	Ile(53)	→ Met	c	cDNA	YABUSAKI et al. (1984)
Rat	c	Identical		c	gene	HINES et al. (1985)
Human	c	Phe(381) Ile(462)	→ Leu → Val	p1 p1	cDNA gene	JAISWAL et al. (1985a) JAISWAL et al. (1986b)
Rat	d	Arg(173) Cys(403)	→ His → Arg	d	gene	SOGAWA et al. (1984)
Rat	d	Arg(31)	→ Lys	d	aa	HANIU et al. (1986)
Mouse	p3	Identical		p3	gene	GONZALEZ et al. (1985a)
Mouse	p3	Ile(384)	→ Met	p2	cDNA	KIMURA et al. (1986)
Human	lm4	Ser(79) Leu(511) Phe(512)	→ Arg → Del. → Arg	p3	cDNA	JAISWAL et al. (1986)
Rat	b	Identical		b	gene	SUWA et al. (1985)
Rat	e	Met(473)	→ Lys	e	aa	YUAN et al. (1983)
Bovine	c21	Ser(14) Tyr(401)	→ Ala → His	c21	gene	CHUNG et al. (1986a)
Bovine	c21	(123/124) Tyr(401) Ser(431)	→ Thr → His → Cys	c21	cDNA	JOHN et al. (1986)
Human	c21	Identical		c21	gene	WHITE et al. (1986)
Human	scc	Cys(16) Leu(274) Ile(301)	→ Tyr → Phe → Met	scc	cDNA	CHUNG et al. (1986b)
Bovine	scc	Asn(57) Asp(106) Asp(197)	→ Asp → Asn → Asn	scc	aa	CHASHCHIN et al. (1986)

same form of P-450. However, rabbit P-450 lm4 sequence analyzed here is a mixture of two sources; residues 2 to 92 were taken from the protein sequence of OZOLS (1986), whereas the rest of the sequence is derived from the cDNA sequence of OKINO et al. (1985). Translation errors were corrected without a notice when genomic or cDNA sequence is available. The fourth residue in rat P-450b was corrected to be Ser (YUAN et al., 1983; SUWA et al., 1985) instead of Thr in the original N-terminal sequence reported by BOTELHO et al., (1979).

Table 3b. Other P-450 Sequences

Closest Sequence Species	Closest Sequence Form	Sequence Species	Sequence Form	Comparable Positions (% of Total)	No. of Difference (% Identity)	Reference
Rat	e	Rat	e(U. C.)	168–491 (66.0)	3 (99.1)	PHILLIPS et al. (1983)
Rat	e	Rat	e	281–491 (43.0)	5 (97.6)	KUMER et al. (1983)
Rat	b	Rat	pb24	295–491 (40.1)	37 (81.2)	AFFOLTER et al. (1986)
Rabbit	lm2	Rabbit	lm2	1–491 (99.6)	17 (96.5)	HEINEMANN and OZOLS (1983)
Rabbit	lm4	Rabbit	lm4	(68.0)	11 (96.9)	FUJITA et al. (1984)
Rabbit	lm3b	Rabbit	pc3	87–490 (82.4)	9 (97.8)	LEIGHTON et al. (1984)
Rat	f	Rat	tf1	87–490 (82.4)	13 (96.5)	FRIEDBERG et al. (1986)
Rat	a	Human	1	162–492 (67.7)	136 (59.0)	PHILLIPS et al. (1985)

Rabbit P-450 lm2 of HEINEMANN and OZOLS and P-450 lm4 of FUJITA et al. were determined by protein sequencing methods. Other sequences were deduced from cDNA sequences.

The N-terminal 50 residues of rabbit P-450 lm6 were taken from MIHARA et al. (1985). The missing part in rabbit P-450 pc2 cDNA (LEIGHTON et al., 1984) was made up by the corresponding gene sequence (GOVIND et al., 1986). Ser(262) of rat P-450d inferred from the cDNA sequence (KAWAJIRI et al., 1984a) was corrected to be Phe as described previously (SOGAWA et al., 1985). The 2498th and 2499th nucleotides in human P-450c 21b gene reported by HIGASHI et al. (1986) were found to be inverted, leading to correction from Pro to Arg at the amino acid site 426 as in the sequence of WHITE et al. (1986). No other change was made, but the initiator Met was added to rabbit P-450 lm4 and rabbit P-450 lm3a to keep consistency in numbering system with those for related sequences.

Throughout this chapter, we use a conventional name of each gene or protein usually given by the author(s) who determined the sequence. We omit hyphens and parentheses, and use lower case letters to identify individual forms of P-450. To facilitate cross references, a universal nomenclature recently proposed by Nebert et al. (1987) is shown together with the conventional names in Table 2.

3.3. P-450 protein families

The terms protein family and superfamily have been used with different meanings (DAYHOFF, 1978; DOOLITTLE, 1981). According to DOOLITTLE, any sequences that are demonstrably related to each other belong to the same family, while a superfamily is composed of two or more families not all of whose members are demonstrably homologous with all the other members in each family. According to this definition, all bacterial and eukaryotic P-450s belong to a single family, while P-450, cytochrome *b*, and some subunits of cytochrome *c* oxidase may possibly compose a superfamily.

On the other hand, more popular notions of protein family and superfamily are those of DAYHOFF (1978). By her definitions, two significantly related sequences belong to a superfamily if they differ at more than 50% positions, belong to a family if they differ at more than 20% but less than 50% positions, and belong to a subfamily if they differ at less than 20% positions. When we apply these definitions to known P-450 sequences, we identify nine families in the unique superfamily. Although the cut-off value discriminating families is more or less arbitrary, we will use this classification of P-450 families until solid criteria based on gene structure or other means become available. The nine family names (Table 2) used in this chapter should be familiar to most readers.

3.4. Sequence alignment

A collection of related protein sequences provides maximal information when they are properly aligned. Thus it is crucially important to get a good alignment. Optimal alignments of two or three sequences are obtained in a rigorous manner (NEEDLEMAN and WUNSCH, 1970; GOTOH, 1982, 1986). Here, an optimal alignment means that the distance value associated with the alignment is minimal (or equivalently, similarity is maximal) among all alternative alignments, where a distance (or similarity) is defined as a sum of weights given to individual matches, replacements, deletions and insertions.

Unfortunately, no practical method is known for aligning many sequences in a mathematically rigorous manner. Our approach is to recurrently apply the pairwise or three-way alignment method to pre-aligned sets of sequences. Namely, we first constructed a preliminary tree that represents relationships among all the P-450 sequences considered. The preliminary tree was obtained on the basis of distance values for selected pairs of the sequences. Inspecting the tree, we chose two or three sequences that were most closely related and obtained their optimal alignment. The aligned sequences were combined and treated as a single sequence. The process was repeated until all the original sequences were combined into a single alignment. For example, we first combined rat P-450c and mouse P-450 p1 sequences. The combined sequence was aligned with human P-450c and rabbit P-450 lm6 by the three-way method. The c-group sequence was aligned with the similarly combined d-group sequence to generate the MC-family sequence. The MC-family sequence was then aligned with the PB-family sequence and the combined C21 — 17α sequence, and so on.

A position in a combined sequence indices a column vector composed of M aligned residues, where M is the number of original sequences included in the combined sequence. A deletion (represented by a dash) is counted as a residue for convention.

In calculation of an optimal alignment, the weight given to a matched pair of positions (column vectors) was evaluated by averaging the mutation data matrix elements (DAYHOFF, 1978) corresponding to all combinations of residues in one column and those in the other. A deletion matched with any residue but a deletion was weighted by 6, whereas a null weight was given to a deletion matched with a deletion in another column. Additional weight of 6 was given to a consecutive run of deletions introduced by the matching algorithms (GOTOH, 1982, 1986).

The alignment thus obtained satisfied several criteria, such as coincidence of predicted secondary structures and hydrophobic patterns in different groups of P-450s. However, sequences in the central region of about 130 residues are extremely variable, so that we could not establish a solid alignment in this region. It was most difficult to align the combined animal sequences with P-450 cam, perhaps because considerable structural rearrangement may have occurred in this variable region since the divergence of prokaryote and eukaryote. From structural considerations (see next section), we imposed a constraint that Arg(187) in P-450 cam should match Lys(256) in rat P-450c and Lys(236) in rat P-450b; without this constraint, Arg(187) in P-450 cam could alternatively match Arg(245) in rat P-450c and Lys(225) in rat P-450b.

Figure 1 shows the final output of the alignment procedure. The residues lost in a mature protein are underlined. Every 20 positions are numbered above the sequences. The residue numbers of individual sequences for the left-most and the right-most amino acids in each row are given on the both sides. Hereafter, we will refer to a position number in square brackets while a residue number in parentheses, e. g., Cys[497] = Cys(461) for rat P-450c and = Cys(436) for rat P-450b. Our residue numbers for rabbit P-450 lm4, rabbit P-450 lm3a and *P. putida* P-450 cam differ by 1 from those in the literature (OZOLS, 1986; KHANI et al., 1987; POULOS et al., 1985), because we regard the initiator Met as residue 1 despite of its absence in purified proteins.

3.5. Construction of a phylogenetic tree

Based on the alignments in Figure 1, we estimated the number of accepted point mutations per 100 residues (PAM) for every pair of the 38 P-450 sequences. Note that PAM is generally greater than the percent difference between

			20 40 60 80 100 120	
Rat	c	1	MPSVYGFPAFTS--ATELLLAVTTFCLGFWVVRVTRTWVPKGLKSPP-GPWGLPF-IGH--VLTL----GKN-PHLSLTKLSQQYGDVLQIRI-GSTPVVVLSGLNTIKQAL--VKQ-GD	105
Mouse	p1	1	MPSMYGLPAFVS--ATELLLAVTVFCLGFWVVRATRTWVPKGLKTPP-GPWGLPF-IGH--MLTV----GKN-PHLSLTRLSQQYGDVLQIRI-GSTPVVVLSGLNTIKQAL--VRQ-GD	105
Human	c	1	ML----FPISMS--ATEFLLASVIFCLVFWVIRASRPQVPKGLKNPP-GPWGWPL-IGH--MLTL----GKN-PHLALSRMSQQYGDVLQIRI-GSTPVVVLSGLDTIRQAL--VRQ-GD	101
Rabbit	6	1	MVSDFGLPTFIS--ATELLLASAVFCLVFWVAGASKPRVPKGLKRLP-GPWGWXX-XXX--MLTL----GKN-PHVALAALSRPYGDVFQIRL-GSTPVVVLSGLDTIKQAL--VRQ-GD	105
Rat	d	1	M----AFSQYISL-APELLLATAIFCLVFWVLRGTRTQVPKGLKSPP-GPWGLPF-IGH--MLTL----GKN-PHLSLTKLSQQYGDVLQIRI-GSTPVVVLSGLNTIKQAL--VKQ-GD	102
Mouse	p3	1	M----AFSQYISL-APELLLATAIFCLVFWMVRASRTQVPKGLKNPP-GPWGLPF-IGH--MLTV----GKN-PHLSLTRLSQQYGDVLQIRI-GSTPVVVLSGLNTIKQAL--VRQ-GD	102
Human	4	1	M----ALSQSVPFSATELLLASAIFCLVFWVLKGLRPRVPKGLKSPP-EPWGWPL-LGH--VLTL----GKN-PHLALSRMSQRYGDVLQISI-GSTPVLVLSRLDTIRQAL--VRQ-GD	103
Rabbit	4	1	M----AMSPAAPLSVTELLLVSAVFSLVFWAVRASRPKVPKGLKRLP-GPWGWPL-LGH--LLTL----GKN-PHVALARLSQRYGDVFQIRL-GSTPVVVLSGLNTIKQAL--VRQ-GD	103
Rat	b	1	M---EP----------SILLLLALLVGFLLLLVRGHPK-SRG-NFPP-GPRPLPL-LGN--LLQL---DRGG-LLNSFMQLREKYGDVFTVHL-GPRPVVMLCGTDTIKEAL--VGQ-AE	93
Rat	e	1	M---EP----------SILLLLALLVGFLLLLVRGHPK-SRG-NFPP-GPRPLPL-LGN--LLQL---DRGG-LLNSFMQLREKYGDVFTVHL-GPRPVVMLCGTDTIKEAL--VGQ-AE	93
Rabbit	2	1	M---EF----------SLLLLLAFLAGLLLLLFRGHPK-AHG-RLPP-GPSPLPV-LGN--LLQM---DRKG-LLRSFLRLREKYGDVFTVYL-GSRPVVVLCGTDAIREAL--VDQ-AE	93
Rat	a	1	M--LDT--------GLLLVVILASLSVMLLVSLW-QQK-IRG-RLPP-GPTPLPF-IGN--YLQL---NTKD-VYSSITQLSERYGPVFTIHL-GPRRVVVLYGYDAVKEAL--VDQ-AE	95
Rabbit	3b	1	M---DL----------LIILGIC-LSCVVLLSLWKKTH-GKG-KLPP-GPTPLPV-VGN--LLQL---ETKD-INKSLSMLAKEYGSIFTLYF-GMKPAVVLYGYEGVIEAL--IYR-GE	92
Rat	f	1	M---DL----------VTFLVLT-LSSLILLSLWRQSS-RRR-KLPP-GPTPLPI-IGN--FLQI---DVKN-ISQSLTKFSKTYGPVFTLYL-GSQPTVILHGYEAIKEAL--IDN-GE	92
Rat	pb1	1	M---DL----------VMLLVLT-LTCLILLSIWRQSS-GRG-KLPP-GPIPLPI-IGN--IFQL---NVKN-ITQSLTSFSKVYGPVFTLYF-GTKPTVILHGYEAVKEAL--IDH-GE	92
Rabbit	pc1	1	X---XX----------XXXXXXX-LSCLLLLSLWKQSY-GGG-KLPP-GPTPFPI-LGN--ILQI---GIQD-ISKSFTKLSEVYGPVFTVYL-GMKPTVVIHGYDAVKEAL--VDL-GE	92
Rabbit	pc2	1	M---DL----------VVVLGLC-LSCLLLLSLWKQSH-GGG-KLPP-GPTPFPI-LGN--VLQL---DFKD-LSKSLTNLSKVYGPVFTVYL-GMKPTVVVHGYEAVKEAL--VDL-GH	92
Rat	m1	1	M---DP----------VLVLVLT-LSSLLLLSLWRQSF-GRG-KLPP-GPTPLPI-IGN--TLQI---YMKD-IGQSIKKFSKVYGPIFTLYL-GMKPFVVLHGYEAVKEAL--VDL-GE	92
Rabbit	1	1	M---DP----------VVVLVLG-LCCLLLLSIWKQNS-GRG-KLPP-GPTPFPI-IGN--ILQI---DAKD-ISKSLTKFSECYGPVFTVYL-GMKPTVVLHGYEAVKEAL--VDL-GE	92
Rat	j	1	M---AV--------LGITIALLVWVATLLVISIWKKIY-NSW-NLPP-GPFPLPI-LGN--IFQL---DLKD-IPKSFTKLAKRFGPVFTLHL-GSRRIVVLHGYKAVKEVL--LNH-KN	95
Human	j	1	M---SA--------LGVTVALLVWAAFLLLVSMWRQVH-SSW-NLPP-GPFPLPI-IGN--LFQL---ELKN-IPKSFTRLAQRFGPVFTLYV-GSQRMVVMHGYKAVKEAL--LDY-KD	95
Rabbit	3a	1	M---AV--------LGITVALLGWMVILLFISVWKQIH-SSW-NLPP-GPFPLPI-IGN--LLQL---DLKD-IPKSFGRLAERFGPVFTVYL-GSRRVVVLHGYKAVREML--LNH-KN	95
Chicken	pb1	1	M---DF--------LGLPTILLLVCISCLLIAAWRSTS-QRG-KEPP-GPTPIPI-IGN--VFQL---NPWD-LMGSFKELSKKYGPIFTIHL-GPKKIVVLYGYDIVKEAL--IDN-GE	95
Mouse	c21a	1	M-----------------LLPGLLLLLLLA---GTRW-LWGQWKLR-KLHLPRL-APG--FLHF---LQPN-LPIYLLGLTQKLGPIYRIRL-GMQDVVVLNSNRTIEEAL--IQK-WV	87
Human	c21b	1	M-----------------LLLG-LLLLPLLA---GARL-LWNWWKLR-SLHLPPL-APG--FLHL---LQPD-LPIYLLGLTQKFGPIYRLHL-GLQDVVVLNSKRTIEEAM--VKK-WA	86
Bovine	c21	1	M-----------------VLAGLLLLLTLLS---GAHL-LWGRWKLR-NLHLPPL-VPG--FLHL---LQPN-LPIHLLSLTQKLGPVYRLRL-GLQEVVVLNSKRTIEEAM--IRK-WV	87
Human	17α	1	M-----------------WELVALLLLTLAYLFWPKRR-CPGAKYPK-SLLSLPL-VGS--LPFL---PRHGHMHNNFFKLQKKYGPIYSVRM-GTKTTVIVGHHQLAKEVL--IKK-GK	91
Bovine	17α	1	M-----------------WLLLAVFLLTLAYLFWPKTK-HSGAKYPR-SLPSLPL-VGS--LPFL---PRRGQQHKNFFKLQEKYGPIYSFRL-GSKTTVMIGHHQLAREVL--LKK-GK	91
Rat	pcn1	1	M---DLLS-ALT--LETWVLLAVVLVLLYGFGTRTHGLFKK--QGIP-GPKPLPF-FGT--VLNY---Y-MG-LWKFDVECHKKYGKIWGLFD-GQMPLFAITDTEMIKNVL--VKECFS	100
Rat	pcn2	1	M---DLLS-ALT--LETWVLLAVILVLLYRLGTHRHGIFKK--QGIP-GPKPLPF-LGT--VLNY---Y-KG-LGRFDMECYKKYGKIWGLFD-GQTPVFAIMDTEMIKNVL--VKECFS	100
Human	1p	1	M---ALIP-DLA--METWLLLAVSLVLLYLYGTHSHGLFKK--LGIP-GPTPLPF-LGN--ILSY---H-KG-FCMFDMECHKKYGKVWGFYD-GQQPVLAITDPDMIKLVL--VKECYS	100
Human	nf	1	M---ALIP-DLA--METWLLLAVSLVLLYLYGTHSHGLFKK--LGIP-GPTPLPF-LGN--ILSY---H-KG-FCMFDMECHKKYGKVWGFYD-GQQPVLAITDPDMIKTVL--VKECYS	100
Rat	laω	1	MSVSALSSTRFTGSISGFLQVASVLGLLLLLVKAVQFYLQRQWLLKAFQQFPSPP-FHW--FFGH-KQFQGDKELQQIMTCVENFPSAFPRWFWGSKAYLIVYDPDYMKVIL--GRS-DP	113
Rabbit	pgω	1	----ALSPTRLPGSLSGLLQVAALLGLLLLLLKAAQLYLHRQWLLRALQQFPCPP-FHW--LLGHSREFQNDQELERIQKWVEKFPGACPWWLSGNKARLLVYDPDYLKVIL--GRS-DP	110
Human	scc	1	MLAKGLPPRSVLVKGCQTFLSAPREGLGRLRVPTGEGAGISTRSPRPFNEIPSPGDNGWLNLYHFWRETGTHKVHLHHVQNFQKYGPIYREKL-GNVESVYVIDPEDVALLFKSEGPNPE	119
Bovine	scc	1	MLARGLPLRSALVKACPPILSTVGEGWGHHRVGTGEGAGISTKTPRPYSEIPSPGDNGWLNLYHFWREKGSQRIHFRHIENFQKYGPIYREKL-GNLESVYIIHPEDVAHLFKFEGSYPE	119
Bovine	11β	1	M---AL-----WAKA-RVRMAGPWLSLHEARLLGTRGAA-APKAVLPFEAMPRCPGNKWMRMLQIWKEQSSENMHLDMHQTFQELGPIFRYDV-GGRHMVFVMLPEDVERLQQADSHHPQ	109
P.putida	cam	1	M-------------TTETIQSNANLAPL-----------------PPHVPEHLVFDFD---MYNP--SNLSAGVQEAWAVLQESNVPDL-VWTRCNGGHWIATRGQLIREAY---ED-YR	80

Fig. 1. Alignment of P-450 sequences. Every 20 positions are numbered above the sequences. The residue numbers of each protein for the left-most and the right-most amino acids in a row are shown on the both sides. A symbol below each position indicates the number of different kinds of residues at the position: 1 (*), 2 (:), 3 (.), or ≧ 4 (blank). Underlined residues are known to be lost in the mature form of a protein.

```
                                   140                 160                 180                 200                 220                 240
Rat       c     106  DFKGRPDLYSFTLIANGQSMTFNPDSGPLWAARRRLAQNALKSFSIASDPTLASSCYLEEHVSKEAEYLISKFQKLMAEVGHFDP-FKYLVVSVANVICAICFGRRYDHDDQEL-LSIVN  223
Mouse     p1    106  DFKGRPDLYSFTLITNGKSMTFNPDSGPVWAARRRLAQNALKSFSIASDPTSASSCYLEEHVSKEANYLVSKLQKVMAEVGHFDP-YKYLVVSVANVICAICFGQRYDHDDQEL-LSIVN  223
Human     c     102  DFKGRPDLYTFTLISNGQSMSFSPDSGPVWAARRRLAQNGLKSFSIASDPASSTSCYLEEHVSKEAEVLISTLQELMAGPGHFNP-YRYVVVSVTNVICAICFGRRYDHNHQEL-LSLVN  219
Rabbit    6     106  DFKGRPDLYSFSFVTKGQSMIFGSDSGPVWAARRRLAQNARNSFSVASDPASSSSCYLEEHVSQEAENLIGRFQELMAAVGHFDP-YRYVVMSVANVICAMCFGRRYDHDDQEL-LSLVN  223
Rat       d     103  DFKGRPDLYSFTLITNGKSMTFNPDSGPVWAARRRLAQDALKSFSIASDPTSVSSCYLEEHVSKEANHLISKFQKLMAEVGHFEP-VNQVVESVANVIGAMCFGKNFPRKSEEM-LNLVK  220
Mouse     p3    103  DFKGRPDLYSFTLITNGKSMTFNPDSGPVWAARRRLAQDALKSFSIASDPTSASSCYLEEHVSKEANHLVSKLQKAMAEVGHFEP-VSQVVESVANVIGAMCFGKNFPRKSEEM-LNIVN  220
Human     4     104  DFKGRPDLYTSTLITDGQSLTFSTDSGPVWAARRRLAQNALNTFSIASDPASSSSCYLEEHVSKEAKALISRLQELMAGPGHFDP-YNQVVVSVANVIGAMCFGQHFPESSDEM-LSLVK  221
Rabbit    4     104  DFKGRPDLYSSSFITEGQSMTFSPDSGPVWAARRRLAQDSLKSFSIASNPASSSSCYLEEHVSQEAENLIGRFQELMAAVGRFDP-YSQLVVSAARVIGAMCFGRHFPQGSEEM-LDVVR  221
Rat       b      94  DFSGRGTIAVIEPIFKEYGVIFA--NGERWKALRRFSLATMRDFGMGKRS-------VEERIQEEAQCLVEELRK--SQGAPLDP-TFLFQCITANIICSIVFGERFDYTDRQF-LRLLE  200
Rat       e      94  DFSGRGTIAVIEPIFKEYGVIFA--NGERWKALRRFSLATMRDFGMGKRS-------VEERIQEEAQCLVEELRK--SQGAPLDP-TFLFQCITANIICSIVFGERFDYTDRQF-LRLLE  200
Rabbit    2      94  AFSGRGKIAVVDPIFQGYGVIFA--NGERWRALRRFSLATMRDFGMGKRS-------VEERIQEEARCLVEELRK--SKGALLDN-TLLFHSITSNIICSIVFGKRFDYKDPVF-LRLLD  200
Rat       a      96  EFSGRGEQATYNTLFKGYGVAFS--SGERAKQLRRLSIATLRDFGVGKRG-------VEERILEEAGYLIKMLQG--TCGAPIDP-TIYLSKTVSNVISSIVFGERFDYEDTEF-LSLLQ  202
Rabbit    3b     93  EFSGRGIFPVFDRVTKGLGIVFS--SGEKWKETRRFSLTVLRNLGMGKKT-------IEERIQEEALCLIQALRK--TNASPCDP-TFLLFCVPCNVICSVIFQNRFDYDDEKF-KTLIK  199
Rat       f      93  KFSGRGSYPMIENVTKGFGIVFS--NGNRWKEMRRFTIMNFRNLGIGKRN-------IEDRVQEEAQCLVEELRK--TKGSPCDP-SLILNCAPCNVICSITFQNHFDYKDKEM-LTFME  199
Rat       pb1    93  EFAERGSFPVAEKINKDLGIVFS--HGNRWKEIRRFTLTTLRNLGMGKRN-------IEDRVQEEARCLVEELRK--TNGSPCDP-TFILGCAPCNVICSIIFQNRFDYKDQDF-LNLME  199
Rabbit    pc1    93  EFSGRIVFPLTAKINKGYGIVFS--NGKRWKETRRFSLMTLRDFGMGKRS-------IEDRVQEEARCLVEELRK--TNGSPCNP-TFILGAAPCNVICSVIFQNRFDYTDQDF-LSLMG  199
Rabbit    pc2    93  ELSGRSRFLVTAKLNKGFGVIFS--NGKRWTETRRFSLMTLRNFGMGKRS-------IEERVQEEAHCLVEELRK--TNASPCDP-TFILGAAPCNVICSVIFQNRFDYTDQDF-LSLMG  199
Rat       m1     93  EFSGRGSFPVSERVNKGLGVIFS--NGMQWKEIRRFSIMTLRTFGMGKRT-------IEDRIQEEAQCLVEELRK--SKGAPFDP-TFILGCAPCNVICSIIFQNRFDYKDPTF-LNLMH  199
Rabbit    1      93  EFAGTGSVPILEKVSKGLGIAFS--NAKTWKEMRRFSLMTLRNFGMGKRS-------IEDRIQEEARCLVEELRK--TNASPCDP-TFILGCAPCNVICSVIFHNRFDYKDEEF-LKLME  199
Rat       j      96  EFSGRGDIPVFQE-YKNKGIIFN--NGPTWKDVRRFSLSILRDWGMGKQG-------NEARIQREAQFLVEELKK--TKGQPFDP-TFLIGCAPCNVIADILFNKRFDYNDKKC-LRLMS  201
Human     j      96  EFSGRGDLPAFHA-HRDRGIIFN--NGPTWKDIRRFSLTTLRNYGMGKQG-------NESRIQREAHFLLEALRK--TQGQPFDP-TFLIGCAPCNVIADILFRKHFDYNDEKF-LRLMY  201
Rabbit    3a     96  EFSGRGEIPAFRE-FKDKGIIFN--NGPTWKDTRRFSLTTLRDYGMGKQG-------NEDRIQKEAHFLLEELRK--TQGQPFDP-TFVIGCTPFNVIAKILFNDRFDYKDKQA-LRLMS  201
Chicken   pb1    96  AFSGRGILPLIEKLFKGTGIVTS--NGETWRQLRRFALTTLRDFGMGKKG-------IEERIQEEAHFLVERIRK--THEEPFNPGKFLIH-AVANIICSIVFGDRFDYEDKKF-LDLIE  202
Mouse     c21a   88  DFAGRPHMLNGKM----DLDLSLGDYSLMWKAHKKLSRSAL-MLGM-RDS-------MEPLIEQLTQEFCERMRA--QAGTPVAI-HKEFSFLTCSIISCLTFGDK---DSTLV-QTLHD  187
Human     c21b   87  DFAGRPEPLTYKLVSKNYPDLSLGDYSLLWKAHKKLTRSAL-LLGI-RDS-------MEPVVEQLTQEFCERMRA--QPGTPVAI-EEEFSLLTCSIICYLTFGDKIK-DDNLM-PAYYK  192
Bovine    c21    88  DFAGRPQIPSYKLVSQRCQDISLGDYSLLWKAHKKLTRSAL-LLGT-RSS-------MEPWVDQLTQEFCERMRV--QAGAPVTI-QKEFSLLTCSIICYLTFGNK---EDTLV-HAFHD  191
Human     17α    92  DFSGRPQMATLDIASNNRKGIAFADSGAHWQLHRRLAMATFALFKDGDQK-------LEKIICQEISTLCDMLAT--HNGQSIDI-SFPVFVAVTNVISLICFNTSYKNGDPEL-NVIQN  200
Bovine    17α    92  EFSGRPKVATLDILSDNQKGIAFADHGAHWQLHRKLALNAFALFKDGNLK-------LEKIINQEANVLCDFLAT--QHGEAIDL-SEPLSLAVTNIISFICFNFSFKNEDPAL-KAIQN  200
Rat       pcn1  101  VFTNRRDFGPVGIMGKAVSVA----KDEEWKRYRALLSPTF-TSGRLKE--------MFPIIEQYGDILVKYLKQEAETGKPVTM-KKVFGAYSMDVITSTSFGVNVDSLNNPK-DPFVE  205
Rat       pcn2  101  VFTNRRDFGPVGIMGKAVSVA----KDEEWKRYRALLSPTF-TSGRLKE--------MFPIIEQYGDILVKYLKQEAETGKPVTM-KKVFGAYSMDVITSTSFGVNVDSLNNPK-DPFVE  205
Human     lp    101  VFTNREPFGPVGFMKSAISIA----EDEEWKRLRSLLSPTF-TSGKLKE--------MVPIIAQYGDVLVRNLRRERETGKPVTL-KDVFGAYSMDVITSSSFGVNVDSLNNPQ-DPLVE  205
Human     nf    101  VFTNRRPFGPVGFMKSAISIA----EDEEWKRLRSLLSPTF-TSGKLKE--------MVPIIAQYGDVLVRNLRREAETGKPVTL-KDVFGAYSMDVITSTSFGVNIDSLNNPQ-DPFVE  205
Rat       laω   114  KANGVYRLLA-PWIGYGLLLL----NGQPWFQHRRMLTPAF-HYDILKP--------YVKNMADSIRLMLDKWEQLAGQDSSIEI-FQHISLMTLDTVMKCAFSHNGSVQVDGNYKSYIQ  218
Rabbit    pgω   111  KAPRNYKLMT-PWIGYGLLLL----DGQTWFQHRRMLTPAF-HYDILKP--------YVGLMVDSVQIMLDRWEQLISQDSSLEI-FQHVSLMTLDTIMKCAFSYQGSVQLDRNSHSYIQ  215
Human     scc   120  RFLIPPWVAYHQYYQRPIGVLLK--KSAAWKKDRVALNQEVMAPEATKNFLP-----LLDAVSRDFVSVLHRRIKKAGSGNYSGDISDDLFRFAFESITNVIFGERQGMLEEVV-NPEAQ  231
Bovine    scc   120  RYDIPPWLAYHRYYQKPIGVLFK--KSGTWKKDRVVLNTEVMAPEAIKNFIP-----LLNPVSQDFVSLLHKRIKQQGSGKFVGDIKEDLFHFAFESITNVMFGERLGMLEETV-NPEAQ  231
Bovine    11β   110  RMILEPWLAYRQARGHKCGVFLL--NGPQWRLDRLRLNPDVLSLPALQKYTP-----LVDGVARDFSQTLKARVLQNARGSLTLGHRAQLFRYTIEASTLVLYGERLGLLTQQP-NPDSL  221
P.putida  cam    81  HFSSECPFIPREAGEAYDFIPTSM-DPPEQRQFRALANQVVGMPVVDK---------LENRIQELACSLIESLR---PQGQ------------CNFTED--YAEPFPIRI----FMLLA  168
                                                         .     :                   .                                     :
```

Fig. 1. — Continued

			260 280 300 320 340 360	
Rat	c	224	LSNEFGEVTG--SGYPADFIP-ILRYLPNSSLDAFKDLNKKFYS--FM--KKLIKEHYRTFEKGHI--RDITDSLIEHCQDRRLDENANVQ----LSDDKVITIVFDLFGAGFDTITTAI	330
Mouse	p1	224	LSNEFGEVTG--SGYPADFIP-VLRYLPNSSLDAFKDLNDKFYS--FM--KKLIKEHYRTFEKGHI--RDITDSLIEHCQDRKLDENANVQ----LSDDKVITIVLDLFGAGFDTVTTAI	330
Human	c	220	LNNNFGEVVG--SGNPADFIP-ILRYLPNPSLNAFKDLNEKFYS--FM--QKMVKEHYKTFEKGHI--RDITDSLIEHCQEKQLDENANVQ----LSDEKIINIVLDLFGAGFDTVTTAI	326
Rabbit	6	224	LNDEFGKVAA--SGSPADFFL-ILRYLPNPALDTFKDLNERFYS--FT--QERVKEHCRSFEKGHI--RDITDSLIKHYRVDRLDENANVQ----VSDEKTVGIVLDLFGAGFDTVTTAI	330
Rat	d	221	SSKDFVENVT--SGNAVDFFP-VLRYLPNPALKRFKNFNDNFVL--FL--QKTVQEHYQDFNKNSI--QDITGALFKH-SEN-YKDNGGL-----IPQEKIVNIVNDIFGAGFETVTTAI	324
Mouse	p3	221	NSKDFVENVT--SGNAVDFFP-VLRYLPNPALKRFKTFNDNFVL--FL--QKTVQEHYQDFNKNSI--QDITSALFKH-SEN-YKDNGGL-----IPEEKIVNIVNDIFGAGFDTVTTAI	324
Human	4	222	NTHEFVETAS--SGNPLDFFP-ILRYLPNPALQRFKAFNQRFLW--FL--QKTVQEHYQDFDKNSV--RDITGALFKH-SKKGPRASGNL-----IPQEKIVNLVNDIFGAGFDTVTTAI	326
Rabbit	4	222	NSSKFVETAS--SGSPVDFFP-ILRYLPNRPLQRFKDFNQRFLR--FL--QKTVREHYEDFDRNSI--QDITGALFKH-SEKNSKANGGL-----IPQEKIVNLVNDIFGAGFDTITTAL	326
Rat	b	201	LFYRTFSLLSSFSSQVFEFFSGFLKYFPGAHRQISKNLQE-ILD--YI--GHIVEKHRATLDPSAP--RDFIDTYLL--RMEKEKSNHHTE----FHHENLMISLLSLFFAGTETSSTTL	307
Rat	e	201	LFYRTFSLLSSFSSQVFEFFSGFLKYFPGAHRQISKNLQE-ILD--YI--GHIVEKHRATLDPSAP--RDFIDTYLL--RMEKEKSNHHTE----FHHENLMISLLSLFFAGTETGSTTL	307
Rabbit	2	201	LFFQSFSLISSFSSQVFELFPGFLKHFPGTHRQIYRNLQE-INT--FI--GQSVEKHRATLDPSNP--RDFIDVYLL--RMEKDKSDPSSE----FHHQNLILTVLSLFFAGTETTSTTL	307
Rat	a	203	MMGQMNRFAASPTGQLYDMFHSVMKYLPGPQQQIIKVTQK-LED--FM--IEKVRQNHSTLDPNSP--RNFIDSFLI--RMQEEKNGNSE----FHMKNLVMTTLSLFFAGSETVSSTL	308
Rabbit	3b	200	YFHENFELLGTPWIQLYNIFPILGHYLPGSHRQLFKNIDG-QIK--FI--LEKVQEHQESLDSNNP--RDFVDHFLI--KMEKEKHKKQSE----FTMDNLITTIWDVFSAGTDTTSNTL	306
Rat	f	200	KVNENLKIMSSPWMQVCNSFPSLIDYFPGTHHKIAKNINY-MKS--YL--LKKIEEHQESLDVTNP--RDFVDYYLI--KQKQANNIEHSE----YSHENLTCSIMDLIGAGTETMSTTL	306
Rat	pb1	200	KLNENMKILSSPWTQFCSFFPVLIDYCPGSHTTLAKNVYH-IRN--YL--LKKIKEHQESLDVTNP--RDFIDYYLI--KWKQENHNPHSE----FTLENLSITVTDLFGAGTETTSTTL	306
Rabbit	pc1	200	KLNENFKILNSPWVQMCNNFPILIDYLPGSHNKILRNNIY-IRN--YV--LEKIKEHQETLDINNP--RDFIDCFLI--KMEQEKDNQQSE----FTIENLMTTLSDVFGAGTETTSTTL	306
Rabbit	pc2	200	KFNENFKILNSPWVQFCNCFPILFDYFPGSHRKAVKNIFY-VKN--YI--TEQIKEHQKSLDINNP--RDFIDCFLI--KMEQEKCNQQSE----FTIENLLTTVSDVFMAGTETTSTTL	306
Rat	m1	200	RFNENFRLFSSPWLQVCNTFPAIIDYFPGSHNQVLKNFFY-IKN--YV--LEKVKEHQESLDKDNP--RDFIDCFLN--KMEQEKHNPQSE----FTLESLVATVTDMFGAGTETTSTTL	306
Rabbit	1	200	SLNENVRILSSPWLQVYNNFPALLDYFPGIHKTLLKNADY-IKN--FI--MEKVKEHQKLLDVNNP--RDFIDCFLI--KMEQEN---NLE----FTLESLVIAVSDLFGAGTETTSTTL	303
Rat	j	202	LFNENFYLLSTPWIQLYNNFADYLRYLPGSHRKIMKNVSE-IKQ--YT--LEKAKEHLQSLDINCA--RDVTDCLLI--EMEKEKHSQEPM----YTMENVSVTLADLFFAGTETTSTTL	308
Human	j	202	LFNENFHLLSTPWLQLYNNFPSFLHYLPGSHRKVIKNVAE-VKE--YV--SERVKEHHQSLDPNCP--RDLTDCLLV--EMEKEKHSAERL----YTMDGITVTVADLFFAGTETTSTTL	308
Rabbit	3a	202	LFNENFYLLSTPWLQVYNNFSNYLQYMPGSHRKVIKNVSE-IKE--YT--LARVKEHHKSLDPSCP--RDFIDSLLI--EMEKDKHSTEPL----YTLENIAVTVADMFFAGTETTSTTL	308
Chicken	pb1	203	MLEENNKYQNRIQTLLYNFFPTILDSLPGPHKTLIKNTET-VDD--FI--KEIVIAHQESFDASCP--RDFIDAFIN--KMEQEK--ENSY----FTVESLTRTTLDLFLAGTGTTSTTL	307
Mouse	c21a	188	CVQDLLQAWNHWSIQILTIIP-LLRFLPNPGLQKLKQIQESRDH--IV--KQQLKQHKESLVAGQW--KDMIDYMLQG-VEKQRDGKDEER----LHEGHVHMSVVDLFIGGTETTATTL	295
Human	c21b	193	CIQEVLKTWSHWSIQIVDVIP-FLRFFPNPGLRRLKQAIEKRDH--IV--EMQLRQHKESLVAGQW--RDMMDYMLQG-VAQPSMEEGSGQ----LLEGHVHMAAVDLLIGGTETTANTL	300
Bovine	c21	192	CVQDLMKTWDHWSIQILDMVP-FLRFFPNPGLWRLKQAIENRDH--MV--EKQLTRHKESMVAGQW--RDMTDYMLQG-VGRQRVEEGPGQ----LLEGHVHMSVVDLFIGGTETTASTL	299
Human	17α	201	YNEGIIDNLS--KDSLVDLVP-WLKIFPNKTLEKLKSHVKIRND--LL--NKILENYKEKFRSDSI--TNMLDTLMQAKMNSDNGNAGPDQDSELLSDNHILTTIGDIFGAGVETTTSVV	311
Bovine	17α	201	VNDGILEVLS--KEVLLDIFP-VLKIFPSKAMEKMKGCVQTRNE--LL--NEILEKCQENFSSDSI--TNLLHILIQAKVNADNNNAGPDQDSKLLSNRHMLATIGDIFGAGVETTTSVI	311
Rat	pcn1	206	KTKKLLRFDF--FDPLF-LSVVLFPFLTPIYEMLNICMFPKDSIE-FF--KKFVYRMKETRLDSVQ--KHRVDFLQL--MMNAHNDSKDKESHTALSDMEITAQSIIFIFAGYEPTSSTL	315
Rat	pcn2	206	KTKKLLRFDF--FDPLF-LSVVLFPFLTPIYEMLNICMFPKDSIA-FF--QKFVHRIKETRLDSKH--KHRVDFLQL--MLNAHNNSKDEVSHKALSDVEIIAQSVIFIFAGYETTSSTL	315
Human	lp	206	NTKKLLRFDF--LDPFF-LSITVFPFLIPILEVLNICVFPREVTN-FL--RKAVKRMKESRLEDTQ--KHRVDFLQL--MIDSHKNSKETESHKALSDLELVAQSIIFIFAGYETTSSVL	315
Human	nf	206	NTKKLLRFDF--LDPFF-LSITVFPFLIPILEVLNICVFPREVTN-FL--RKSVKRMKESRLEDTQ--KHRVDFLQL--MIDSQ-NSKETESHKALSDLELVAQSIIFIFAGYETTSSVL	314
Rat	laω	219	AIGNLNDLFHSRVRNIFHQNDTIYNFSSNGHLFNRACQLAHDHTDGVIKLRKDQLQNAGELEKVKK--KRRLDFLDI--LLLARMENGDS-----LSDKDLRAEVDTFMFEGHDTTASGV	329
Rabbit	pgω	216	AINDLNNLVFYRARNVFHQSDFLYRLSPEGRLFHRACQLAHEHTDRVIQQRKAQLQQEGELEKVRR--KRRLDFLDV--LLFAKMENGSS-----LSDQDLRAEVDTFMFEGHDTTASGV	326
Human	scc	232	RFIDAIYQMFHTSVPMLNLPPDLFRLFRTKTWKDHVAAWDVILS--KA--DIYTQNFYWELRQKGSVHHDYRGILYR--LLGDSK----------MSFEDIKANVTEMLAGGVDTTSMTL	335
Bovine	scc	232	KFIDAVYKMFHTSVPLLNVPPELYRLFRTKTWRDHVAAWDTIFN--KA--EKYTEIFYQDLRRKTEF-RNYPGILYC--LLKSEK----------MLLEDVKANITEMLAGGVNTTSMTL	334
Bovine	11β	222	NFIHALEAMLKSTVQLMFVPRRLSRWMSTNMWREHFEAWDYIFQ--YA--NRAIQRIYQELALGHP--WHYSGIVAE--LLMRAD----------MTLDTIKANTIDLTAGSVDTTAFPL	323
P.putida	cam	169	GLPE------------EDIP-HLKYLTD---QMTRPDGSMTFA--EA--KEALYDYLIPIIEQR---RQKPGTDAIS-IVANGQVNGRP-----ITSDEAKRMCGLLLVGGLDTVVNFL	258
			.: :. .	

Fig. 1. — Continued

```
                           380                 400                 420                 440                 460                 480
Rat       c     331  SWSLMYLVTNPRIQR-KIQEELDTVIGRD---RQPRLSDRPQLPYLEAFILETFRHSSFVPFTIPHSTIRDTSL-NGFYIPKGHCVFVNQWQVNHDQELWGDPNEFRPERFLTSSGT-LD  444
Mouse     p1    331  SWSLMYLVTNPRVQR-KIQEELDTVIGRD---RQPRLSDRPQLPYLEAFILETFRHSSFVPFTIPHSTTRDTSL-NGFYIPKGCCVFVNQWQVNHDRELWGDPNEFRPERFLTPSGT-LD  444
Human     c     327  SWSLMYLVMNPRVQR-KIQEELDTVIGRS---RRPRLSDRSHLPYMEAFILETFRHSSFVPFTIPHSTTRDTSL-KGFYIPKGRCVFVNQWQINHDQKLWVNPSEFLPERFLTPDGA-ID  440
Rabbit    6     331  SWSLMYLVTKRQHTREKIQEELDAVVGRA---RRPRFSDRPQLPYLEAVIMETFRHTSFLPFTIPHSTTRDTSL-GGFYIPKGRCVFVNQWQNNHDRELWGVPRSFRPERFLTPSGT-LD  445
Rat       d     325  FWSILLLVTEPKVQR-KIHEELDTVIGRD---RQPRLSDRPQLPYLEAFILEIYRYTSFVPFTIPHSTTRDTSL-NGFHIPKECCIFINQWQVNHDEKQWKDPFVFRPERFLTNDNTAID  439
Mouse     p3    325  TWSILLLVTWPNVQR-KIHEELDTVVGRD---RQPRLSDRPQLPYLEAFILEIYRYTSFVPFTIPHSTTRDTSL-NGFHIPKERCIYINQWQVNHDEKQWKDPFVFRPERFLTNNNSAID  439
Human     4     327  SWSLMYLVTKPEIQR-KIQKELDTVIGRE---RRPRLSDRPQLPYLEAFILETFRHSSFLPFTIPHSTTRDTTL-NGFYIPKKCCVFVNQWQVNHDPELWEDPSEFRPERFLTADGTAIN  441
Rabbit    4     327  SWSLMYLVTNPRRQR-KIQEELDAVVGRA---RQPLLSDRPQLPYLEAFILELFRHTSFVPFTIPHSTTRDTTL-NGFHIPKECCIFINQWQINHDPQLWGDPEEFRPERFLTADGAAIN  441
Rat       b     308  RYGFLLMLKYPHVAE-KVQKEIDQVIGSH---RLPTLDDRSKMPYTDAVIHEIQRFSDLVPIGVPHRVTKDTMF-RGYLLPKNTEVYPILSSALHDPQYFDHPDSFNPEHFLDANGA-L-  420
Rat       e     308  RYGFLLMLKYPHVTV-KVQKEIDQVIGSH---RPPSLDDRTKMPYTDAVIHEIQRFADLAPIGLPHRVTKDTMF-RGYLLPKNTEVYPILSSALHDPQYFDHPDTFNPEHFLDADGT-L-  420
Rabbit    2     308  RYGFLLMLKYPHVTE-RVQKEIEQVIGSH---RPPALDDRAKMPYTDAVIHEIQRLGDLIPFGVPHTVTKDTQF-RGYVIPKNTEVFPVLSSALHDPRYFETPNTFNPGHFLDANGA-L-  420
Rat       a     309  RYGFLLLMKHPDVEA-KVHEEIEQVIGRN---RQPQYEDHMKMPYTQAVINEIQRFSNLAPLGIPRRIIKNTTF-RGFFLPKATDVFPILGSLMTDPKFFPSPKDFDPQNFLDDKGQ-L-  421
Rabbit    3b    307  KFALLLLLKHPEITA-KVQEEIEHVIGRH---RSPCSQDRSRMPYTDAVMHEIQRYVDLVPTSLPHAVTQDIEF-NGYLIPKGTDIIPSLTSVLYDDKEFPNPEKFDPGHFLDESGN-F-  419
Rat       f     307  RYALLLLMKYPHVTA-KVQEEIDRVIGRH---RSPCMQDRKHMPYTDAHDHEVQRFINFVPTNLPHAVTCDIKF-RNYLIPKGTKVLTSLTSVLHDSKEFPNPEMFDPGHFLDENGN-F-  419
Rat       pb1   307  RYALLLLLKCPEVTA-KVQEEIDRVVGKH---RSPCMQDRSRMPYTDAHDHEVQRFIDLIPTNLPHAVTCDIKF-RNYLIPKGTTIITSLSSVLHDSKEFPDPEIFDPGHFLDGNGK-F-  419
Rabbit    pc1   307  RYGLLLLMKHPEVIA-KVQEEIERVIGRH---RSPCMQDRSRMPYTDATVHEIQRYINLIPNNVPRATTCNVKF-RSYLIPKGTAVITSLTSMLYNDKEFPNPDRFDPGHFLDASGK-F-  419
Rabbit    pc2   307  RYGLLLLMKHPEVIA-KVQEEIERVIGRH---RSPCMQDRSRMPYTDATVHEIQRYINLIPNNVPHTTICNLKF-RNYLIPKGTDVLTSLSSVLHDDKEFPNPDRFDPGHFLDASGN-F-  419
Rat       m1    307  RYGLLLLLKHVDVTA-KVQEEIERVIGRN---RSPCMKDRSQMPYTDAVVHEIQRYIDLVPTNLPHLVTLDIKF-RNYFIPKGTNVIVSLSSILHDDKEFPNPEKFDPGHFLDERGN-F-  419
Rabbit    1     304  RYSLLLLLKHPEVAA-RVQEEIERVIGRH---RSPCMQDRSRMPYTDAVIHEIQRFIDLLPTNLPHAVTRDVRF-RNYFIPKGTDIITSLTSVLHDEKAFPNPKVFDPGHFLDESGN-F-  416
Rat       j     309  RYGLLILMKYPEIEE-KLHEEIDRVIGPS---RVPAVRDRLDMPYMDAVVHEIQRFINLVPSNLPHEATRDTVF-QGYVIPKGTVVIPTLDSLLYDSHEFPDPEKFKPEHFLNENGK-F-  421
Human     j     309  RYGLLILMKYPEIEE-KLHEEIDRVIGPS---RIPAIKDRQEMPYMDAVVHEIQRFITLVPSNLPHEATRDTIF-RGYLIPKGTVVVPTLDSVLYDNQEFPDPEKFKPEHFLNENGK-F-  421
Rabbit    3a    309  RYGLLILLKHPEIEE-KLHEEIDRVIGPS---RMPSVRDRVQMPYMDAVVHEIQRFIDLVPSNLPHEATRDTTF-QGYVIPKGTVVIPTLDSLLYDKQEFPDPEKFKPEHFLNEEGK-F-  421
Chicken   pb1   308  RYGLLILLKHPEIEE-KMHKEIDRVVGRD---RSPCMADRSQLPYTDAVIHEIQRFIDFLPLNVPHAVIKDTKL-RDYFIPKDTMIFPLLSPILQDCKEFPNPEKFDPGHFLNANGT-F-  420
Mouse     c21a  296  SWAVAFLLHHPEIQK-RLQAELDLKLGPG---SQLLYRNRMQLPLLMATIAEVLRLRPVVPLALPHRATRASSI-SGYDIPKDMVIIPNIQGANLDEMVWELPSKFWPDRFLEP------  404
Human     c21b  301  SWAVVFLLHHPEIQQ-RLQEELDHELGPGASSSRVPYKDRARLPLLNATIAEVLRLRPVVPLALPHRTTRPSSI-SGYDIPEGTVIIPNLQGAHLDETVWERPHEFWPDRFLEP------  412
Bovine    c21   300  SWAVAFLLHHPEIQR-RLQEELDRELGPGASCSRVTYKDRARLPLLNATIAEVLRLRPVVPLALPHRTTRPSSI-FGYDIPEGMVVIPNLQGAHLDETVWEQPYEFRPDRFLEP------  411
Human     17α   312  KWTLAFLLHNPQVKK-KLYEEIDQNVGFS---RTPTISDRNRLLLLEATIREVLRLRPVAPMLIPHKANVDSSI-GEFAVDKGTEVIINLWALHHNEKEWHQPDQFMPERFLNPAGTQLI  426
Bovine    17α   312  KWIVAYLLHHPSLKK-RIQDDIDQIIGFN---RTPTISDRNRLVLLEATIREVLRIRPVAPTLIPHKAVIDSSI-GDLTIDKGTDVVVNLWALHHSEKEWQHPDLFMPERFLDPTGTQLI  426
Rat       pcn1  316  SFVLHSLATHPDTQK-KLQEEIDRALPNK---APPTYDTVMEMEYLDMVLNETLRLYP-IGNRLERVCKKDVEI-NGVFMPKGSVVMIPSYALHRDPQHWPEPEEFRPERFSKENKG---  426
Rat       pcn2  316  SFVLYFLATHPDIQK-KLQEEIDGALPSK---APPTYDIVMEMEYLDMVLNETLRLYP-IGNRLERVCKKDIEL-DGLFIPKGSVVTIPTYALHHDPQHWPKPEEFHPERFSKENKG---  426
Human     lp    316  SFIMYELATHPDVQQ-KLQEEIDAVLPNK---APPTYDTVLQMEYLDMVVNETLRLFP-IAMRLERVCKKDVEI-NGMFIPKGWVVMIPSYALHRDPKYWTEPEKFLPERFSKKNKD---  426
Human     nf    315  SFIMYELATHPDVQQ-KLQEEIDAVLPNK---APPTYDTVLQMEYLDMVVNETLRLFP-IAMRLERVCKKDVEI-NGMFIPKGWVVMIPSYALHRDPKYWTEPEKFLPERFSKKNKD---  425
Rat       laω   330  SWIFYALATHPKHQQ-RCREEVQSVLGDG---SSITWDHLDQIPYTTMCIKEALRLYP-PVPGIVRELSTSVTFPDGRSLPKGIQVTLSIYGLHHNPKVWPNPEVFDPSRFAPDS-----  439
Rabbit    pgω   327  SWIFYALATHPEHQH-RCREEIQGLLGDG---ASITWEHLDQMPYTTMCIKEALRLYP-PVPSVTRQLSKPVTFPDGRSLPKGVILFLSIYGLHYNPKVWQNPEVFDPFRFAPDS-----  436
Human     scc   336  QWHLYEMARNLKVQD-MLRAEVLAARHQA---QGDMATMLQLVPLLKASIKETLRLHP-ISVTLQRYLVNDLVL-RDYMIPAKTLVQVAIYALGREPTFFFDPENFDPTRWLSKDK----  445
Bovine    scc   335  QWHLYEMARSLNVQE-MLREEVLNARRQA---EGDISKMLQMVPLLKASIKETLRLHP-ISVTLQRYPESDLVL-QDYLIPAKTLVQVAIYAMGRDPAFFSSPDKFDPTRWLSKDK----  444
Bovine    11β   324  LMTLFELARNPEVQQ-AVRQESLVAEARI---SENPQRAITELPLLRAALKETLRLYP-VGITLEREVSSDLVL-QNYHIPAGTLVKVLLYSLGRNPAVFARPESYHPQRWLDRQG----  433
P.putida  cam   259  SFSMEFLAKSPEH-----RQEL---IE----------RPERIP---AACEELLRRFSLVAD--GRILTSDYEF-HGVQLKKGDQILLPQMLSGLDERENACPMHVDFSRQ---------  344
                           :        :                 :  *  *          :     .     .. .              . * . : ..
```

Fig. 1. — Continued

```
                                      500                 520                 540                 560
Rat       c     445  KHLSEKVILFGLGKRKCIGETIGRLEVFLFLAILLQQM-EFNVSPGEK-VDMTPAY--GLTLKHARCEHFQVQMRSSGPQHLQA---  524
Mouse     p1    445  KRLSEKVTLFGLGKRKCIGETIGRSEVFLFLAILLQQI-EFKVSPGEK-VDMTPTY--GLTLKHARCEHFQVQMRSSGPQHLQA---  524
Human     c     441  KVLSEKVIIFGMGKRKCIGETIARWEVFLFLAILLQRV-EFSVPLGVK-VDMTPIY--GLTMKHACCEHFQMQLRS----------  512
Rabbit    6     446  KGPDDKVLLFGLGKRKCIAETIGRLEVFLFLATLLQQV-EFSVSPGTT-VDMTPIY--GLTMKHARCEHFQAKLRFEA--------  519
Rat       d     440  KTLSEKVMLFGLGKRRCIGEIPAKWEVFLFLAILLHQL-EFTVPPGVK-VDLTPSY--GLTMKPRTCEHVQAWPRFSK--------  513
Mouse     p3    440  KTQSEKVMLFGLGKRRCIGEIPAKWEVFLFLAILLQHL-EFSVPPGVK-VDLTPNY--GLTMKPGTCEHVQAWPRFSK--------  513
Human     4     442  KPLSEKMMLFGMGKRRCIGEVLAKWEIFLFLAILLQQL-EFSVPPGVK-VDLTPIY--GLTMKHARCEHVQARLPFSIN-------  516
Rabbit    4     442  KPLSEKVTLFGLGKRRCIGEILARWEVFLFLAILLQRL-EFSVPPGVP-VDLTPIY--GLTMKHPRCEHVQARPRFSDQ-------  516
Rat       b     421  KK-SEAFMPFSTGKRICLGEGIARNELFLFFTTILQNF-SVSSHLAPKDIDLTPKES-GIGKIPPTYQICFSAR------------  491
Rat       e     421  KK-SEAFMPFSTGKRICLGEGIARNELFLFFTTILQNF-SVSSHLAPKDIDLTPMES-GIAKIPPTYQICFSAR------------  491
Rabbit    2     421  KR-NEGFMPFSLGKRICLGEGIARTELFLFFTTILQNF-SIASPVPPEDIDLTPRES-GVGNVPPSYQIRFLAR------------  491
Rat       a     422  KK-NAAFLPFSTGKRFCLGDGLAKMELFLLLTTILQNF-RFKFPMKLEDINESPKPL-GFTRIIPKYTMSFMPI------------  492
Rabbit    3b    420  KK-SDYFMPFSTGKRACVGEGLARMELFLLLTTILQHF-TLKPLVDPKDIDPTPVEN-GFVSVPPSYELCFVPV------------  490
Rat       f     420  KK-SDYFLPFSAGKRACVGEGLARMQLFLFLTTILQNF-NLKSLVHPKDIDTMPVLN-GFASLPPTYQLCFIPS------------  490
Rat       pb1   420  KK-SDYFMPFSAGKRMCAGEGLARMELFLFLTTILQNF-KLKSVLHPKDIDTTPVFN-GFASLPPFYELCFIPL------------  490
Rabbit    pc1   420  RK-SDYFMPFSTGKRVCVGEVLARMELFLFLTAILQNF-TPKPLVDPKDIDTTPLVS-GLGRVPPLYQLSFIPA------------  490
Rabbit    pc2   420  RK-SDYFMPFSTGKRVCVGEALARMELFLFLTAILQNF-TPKPLVNPNNVDENLFSS-GIVRVPPLYRVSFIPV------------  490
Rat       m1    420  KK-SDYFMPFSAGKRICAGEALARTELFLFFTTILQNF-NLKSLVDVKDIDTTPAIS-GFGHLPPFYEACFIPVQRADSLSSHL---  500
Rabbit    1     417  KK-SDYFMPFSAGKRMCVGEGLARMELFLFLTSILQNF-KLQSLVEPKDLDITAVVN-GFVSVPPSYQLCFIPI------------  487
Rat       j     422  KY-SDYFKAFSAGKRVCVGEGLARMELFLLLSAILQHF-NLKSLVDPKDIDLSPVTV-GFGSIPPQFKLCVIPRS-----------  493
Human     j     422  KY-SDYFKPFSTGKRVCAGEGLARMELFLLLCAILQHF-NLKPLVDPKDIDLSPIHI-GFGCIPPRYKLCVIPRS-----------  493
Rabbit    3a    422  KY-SDYFKPFSAGKRVCVGEGLARMELFLLLSAILQHF-NLKPLVDPEDIDLRNITV-GFGRVPPRYKLCVIPRS-----------  493
Chicken   pb1   421  RR-SDYFMPFSAGKRICAGEGLARMEIFLFLTSILQNF-SLKPVKDRKDIDISPIIT-SLANMPRPYEVSFIPR------------  491
Mouse     c21a  405  GK-NPRTPSFGCGARVCLGEPLARLELFVVLARLLQAF-TLLPPPDGTLPSLQPQPYAGINLPIPPFQVRLQPRNL--APQDQGERP  487
Human     c21b  413  GK-NSRALAFGCGARVCLGEPLARLELFVVLTRLLQAF-TLLPSGD-ALPSLQPLPHCSVILKMQPFQVRLQPRGM--GAHSPGQNQ  494
Bovine    c21   412  GA-NPSALAFGCGARVCLGESLARLELFVVLLRLLQAF-TLLPPPVGALPSLQPDPYCGVNLKVQPFQVRLQPRGVEAGAWESASAQ  496
Human     17α   427  SP-SVSYLPFGAGPRSCIGEILARQELFLIMAWLLQRF-DLEVPDDGQLPSLEGIP--KVVFLIDSFKVKIKVRQAWREAQAEGST-  508
Bovine    17α   427  SP-SLSYLPFGAGPRSCVGEMLARQELFLFMSRLLQRF-NLEIPDDGKLPSLEGHA--SLVLQIKPFKVKIEVRQAWKEAQAEGSTP  509
Rat       pcn1  427  SIDPYVYLPFGNGPRNCIGMRFALMNMKLALTKVLQNF-SFQPCKETQIPLKLSRQ--GLLQPTKPIILKVVPRDEIITGS------  504
Rat       pcn2  427  SIHPYVYLPFGNGPRNCIDMRFALMNMKLALTKVLQNF-SFQPCKETQIPLKLSRQ--AILEPEKPIVLKVLPRDAVINGA------  504
Human     1p    427  NIDPYIYTPFGSGPRNCIGMRFALMNMKLALIRVLQNF-SFKPCKETQIPLKLSLG--GLLQPEKPVVLKVESRDGTVSGA------  504
Human     nf    426  NIDPYIYTPFGSGPRNCIGMRFALMNMKLALIRVLQNF-SFKPCKETQIPLKLSLG--GLLQPEKPVVLKVESRDGTVSGA------  503
Rat       1aω   440  PRHSHSFLPFSGGARNCIGKQFAMSEMKVIVALTLLRF-ELLP-DPTKVPIPLPRL---VLKSKNGIYLYLKKLH-----------  509
Rabbit    pgω   437  AYHSHAFLPFSGGARNCIGKQFAMRELKVAVALTLVRF-ELLP-DPTRIPIPIARV---VLKSKNGIHLRLRKLH-----------  506
Human     scc   446  NITYFRNLGFGWGVRQCLGRRIAELEMTIFLINMLENF-RVEIQHLSDVGTTFNLI--LMPEKPISFTFWPFNQEATQQ--------  521
Bovine    scc   445  DLIHFRNLGFGWGVRQCVGRRIAELEMTLFLIHILENF-KVEMQHIGDVDTIFNLI--LTPDKPIFLVFRPFNQDPPQA--------  520
Bovine    11β   434  SGSRFPHLAFGFGVRQCLGRRVAEVEMLLLLHHVLKNF-LVETLEQEDIKMVYRFI--LMPSTLPLFTFRAIQ---~----------  503
P.putida  cam   345  ---KVSHTTFGHGSHLCLGQHLARREIIVTLKEWLTRIPDFSIAPGAQIQHKSGIVS-GVQALPLVWDPATTKAV------------  415
                             *: * : * .   : .  .     *
```

Fig. 1. — Continued

present-day sequences, since multiple mutations may have occurred at a locus. DAYHOFF (1978) proposed a table that relates a percent identity to a PAM, Instead of using this table, we adopted a maximum likelihood method (CRAMER, 1946) to estimate a PAM on the assumption that evolutionary changes of amino acids follow a Markov process determined by the transition matrix presented in Figure 82 in DAYHOFF (1978). Briefly, we first count the individual numbers of the 20×20 types of amino-acid changes associated with the conversion from one sequence to the other. The diagonal elements of this table represent the numbers of unchanged residues. The PAM value that maximizes the probability of observing these numbers of amino-acid changes and identities is defined as the maximum likelihood estimate of PAM. We omitted deletions or insertions from the present analysis, since the frequency of evolutionary events that change sequence length is not well understood.

Using the 703 ($= 38 \times 37/2$) PAM values, we constructed a phylogenetic tree of P-450 protein sequences (Fig. 2) by UPGMA (unweighted pair group method of averaging) (SOKAL and SNEATH, 1963). Although UPGMA assumes a constant evolutionary change rate, a constant rate is not realized by P-450 sequences as discussed below. However, the tree topology may not be significantly affected even if other tree reconstruction methods are used, since the clusters are well separated from one another.

3.6. Rates of amino-acid change

The PAM values between orthologous proteins in different mammalian species (Table 4) show that no pair of the four species, human, rat or mouse, rabbit and bovine, is apparently closer than other pairs. This observation is consistent with the well-accepted scheme that various orders of mammals differentiated in a short period around 80 million years (My) ago (DICKERSON, 1972; MCLAUGHLIN and DAYHOFF, 1972). If we accept this scheme, the average rate of amino acid replacement is calculated to be 1.8 ± 0.3 (SD) Pauling, or equivalently the unit evolutionary period (UEP) equals 2.8 ± 0.5 (SD), where one Pauling stands for 10^{-9} replacements per site per year (KIMURA, 1969) and the UEP is defined as the time in million years for a 1% change in amino acid sequence to occur (DICKERSON, 1972). As shown in Table 4, the variation in the rates of different lines is significantly greater than that expected from purely statistical variations (approximately ± 0.13 Pauling). Functional and/or structural constraints must have operated differently on one line versus another.

It was rather unexpected that the P-450s such as C21, 17α, and SCC that catalyze specific metabolic reactions would show relatively high evolutionary rates compared to those with broad substrate specificities (members in PB and MC families). The alcohol inducible forms (rat and human P-450j and rabbit P-450 lm3a) show the slowest rates of 1.3 $\sim$ 1.5 Pauling (UEP = 3.4 $\sim$ 3.8). A similar rate (1.3 Pauling or UEP = 3.9) is observed for the divergence between chicken P-450 pb1 and a mammalian group composed of the alcohol inducible forms and the so-called "constitutive" forms.

Table 4. Comparison of orthologous P-450 sequences

Gene	Name	Species 1	Species 2	PAM	Match %	Rate[1] Pauling	UEP[1]
IA1	c	rat	mouse	6.15	93.13	1.80	2.76
		rodent[2]	human	21.73	78.15	1.36	3.68
		rodent	rabbit	30.42	73.66	1.90	2.62
		human	rabbit	29.09	73.93	1.81	2.75
IA2	d	rat	mouse	6.03	93.37	1.77	2.82
		rodent	human	30.92	73.26	1.93	2.59
		rodent	rabbit	30.16	73.94	1.89	2.65
		human	rabbit	24.84	77.71	1.55	3.22
IIB1	b	rat	rabbit	26.32	77.19	1.65	3.04
IIE1	j	rat	human	23.54	78.70	1.47	3.40
		rat	rabbit	20.83	81.14	1.30	3.84
		human	rabbit	23.54	79.11	1.47	3.40
XXI	c21	mouse	human	36.18	70.77	2.26	2.21
		mouse	bovine	35.78	70.97	2.24	2.24
		human	bovine	24.99	78.11	1.56	3.20
XVII	17α	human	bovine	34.51	70.92	2.15	2.32
III[3]	pcn	rat	human	33.90	72.22	2.12	2.36
XII	scc	human	bovine	32.09	72.55	2.00	2.49
IIC/E[4]		mammal	chicken	76.98	52.49	1.28	3.90

1 The dates of branching are assumed to be 17 my for rat and mouse (Miyata et al., 1982), 80 my for various orders of mammals, and 300 my for mammals and birds (Dickerson, 1972; McLaughlin and Dayhoff, 1972).
2 The data obtained with the rat sequence and the mouse sequence are averaged.
3 The values obtained from four comparisons between rat P-450 pcn1 or P-450 pcn2 and human P-450 lp or P-450 nf are averaged.
4 Each value shown is an average of comparisons between chicken P-450 pb1 and each of 10 mammalian P-450s, rat f, pb1, m1, j, rabbit 3b, 1, pc1, pc2, 3a, and human j.

The time scale in Figure 2 is drawn on the assumption of a constant evolutionary rate of 1.76 Pauling, which is the average of the 19 values listed in Table 4. Thus, we estimate that PB and MC families diverged some 470 My ago, their common ancestor branched from C21/17α group about 550 My ago, the divergence between microsomal and mitochondrial P-450s occurred more than 900 My ago, and so on. However, one must interpret these values with

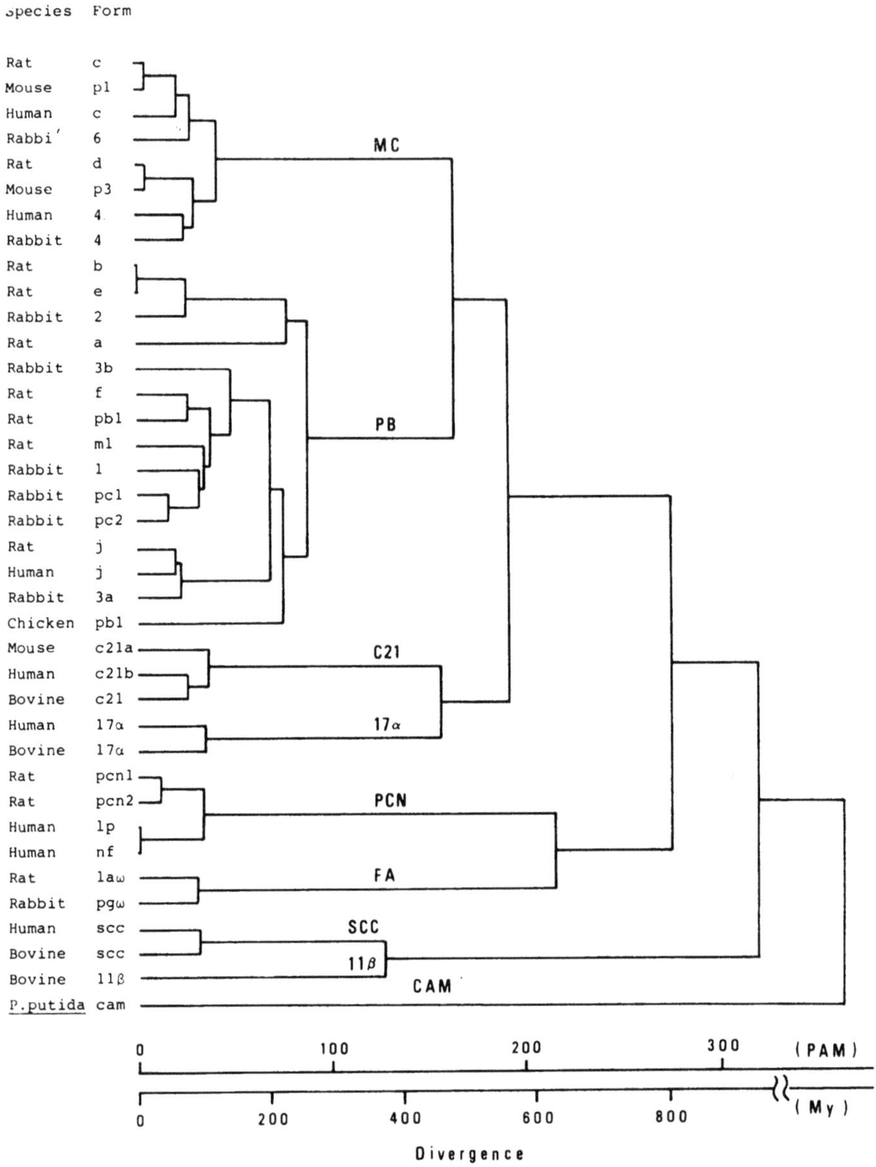

Fig. 2. The phylogenetic tree of 38 P-450 proteins constructed with UPG method. The divergence time below the tree is scaled on the assumption of the constant amino-acid change rate of 1.76 Pauling (UEP = 2.92).

caution, since, as mentioned above, amino acid change rates can vary as much as twofold in different lineages. To get precise dates, we must wait for sequence data of a variety of lower vertebrate, invertebrate, plant, fungal and bacterial P-450s.

3.7. Evolution of P-450 genes

At the present time, we cannot tell the whole story of the evolution of P-450 genes, since our knowledge is almost limited to animal genes. There are many kinds of microbial P-450 besides the well characterized one of *P. putida*, P-450cam (for reviews, SLIGAR and MURRAY, 1986; KÄPPELI, 1986). These microbial P-450s catalyze hydroxylation of diverse lipophilic substrates such as steroids, fatty acids, alkanes and monoterpenes. Some substrates are common to those of animal P-450s. However, a common substrate does not mean a close evolutionary relationship. The phylogenetic tree (Fig. 2) suggests that a number of P-450 molecular species so far identified in higher animals have evolved from an ancestor during the last one billion years.

We suspect that the common ancestor of animal P-450s was involved in steroid metabolism, because steroids are good substrates of at least one member in most families of animal P-450s. The substrate specificity of the ancestral enzyme might have been stringent, as is observed for mitochondrial P-450s, P-450 scc and P-450 11β. The subsequent duplications and mutations produced many P-450 forms with relaxed specificities to steroids, and eventually some species of P-450 with omnivorous properties were generated at the expense of the high affinity to steroid rings.

After the first momentous divergence between microsomal and mitochondrial types, microsomal P-450 has greatly increased its members. The PB family is prominent in its prolific nature. The complicated history of this gene family including gene duplications and gene conversions has been extensively discussed (AFFOLTER and ANDERSON, 1984; ADESNIK and ATCHISON, 1986; ATCHISON and ADESNIK, 1986).

4. High-order structure of animal P-450s

In the previous section, we have shown that all P-450s are evolutionarily related. As is known for various protein families, tertiary structure is much better conserved than primary structure. Keeping this in mind, we can now make a confident prediction of the higher-order structure of animal P-450s and assign functional domains along the amino acid sequences. This is made possible by the wealth of sequence data, careful sequence alignment and various predictive methods now available, in conjunction with the unique X-ray structure of bacterial P-450 cam (POULOS et al., 1985, 1986).

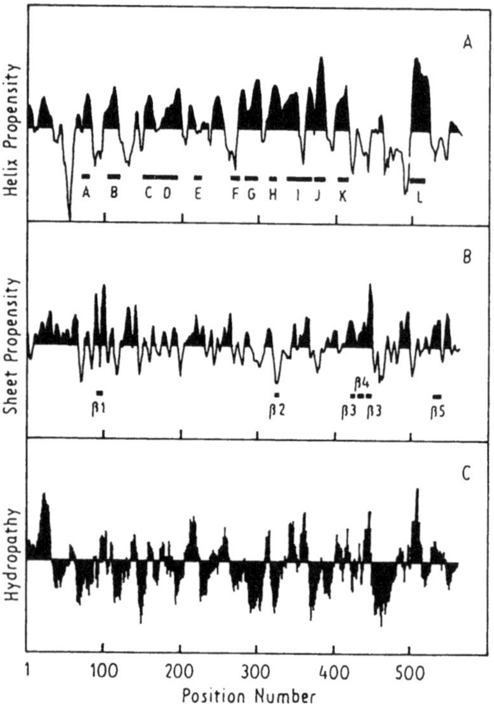

Fig. 3. Prediction of secondary structure and hydropathy of animal P-450s. (A) Helix-forming propensity and (B) sheet-forming propensity are calculated by the method of Garnier et al. (1978) and averaged over the 37 animal P-450s using the weight values listed in Table 2. The locations of 12 helices and 5 β-structures in P-450 cam are indicated by solid bars. (C) Hydropathy profile averaged over 37 animal P-450s using the weight values in Table 2. Hydropathy of each protein is obtained by the method of Kyte and Doolittle (1982) with the window size of 9 residues.

4.1. Prediction of secondary structure

Several reports have predicted secondary structures of P-450 molecules based on amino acid sequences (Haniu et al., 1982b, 1986; Gotoh et al., 1983). Critical assessments of available prediction methods including those used in these studies indicate that the accuracy of prediction is $50 \sim 55\%$ for three states of α-helix, β-sheet or loop (Kabsch and Sander, 1983; Nishikawa, 1983). Comparing the X-ray structure of P-450 cam (Poulos et al., 1985) with our previous prediction (Gotoh et al., 1983), we found that 52% sites were correctly predicted, disregarding the *N*-terminal nine residues with vague crystallographic structure and two residues that were missing in the sequence used for the prediction. It is remarkable that some peculiar structures such as β3/β4 region (repeated units of β-pairs) and the heme-binding region (a loop followed by a helix) were correctly predicted. We demonstrate below that the results of analyses extended to a collection of animal P-450 sequences are highly reliable despite of some suspicions about the existing prediction methods (Black and Coon, 1986).

Figure 3A shows a helix-forming propensity profile calculated by the method of Garnier et al. (1978) and averaged over the 37 animal P-450 sequences. On averaging, we assigned a weight to each sequence so that every piece of sequence information would contribute equally to the result. The weighting values are obtained during the course of construction of the phylogenetic tree (Fig. 2). The details of the calculation procedure will be described elsewhere (O. GOTOH, in preparation). The weighting values listed in Table 2 were also used in averaging sheet-forming propensity (Fig. 3B), turn and random propensities (data not shown), and hydropathy indices (Fig. 3C).

The locations of the 12 helices in P-450 cam (POULOS et al., 1985) are indicated in Fig. 3A according to the alignment (Fig. 1), and these locations fit strikingly well to the regions with high helix propensities. The excellent fits of helix locations obtained from different sources of information strongly suggest that the backbone structure is essentially conserved between animal P-450s and P-450 cam.

Prediction of secondary structure of P-450 cam gave a considerable overestimate of the amount of β-structure (GOTOH et al., 1983). This is probably the case for animal P-450s whose β-sheet profile is abundant in peaks (Fig. 3B). However, the five antiparallel β-pairs in P-450 cam except $\beta 2$ are properly aligned to animal P-450 regions with high β-propensities, further supporting the idea of common tertiary structures between animal and bacterial P-450s.

Although structural conservation between P-450 cam and animal P-450s is now apparent, there are two exceptional regions. The prominent *N*-terminal peak in the helix-propensity profile overlaps with the most hydrophobic region in microsomal P-450s (Fig. 3C). This region is certainly recognized by the signal recognition machinery to be inserted into the endoplasmic reticulum (ER) membrane (SAKAGUCHI et al., 1984) (see below), and hence P-450 cam lacks the corresponding portion. The locations of helix E and $\beta 2$ do not match well with high probability regions. The corresponding regions in animal P-450s are located near the center of a molecule and highly variable in sequence (Fig. 4A). We speculate that not only primary structure but also higher-order structures are somewhat variable in this central region.

Once we accept a common structure of P-450 cam and animal P-450s, then we can refer to individual domains in a P-450 protein in terms of corresponding regions in P-450 cam, e.g., Helix A or $\beta 1$ region, etc. We will start this with examining highly conserved regions in animal P-450s.

4.2. Highly conserved regions

There are ten positions where all animal P-450 sequences share unique amino acids. Amino acids at seven of the ten positions are also common to those in P-450 cam. All these seven strictly conserved positions are located in the C-terminal half of a molecule, and clustered into three conserved domains (Fig. 4A). For reasons explained below, we call the most *N*-proximal conserved cluster the Helix-K region, the second cluster the aromatic region and the third cluster the heme-binding or HR2 region. Undoubtedly, these three conserved domains are relevant to important functions common to all P-450s. We will first discuss functional implication of these three regions, and then proceed to other regions.

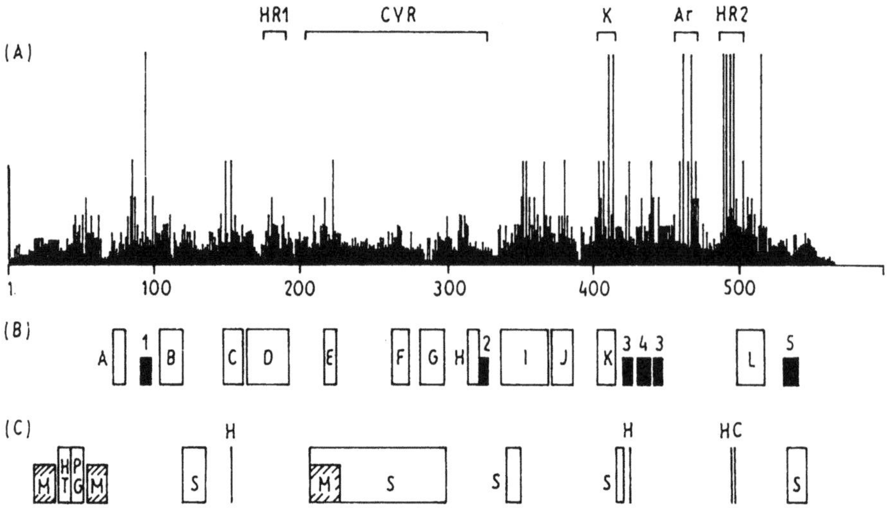

Fig. 4. Amino acid conservativity in animal P-450s. (A) The height of each bar represents the number of residues except for deletions at the position divided by the number of different kinds of residues at that position. HR1, central highly variable region (CVR), and three remarkably conserved regions, Helix K (K), Aromatic (Ar), and HR2 regions, are indicated above the conservativity profile. (B) Locations of 12 helices and 5 β-structures based on the tertiary structure of P-450 cam and alignment in Fig. 1 are indicated by open and closed boxes, respectively. (C) Potential membrane-bound domains (M), substrate-binding regions (S), the halt-transfer signal (HT), proline/glycine cluster (PG) are indicated by boxes. Some substrate binding regions may also be bound to the membrane. The heme-chelating cysteine (C) and the three cationic residues (H) that interact with heme propionates are indicated by bars.

(1) Heme-binding (HR2) and proximal helix (Helix-L) regions

It is now well established that the fifth ligand to the heme of P-450 is a cysteine, Cys[497], located in strongly homologous region 2 (HR2) (Gotoh et al., 1983), about 50 residues upstream from the C-terminus. Identification of the heme-binding domain on determined amino acid sequences took a tortuous course, however.

From several lines of evidence, it was certain that the *b*-type heme of P-450 is bound to the thiolate group of a cysteine (Gunsalus and Sligar, 1978; White and Coon, 1982; Hahn et al., 1982). Based on similarity in amino acid compositions of heme-binding peptides (Dus et al., 1974) and that of a part of rat P-450b, Fujii-Kuriyama et al. (1982a) suspected that Cys(436) located near the C-terminus might be the heme-binding cysteine. The equivalent cysteine in rabbit P-450 lm2 was also assumed to be the heme-binding site by Heinemann and Ozols (1982) based on local sequence homology between P-450 lm2 (or P-450b) and P-450 cam. However, closer inspection (Gotoh et al., 1983) revealed that neither reasoning was conclusive because P-450b (or

P-450 lm2) and P-450 cam share two highly homologous regions containing a cysteine, HR1 and HR2, and because no region showed definitive compositional similarity to the heme-peptide isolated from the same protein species (Dus et al., 1974; Dus, 1980). No other regions around the six cysteines in P-450b and the eight cysteines in P-450 cam are significantly homologous, indicating that either HR1 or HR2 is the heme-binding site.

HR1 is located in the N-terminal half of P-450, while HR2 is the C-proximal region pointed out by Fujii-Kuriyama et al. (1982a) and Heinemann and Ozols (1982). HR2 is more tightly conserved than HR1 and provides the heme with more hydrophobic and spacious environment. Although these features suggested that HR2 was the better candidate for the heme-binding region (Gotoh et al., 1983), there was abundant discussion in favor of HR1 as the heme-binding domain (Akhrem et al., 1980; Black et al., 1982; Haniu et al., 1982c, 1983; Tarr et al., 1983; Poulos, 1984).

A significant advance toward the settlement of this problem was obtained from the complete sequence of rat P-450d deduced from a cDNA sequence by Kawajiri et al. (1984). On the basis of an optimal alignment between rat P-450d and P-450b, Kawajiri et al. showed that the cysteine in HR1 of P-450b is replaced by a histidine in P-450d, while the cysteine as well as its surrounding residues in HR2 is well conserved (Fig. 1). This led Kawajiri et al. to the conclusion that HR2 is most likely to be the heme binding domain. Essentially the same conclusion was drawn later from the sequences of rat P-450c (Sogawa et al., 1984) and mouse P-450 p1 and P-450 p3 (Kimura et al., 1984a, 1984b; Gonzalez et al., 1985a).

In spite of this evidence, some investigators persisted in favoring HR1 with the heme-binding domain relying on an alternative alignment between P-450b (or P-450 lm2) and P-450d (or P-450 lm4) (Fujita et al., 1984; Haniu et al., 1984; Adesnik and Atchison, 1986). However, the conclusion of HR2 being the heme-binding domain became undoubted when the sequence of a mitochondrial P-450, bovine P-450 scc, was determined by Morohashi et al. (1984). The mature form of this P-450 contains only two cysteines, both of which are located in the C-terminal half of the molecule. Similarly, Gonzalez et al. (1985b) found that rat P-450 pcn also lacks a cysteine residue in its N-proximal half. Morohashi et al. (1987a) recently found that the mature form of human P-450 scc contains only one cysteine, which is located in HR2 and ought to be the heme-binding residue. Needless to say, X-ray crystallography of P-450 cam (Poulos et al., 1985) has finally confirmed the conclusion drawn from sequence data.

Besides the heme-chelating cysteine Cys[497], Phe[490] and Gly[493] are strictly conserved throughout all known P-450s. Other well conserved positions include Gly/Ser[491], Arg/His[495], Gly/Ala/Asp[499], Gly/Ala[503], and Glu/Gln/Asn[506]. These residues must be important for maintaining a proper heme-binding environment.

The hydrophobic patch [507—518] which contains invariant Leu[515] lies at the C-terminus of Helix L. Although the hydrophobic feature is well

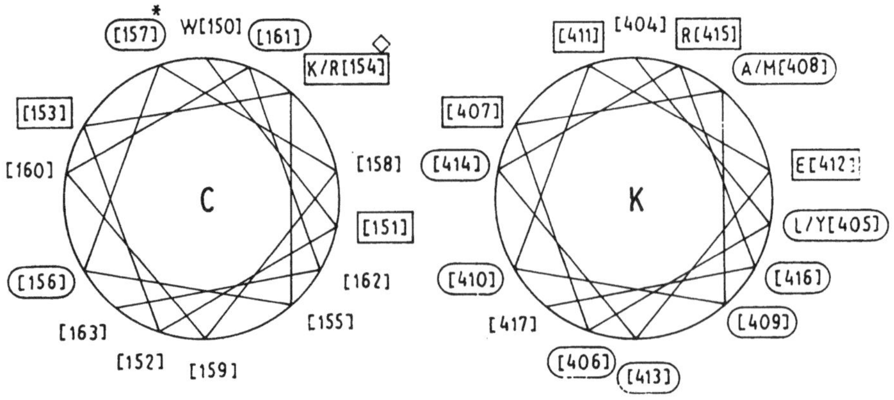

Fig. 5. Axial helical projection (SHIFFER and EDMUNDSON, 1967) of Helix C (left), and Helix K (right). Positions with single-letter amino acid codes are conserved throughout animal P-450s with few exceptions. A hydrophobic position (averaged hydropathy index of KYTE and DOOLITTLE (1982) is greater than 1.0) is enclosed in a box with round edges, while a hydrophilic position (averaged hydropathy index is less than −2.0) is enclosed in a rectangular box. The asterisk indicates the site where Ser in rabbit P-450 lm2 is phosphorylated (MÜLLER et al., 1985; PYERIN et al., 1986). Arg or Lys at [154] probably interacts with a propionate group of the heme (◇).

conserved (Fig. 3C), the C-terminal half of Helix L does not seem to face the heme plane. The functional importance of this portion remains to be elucidated.

(2) Helix-K region

The second best conserved region in all P-450 sequences coincides with Helix K in P-450 cam. This conserved region also overlaps with the "conserved tridecapeptide" found by OZOLS et al. (1981) in two members of PB-family P-450s. There are two invariable charged residues, Glu[412] and Arg[415]. Position [408] is Ala in all the P-450s but the members in PCN and FA families which have Met at this position. Either Leu or Tyr appears at position [405] of animal P-450s, while this residue is deleted in P-450 cam. Interestingly, the residues at the four well conserved positions lie on the same side of the helix as illustrated by a helical wheel (Fig. 5). The opposite side of the helix is predominantly occupied by hydrophobic residues, showing that the helix has amphiphilic nature. The helix (Helix K) in an animal P-450 is probably longer by one turn than that in P-450 cam as indicated in the alignment (Fig. 1).

(3) Aromatic region

The third well conserved region is located between HR2 and Helix-K region, and characterized by its peculiar amino acid sequence. An animal sequence at [460—471] takes the form, $A_1-X-X-P-X-X-A_2-X-P-X-B-A_3$, where A_i represents an amino acid with an aromatic side chain, P means Pro, B means either Arg or His, and X indicates a weakly conserved position. More precisely, A_1 is either Trp or Phe, A_2 is Phe in all animal P-450s but P-450 11β (which has Tyr at this position) and A_3 is Phe in all microsomal P-450s but Trp in all mitochondrial P-450s. Position B in rat P-450a is exceptionally Asn. P-450 cam shares only the first Pro and B (Arg) positions with animal P-450s.

Interestingly, Helix K and this aromatic region are spatially close to each other and located in the proximal side of the heme plane. Most probably, they cooperate in some important function common to all P-450 proteins. We propose that they may be involved in interaction with an electron-donor protein. The three charged residues, Glu[412], Arg[415] and Arg/His[470], may contribute to specific intermolecular ion-pairing and may also function as a channel of electron transfer. On the other hand, the three conserved aromatic residues can constitute a $\pi - \pi$ electron bridge (Poulos and Kraut, 1980), or take part in the interaction as signals for specific recognition. The divergence at A_2 and A_3 positions among microsomal P-450s, mitochondrial P-450s and P-450 cam may reflect the difference in the electron-donor proteins, cytochrome P-450 reductase for microsomal enzymes, adrenodoxin for mitochondrial enzymes, and putidaredoxin for P-450 cam. The above hypothesis will be testable by site-directed modification, mutagenesis or immunological methods.

(4) Basic residues that interact with heme propionates

In P-450 cam three basic residues, Arg(113), Arg(300) and His(356), form ion pairs with heme propionates. All these positively charged residues are conserved in other P-450s; Arg or Lys at position [154], His or Arg at [426], and Arg at [495]. This observation strongly supports the idea that animal P-450s have essentially the same tertiary structure as P-450 cam.

(5) Distal helix (Helix I) region

The heme of P-450 cam is sandwiched between two long helices, Helix L in the proximal side and Helix I in the distal side. Poulos and co-workers (Poulos et al., 1985; Poulos, 1986) were the first to suggest an important functional role of an apolar patch in the distal helix [346—362], in connection with contact to the heme surface and binding of the oxygen and substrate molecules. The best conserved position in this region is Thr[355]; only rat P-450 pcn1 has a Pro at this position. However, the significance of Pro[355] is questionable because rat P-450 pcn2, which is nearly 90% homologous to P-450 pcn1, has Thr at this position, and because a proline must break the α-helical conformation probably taken by all the other P-450s. We consider that Thr[355] is invariable and plays a crucial role in catalytic function. Thr[355] in P-450 cam is also essential for maintenance of the deformed helical structure with an unusual hydrogen-bonding pattern

(POULOS et al., 1985). The deep depression in helix propensities at the middle of Helix-I region (Fig. 3A) indicates that deformation in helix-I structure is also shared by animal P-450 proteins.

The residues at positions [351] and [352] are closest to the heme plane, and only small amino acids are allowed at these positions due to steric constraints. The constraints to [352] appear to be more severe than those to [351]; except for Ser in bovine P-450 11β, all P-450s have Gly at [352], while several amino acids are found at [351].

Position [354] is occupied by either Glu or Asp, except for chicken P-450 pb1 which has Gly at this position. By analogy to P-450 cam, the acidic side chain may form an internal ion pair with a basic residue located in the central highly variable region (Fig. 4A). The partner cationic residue must be conserved, which we took advantage of, to align P-450 cam with animal P-450s. Further details will be discussed later together with the residues responsible for substrate binding.

(6) Helix J region

Helix J region is rich in charged residues among which Glu at [381] is most remarkably conserved with a single conservative variation, Asp, in bovine P-450 17α. Positive charge at [377] and negative charge at [383] are also well conserved among microsomal P-450s. Like Helix K, Helix J of animal P-450s may be longer than that in P-450 cam by one or two turns (Fig. 1). Helix J is antiparallel to Helix K, and can be incorporated together with Helix K in the specific intermolecular interaction with a component of the electron-transfer system.

(7) Helix D (HR1) region

As mentioned previously, HR1 was once considered to be a favorite candidate for the heme-binding site. It is now clear that this region is conserved only marginally. Thus, the marked resemblance of P-450b and P-450 cam sequences in this region must be due to a fortuitous convergence. HR1 in P-450 cam overlaps with Helix D, which does not seem to be involved in catalytic function. Possibly, Helix D may be an important structural component in the protein architecture.

(8) N-terminal region

The N-terminal region in P-450 is concerned with cellular localization of the protein. Thus, microsomal, mitochondrial and soluble P-450s have individually distinct sequence organizations for initial 15 ~ 45 residues.

The N-terminal region of microsomal P-450s may be divided into five parts: the extreme N-terminal part, the extremely hydrophobic stretch, a cluster of basic residues, a cluster of prolines and glycines, and the second hydrophobic stretch. The extreme N-terminal part varies in size, and sometimes contains acidic residues but rarely contains basic residues. The following hydrophobic stretch is fairly constant in length (16 ~ 20 residues), and probably in helical conformation (Fig. 3A). There is little doubt that this hydrophobic stretch is inserted into ER membrane, because the N-terminal hydrophobic sequence of P-450 lm2 has been shown to be recognized by the signal recognition machinery (SAKAGUCHI, et al., 1984), similar to the case with most secretory proteins. However, it is not clear whether it penetrates the ER membrane or takes a fold-back structure, and in which side of the ER membrane the N-terminal residue resides.

The cluster of basic amino acids that follows the hydrophobic stretch is considered to function as a halt-transfer signal (SABATINI et al., 1982), and discriminates microsomal P-450s from secretory proteins. The positive charges can also interact with carboxyl groups of membrane lipid to stabilize the protein body on the membrane.

The proline/glycine cluster first noted by BLACK et al. (1982) is present in all P-450s including microsomal, mitochondrial and bacterial proteins. Thus, we can speculate that the higher-order structure common to all P-450 proteins starts from this region. In a microsomal protein, the proline/glycine cluster may link the N-terminal membrane-bound segment to the globular portion of the protein.

The second hydrophobic stretch is short and contains a few positively charged residues. It is unlikely that this portion penetrates the microsomal membrane but it may dip into it. It is noteworthy that there is no negatively charged residue between or within the first and the second hydrophobic stretches in microsomal P-450s.

Mitochondrial P-450s, as other mitochondrial components, are synthesized as larger precursors in the cytoplasm and then translocated to mitochondria. The N-terminal ~40 residue chain of a precursor is cleaved off after its tagging function has been completed. The extra peptides of P-450 scc and P-450 11β have diverged considerably from each other. Hence their tagging function does not require strong sequence homology. The secondary or tertiary structure should be essential for the tagging function, as suggested for other mitochondrial proteins (ROISE et al., 1986; VON HEIJNE, 1986).

(9) Other conserved positions

Two well conserved glycines are found in animal P-450s between Helix A and Helix B regions. The sequence alignment (Fig. 1) suggests that Gly[95] lies at the turn position of β1. Gly/Pro[86] is also a component of a turn located between Helix A and β1. Neither position [86] nor [95] is conserved in P-450 cam.

All animal P-450s except for rat P-450a share Trp at [150]. P-450a has, instead, an Ala at this position, and P-450 cam has a Gln from which Helix C starts. This Gln makes a hydrogen bond with Arg(113). As mentioned above, its side chain interacts with a heme propionate and the corresponding position

[154] is well conserved in all P-450 s. Therefore, the conserved Trp[150] must be close to the heme (Fig. 5). The indole ring of the Trp may stack on the heme plane to stabilize protein-to-heme interaction.

4.3. Substrate binding sites

Comparing temperature factors of substrate-bound and -free crystals of P-450 cam, Poulos et al. (1986) found that substrate binding decreases the flexibility in three separate regions of the protein; One of the three regions is located between Helix B and Helix C, spanning from [121] to [137]. In P-450cam, Phe(88) and Tyr(97) directly contact the camphor molecule. Position [128] corresponding to Phe(88) is exclusively occupied by hydrophobic amino acids, whereas position [137] corresponding to Tyr(97) is more variable. This is not surprising because the hydrogen bond between Tyr(97) and the camphor is specific to P-450 cam. The high variability in this region (Fig. 4A) partly accounts for the different substrate specificities of different enzymes.

The region spanning from [260] to [300] corresponds to the P-450 cam segment of which the motion is most strongly affected by substrate binding. Thr(186) in the loop that joins Helix F and Helix G is closest to the substrate camphor. Sequence alignment indicates that the region showing the greatest flexibility shift upon camphor binding corresponds to the highly variable domain near the center of animal P-450 sequences (Fig. 4A). This domain includes P-450 scc sequence (210—221) which we previously proposed to be a substrate-binding site based on sequence similarity to a prostatic steroid-binding protein (Gotoh et al., 1985). It is very likely, therefore, that this highly variable domain provides animal P-450s with the major substrate-binding capacity and/or the channel for access of a substrate and release of the product.

An experimental observation supporting the above idea has recently been obtained by Onoda et al. (1987). They identified the residues of a porcine testis P-450 (probably the same as or closely related to P-450 17α) which were labelled by a reactive substrate analogue. Binding to one of the two target cysteine residues is competitively inhibited by the presence of the authentic substrate, implying that the competitively masked cysteine, Cys(235), contacts the substrate. Although Onoda et al. proposed that Cys(235) is located in HR1 (Helix D region), our alignment shows that it is at position [278] in the highly variable region under consideration.

The high variability in sequence is consistent with the extraordinarily wide range of substrate specificities exhibited by the multiple forms of P-450. For the same reason, however, it is very difficult to establish sequence alignment on the basis of structural and functional equivalence. The X-ray structure of P-450 cam indicates an internal ion pair between Arg(187) and Asp(252). Asp(252) is in the distal helix region, and lies at the center of the third region with significant substrate-induced flexibility shift (Poulos et al., 1986). The corresponding position [354] is occupied predominantly by either Asp or Glu. Hence it is likely that an ion pair between the negative charge at [354] and a positive charge in the highly variable region exists also in animal P-450s. The best candidate for the partner is Lys or Arg at [276] in MC, PB, C21 and 17α family P-450s. The alignment shown in Fig. 1 was obtained on the assumption that Arg(187) of P-450 cam matches these cationic residues at [276]. If this is true, Helix F and possibly Helix G in these P-450s would be shorter than those in P-450 cam. The partner cationic residue in PCN, FA, SCC and 11β families has not been identified. Further investigations are needed to provide a convincing image about the topology of substrate-binding sites of individual P-450 proteins.

Another residue in P-450 cam, Val(296), intimately contacts the camphor molecule. Only apolar residues are found at the corresponding position [420] in animal P-450s, suggesting that the residues at this and surrounding positions contribute to substrate binding. Moreover, a short stretch of variable regions found in closely related forms of rat P-450b and P-450e (FUJII-KURIYAMA et al., 1982a; YUAN et al., 1983) covers this residue, consistent with a small but definite difference in catalytic activity of the two enzymes. However, the freedom of variation at this and surrounding positions is rather limited. As opposed to the postulate of POULOS et al. (1985), we think that it is not this region but the highly variable region mentioned above that primarily controls substrate specificity.

The fifth region possibly responsible for substrate binding overlaps with β5 region. Ile(396) in P-450 cam contacts the substrate, and forms a part of the potential substrate entrance channel (POULOS et al., 1985). Like other possible substrate-binding regions, this region is also variable in sequence (Fig. 4A). The locations of the five separate putative substrate binding regions discussed above are illustrated in Fig. 4C.

4.4. Sites of modification

(1) Phosphorylation

MÜLLER et al. (1985) and PYERIN et al. (1986) found that purified or microsomal rabbit P-450 lm2 is phosphorylated at a single site, Ser(128), in the presence of exogenously added cAMP-dependent protein kinase and ATP. The corresponding position [157] is in Helix C region and downstream by three residues from the conserved Arg/Lys[154] (Fig. 5). As mentioned previously, Arg/Lys[154] forms an ion pair with a heme propionate. It is reasonable to postulate that phosphorylation at Ser[157] alters the electrostatic field nearby, weakens the interaction between the heme and protein, and finally causes the conversion from P-450 to P-420 (TANIGUCHI et al., 1985).

(2) Chemical modification

Active site-directed modification of P-450 has been undertaken by several groups (GIBSON and TAMBURINI, 1986; JOHNSON et al., 1986; FREY et al., 1986;

Onoda et al., 1987). Except for the work of Onoda et al. introduced previously, identification of the active site or substrate-binding site has not been accomplished by this approach.

Bernhardt et al. (1984) reported that selective modification of a lysine residue in rabbit P-450 lm2 by fluorescein isothiocyanate impaired electron transfer from the cytochrome P-450 reductase. The implied position, Lys[442], is located just between $\beta 4$ and the returning part of $\beta 3$. Lys[442] is conserved in most P-450s but human and bovine P-450 c21 have an opposite charge, Glu, at this position. No charged residue is found in the immediate vicinity of this position in bovine P-450 11β. Thus it is not certain whether Lys[442] directly contributes to an intermolecular electrostatic interaction.

Parkinson et al. (1986a, 1986b) recently showed that an alkylating agent, 2-bromo-4'-nitroacetophenone (BrNAP) covalently binds to rat P-450 c leading to substantial inhibition of the catalytic activity. Binding to other forms of rat microsomal P-450 (a — j) did not significantly affect the enzymatic activity. The major site of modification in P-450c, Cys(293), is unique to c-group P-450s, and other forms lack the corresponding residue (Fig. 1). Cys(293) = [319] is located within Helix H or $\beta 3$, but this location does not provide a hint as to the mechanism by which BrNAP binding uncouples oxygen consumption from product formation.

4.5. Membrane topology

Based on the amino acid sequences of rabbit P-450 lm2 and P-450 lm3b, Ozols and his associates (Heinemann and Ozols, 1982; Ozols et al., 1985) and Tarr et al. (1983) proposed several models of membrane binding topology for these proteins. These models show eight or ten trans-membrane segments with relatively small polar segments exposed on both the cytoplasmic and luminal faces of the ER membrane. The accumulated sequence data and recent experimental findings are not consistent with these models with multiple transmembrane segments.

First of all, the averaged hydropathy profile (Fig. 3C) shows few hydrophobic stretches long enough to span the membrane, except for that near N-terminal. Second, it is very unlikely that a membranous protein would take essentially the same tertiary structure as a soluble counterpart (P-450 cam) under totally different environmental conditions, either in lipid or in water, while we are convinced of the universality of P-450 tertiary structure. Finally, immunochemical studies by Matsuura et al. (1979, 1981, 1983) with antibodies raised against rat P-450b clearly showed the absence of antigenic sites on the luminal surface of the ER membrane, while extensive binding of antibodies to the cytoplasmic surface was observed. Recent study by De Lemos-Chiarandini et al. (1987) with site-specific antibodies has reinforced the observation of Matsuura et al. and provided further information about the membrane topology. Eight of 15 discrete segments in rat P-450b examined were accessible from the cytoplasmic side by the specific antibodies raised against corresponding synthetic oligopeptides, whereas other seven segments were not or only weakly recognized. Four of the seven weakly recognized segments (Nos. 5, 8, 10, and 12) are mapped within hypothetical substrate-binding regions discussed above, and also match with hydrophobic regions. These observations might imply that substrate-binding sites (or entrance channel) face to (or are partially buried in) the membrane bilayer so that a substrate enters from the lipid phase (KOMINAMI et al., 1986).

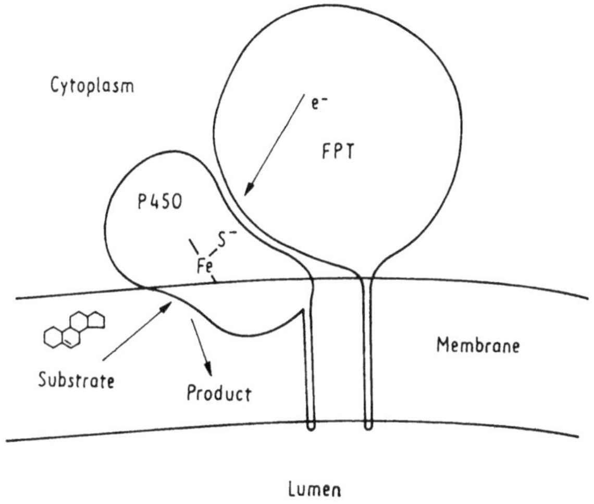

Fig. 6. A schematic model of microsomal P-450 (P-450) and cytochrome P-450 reductase (FPT) on ER membrane. The rationale behind this model is discussed in the text.

The picture shown in Figure 6 is drawn on the basis of the following findings and assumptions. (1) The bulk of cytochrome P-450 reductase including the active center is exposed on the cytoplasmic face of the ER membrane with a small portion of the protein inserted in the membrane (MASTERS and OKITA, 1980; BLACK and COON, 1982). (2) Only an N-terminal segment of a P-450 protein spans the membrane bilayer. (3) The P-450 region(s) that mediates electron transfer from the reductase is located in the proximal side of the heme plane, and faces the cytoplasmic side of the membrane to interact with the reductase. (4) The substrate binding site(s) (or entrance channel) faces the membrane so that access of a substrate and release of the product occur in the lipid phase (KOMINAMI et al., 1986). (5) The heme plane is tilted by 55° from the membrane (GUT et al., 1983). The rough model depicted in Figure 6 can be a base for future refinement.

4.6. Summary of P-450 structure

What we wish to primarily emphasize is the universality rather than the individuality in higher-order structure of all P-450 proteins. As illustrated in

Figure 4, more than 10 positions critical for structural maintenance are strictly conserved or variable only to conservative residues. The good fit of the locations of predicted helices and β-structure in animal P-450s to those expected from the structure of P-450 cam (Fig. 3) is further strong evidence of common secondary and tertiary structures of animal and bacterial P-450s. These findings imply that the difference in cellular localization and the difference in electron transfer system do not involve large-scale structural reorganization in divergent family of P-450s.

In contrast to the constancy in the basic higher-order structure, the five putative substrate-binding domains (Fig. 4C) are highly variable in sequence (Fig. 4A). It will be an important subject of future studies to disclose the role of each residue in these domains in determining the substrate specificities of individual enzymes.

5. Gene structure of P-450

5.1. Gene numbers

Cot analysis and Southern blot analysis of rat total genomic DNA have established 6 to 11 gene sequences which cross-hybridize with P-450b cDNA sequence (Mizukami et al., 1983a; Atchison and Adesnik, 1983; Leighton et al., 1984), although it is not known whether all of the gene sequences are functional or not. The cDNAs for rat P-450 pb1, P-450 f (Gonzalez et al., 1986a), P-450 m1 (Yoshioka et al., 1987), and P-450j (Song et al., 1986) do not cross-hybridize with P-450b under usual hybridization conditions, although they belong to the same gene family (Table 2). Each of these cDNAs detected one or a few genes which hybridized with the respective cDNA in Southern blot analysis. Accordingly, the gene family including P-450b, e, m1, f, pb1 and j may be estimated to have 20 (or even more) family members.

The gene family of MC-inducible P-450c and d appears to be composed of the two members on the basis of the analysis of the isolated genomic clones and Southern blot of the genomic DNA (Sogawa et al., 1984, 1985; Gonzalez et al., 1985a). Because of the simple genomic blotting pattern even under relaxed hybridization conditions, the MC-inducible P-450 family may contain few, if any, additional cross-hybridizing members.

Southern blot analysis with a PCN-family cDNA suggested that there are three to five gene or gene-like sequences in rat and human that anneal with the cDNA (Hardwick et al., 1983; Molowa et al., 1986; Beaune et al., 1986). Hardwick et al. (1987) recently suggested the presence of two or three genes (or pseudogenes) related to P-450 laω in the rat genome. One of the genes may code for the enzyme equivalent to rabbit P-450 pgω (Matsubara et al., 1987).

The numbers of genes in C21, 17α, SCC and 11β families are small. There are two C21 genes in a human or mouse genome, one of which is known to be a pseudogene (Higashi et al., 1986; White et al., 1986; Chaplin et al., 1986).

Pigs may have two distinct enzymes related to P-450 17α (NAKAJIN et al., 1984), and humans also have two or more 17α-family genes (MATTESON et al., 1986). Human P-450 scc gene is possibly a single copy gene (MOROHASHI et al., 1987a), whereas two or more members may exist in the 11β gene family (MOROHASHI et al., 1987b).

Judging from all the data so far obtained, it is plausible to speculate that the total number of P-450 superfamily genes in a mammal could easily go over 50.

5.2. Gene structure

The gene structures of 11 species of cytochrome P-450 have been elucidated. These include five genes in MC family (SOGAWA et al., 1984, 1985; GONZALEZ et al., 1985a; JAISWAL et al., 1985b; HINES et al., 1985; KAWAJIRI et al., 1986), three genes in the PB family, P-450b and P-450e (ATCHISON and ADESNIK, 1983; MIZUKAMI et al., 1983b; SUWA et al., 1985) and P-450 pc2 (GOVIND et al., 1986), two genes in the C21 family (HIGASHI et al., 1986; WHITE et al., 1986; CHAPLIN et al., 1986), and the gene of human P-450 scc (MOROHASHI et al., 1987a). The outlines of these gene structures are summarized in Fig. 7. All these genes show split gene structures of seven exons for MC family genes, nine exons for PB family genes, ten exons for C21 family genes and nine exons for human P-450 scc.

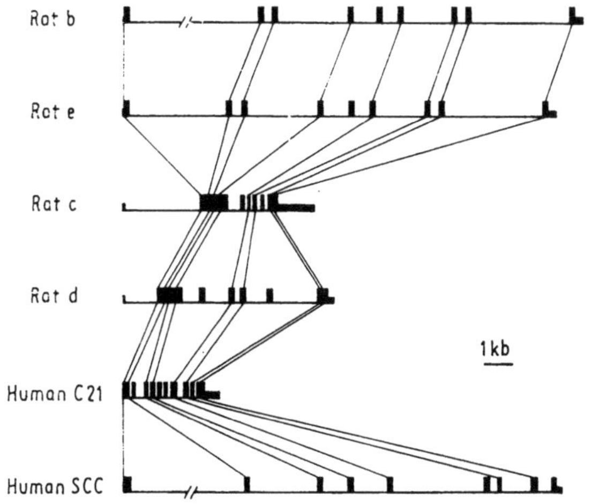

Fig. 7. Outline of the exon/intron structure of six P-450 genes. Thin lines are drawn between corresponding codons in different genes for several conserved residues. The length of the first intron of human P-450 scc gene is not known.

All the exon/intron junctions but one follow the canonical GT/AG rules. The sequence of the sixth intron of human P-450 scc gene begins unusually with GC at the 5' end. Only a few genes have so far been reported to possess this unusual GC sequence at the 5' end of the intron boundary (MOROHASHI et al., 1987a). Table 5 summarizes the boundary sequences of this particular type of the splicing junction. The boundary sequence with the unusual GC sequence at

Table 5. Genes with unusual boundary sequence (GC) at the 5' end of an intron

Species	Gene	Intron	Exon/Intron	Reference
Human	P-450 scc	6	CTAAG/GCAAGCC	MOROHASHI et al. (1987a)
Chicken	α-Globin (α^D)	2	TCAAG/GCAAGCA	DODGSON and ENGEL (1983)
Mouse	α-A_2 Crystallin	2	TCAAG/GCAAGTT	KING and PIATIGORSKY (1983)
Human	Cu, Zn-Superoxide dismutase	1	AGAAG/GCAAGGG	LEVANON et al. (1985)
Human	MAR[1] γ Subunit	9	GCAAG/GCAAGGA	SHIBAHARA et al. (1985)
	Consensus		– AAG/GCAAG –	

[1] Muscle acetylcholine receptor.

the 5' end of the intron has a longer conserved sequence, AAG/GCAAG, as pointed out by SHIBAHARA et al. (1985) than the canonical junctional sequence. In this sequence, the first three nucleotides belong to the exon and the remaining five at the 3' side are included in the intron. A physiological function of this unusual junctional sequence has not yet been identified.

The exon/intron arrangements of P-450 genes so far elucidated are conserved within each family; no discrepancy in the positions of introns in relation to the amino acid sequences has been found within PB, MC or C21 family genes. When the gene structures are compared between families, however, the insertion sites of introns in relation to amino acid sequence are very different (Fig. 8). It seems difficult to explain these variations in the gene structure of the four P-450 families only by the assumption of the random deletion of introns from an ancestral gene which contained all the introns in the four families, because too few introns are conserved for their insertion site among these four families.

This situation is not limited to the P-450 superfamily, but extended to other gene families including actin (GALLWITZ and SURES, 1980; NG and ABELSON, 1980; FIRTEL et al., 1979; FYRBERG et al., 1981; DURICA et al., 1980), myosin (KARN et al., 1983), serine protease (ROGERS, 1985), and ovalbumin-antitrypsin superfamilies (LEICHT et al., 1982; TANAKA et al., 1984). As suggested for the gene families of actin and serine protease, another possible mechanism for the generation of these variations in gene structure, i. e., the addition of introns after divergence of P-450 families should be considered (ROGERS, 1985; CAVALIER-SMITH, 1985). It could be that some introns are

```
  1 MPSVYGFPAF TS--ATELLL AVTTFCLGFW VVRVTRTWVP KGLKSPP-GP WGLPF-IGH- -VLTL----G KN-PHLSLTK LSQQYGDVLQ IRI-GSTPVV VLSGLNTIKQ AL--VKQ-GD  105
  1 M---EP---- ------SILL LLALLVGFLL LLVRGHPK-S RG-NFPP-GP RPLPL-LGN- -LLQL---DR GG-LLNSFMQ LREKYGDVFT VHL-GPRPVV MLCGTDTIKE AL--VGQ-AE   93
  1 M--------- --------LL LG-LLLLPLL A---GARL-L WNWWKLR-SL HLPPL-APG- -FLHL---LQ PD-LPIYLLG LTQKFGPIYR LHL-GLQDVV VLNSKRTIEE AM--VKK-WA   86
  1 MLAKGLPPRS VLVKGCQTFL SAPREGLGRL RVPTGEGAGI STRSPRPFNE IPSPGDNGWL NLYHFWRETG THKVHLHHVQ NFQKYGPIYR EKL-GNVESV YVIDPEDVAL LFKSEGPNPE  119

106 DFKGRPDLYS FTLIANGQSM TFNPDSGPLW AARRRLAQNA LKSFSIASDP TLASSCYLEE HVSKEAEYLI SKFQKLMAEV GHFDP-FKYL VVSVANVICA ICFGRRYDHD DQEL-LSIVN  223
 94 DFSGRGTIAV IEPIFKEYGV IFA--NGERW KALRRFSLAT MRDFGMGKRS -------VEE RIQEEAQCLV EELRK--SQG APLDP-TFLF QCITANIICS IVFGERFDYT DRQF-LRLLE  200
 87 DFAGRPEPLT YKLVSKNYPD LSLGDYSLLW KAHKKLTRSA L-LLGI-RDS -------MEP VVEQLTQEFC ERMRA--QPG TPVAI-EEEF SLLTCSIICY LTFGDKIK-D DNLM-PAYYK  192
120 RFLIPPWVAY HQYYQRPIGV LLK--KSAAW KKDRVALNQE VMAPEATKNF LP-----LLD AVSRDFVSVL HRRIKKAGSG NYSGDISDDL FRFAFESITN VIFGERQGML EEVV-NPEAQ  231

224 LSNEFGEVTG --SGYPADFI P-ILRYLPNS SLDAFKDLNK KFYS--FM-- KKLIKEHYRT FEKGHI--RD ITDSLIEHCQ DRRLDENANV Q----LSDDK VITIVFDLFG AGFDTITTAI  330
201 LFYRTFSLLS SFSSQVFEFF SGFLKYFPGA HRQISKNLQE -ILD--YI-- GHIVEKHRAT LDPSAP--RD FIDTYLL--R MEKEKSNHHT E----FHHEN LMISLLSLFF AGTETSSTTL  307
193 CIQEVLKTWS HWSIQIVDVI P-FLRFFPNP GLRRLKQAIE KRDH--IV-- EMQLRQHKES LVAGQW--RD MMDYMLQG-V AQPSMEEGSG Q----LLEGH VHMAAVDLLI GGTETTANTL  300
232 RFIDAIYQMF HTSVPMLNLP PDLFRLFRTK TWKDHVAAWD VILS--KA-- DIYTQNFYWE LRQKGSVHHD YRGILYR--L LGDSK----- -----MSFED IKANVTEMLA GGVDTTSMTL  335

331 SWSLMYLVTN PRIQR-KIQE ELDTVIGRD- --RQPRLSDR PQLPYLEAFI LETFRHSSFV PFTIPHSTIR DTSL-NGFYI PKGHCVFVNQ WQVNHDQELW GDPNEFRPER FLTSSGT-LD  444
308 RYGFLLMLKY PHVAE-KVQK EIDQVIGSH- --RLPTLDDR SKMPYTDAVI HEIQRFSDLV PIGVPHRVTK DTMF-RGYLL PKNTEVYPIL SSALHDPQYF DHPDSFNPEH FLDANGA-L-  420
301 SWAVVFLLHH PEIQQ-RLQE ELDHELGPGA SSSRVPYKDR ARLPLLNATI AEVLRLRPVV PLALPHRTTR PSSI-SGYDI PEGTVIIPNL QGAHLDETVW ERPHEFWPDR FLEP------  412
336 QWHLYEMARN LKVQD-MLRA EVLAARHQA- --QGDMATML QLVPLLKASI KETLRLHP-I SVTLQRYLVN DLVL-RDYMI PAKTLVQVAI YALGREPTFF FDPENFDPTR WLSKDK----  445

445 KHLSEKVILF GLGKRKCIGE TIGRLEVFLF LAILLQQMEF NVSPGEK-VD MTPAY--GLT LKHARCEHFQ VQMRSSGPQH LQA---       Rat    P450c       524
421 KK-SEAFMPF STGKRICLGE GIARNELFLF FTTILQNFSV SSHLAPKDID LTPKES-GIG KIPPTYQICF SAR------- ------     Rat    P450b       491
413 GK-NSRALAF GCGARVCLGE PLARLELFVV LTRLLQAFTL LPSGD-ALPS LQPLPHCSVI LKMQPFQVRL QPRGM--GAH SPGQNQ     Human  P450c21b    494
446 NITYFRNLGF GWGVRQCLGR RIAELEMTIF LINMLENFRV EIQHLSDVGT TFNLI--LMP EKPISFTFWP FNQEATQQ-- ------     Human  P450scc     521
```

Fig. 8. Locations of introns along four P-450 protein sequences extracted from Figure 1. The sites of introns are indicated by triangles with intron numbers: ▽, intron lies between the codons for flanking amino acids; ▼, intron lies between the first and the second nucleotides of the codon for the following amino acid; ▼, intron lies between the second and the third nucleotides of the codon for the preceding amino acid. The first intron of rat P-450c gene is localized 14 bp upstream from the initiation codon.

vestiges of transposon-like sequences that have been inserted into genes, become fixed and then diverged in nucleotide sequence. In the case of P-450 genes, however, we have not yet found DNA sequences reminiscent of the insertion of transposon-like sequences in the introns. The DNA sequences in the introns may have diverged so rapidly after insertion into the preexisting gene that their characteristic features cannot be recognized.

5.3. Chromosomal localization

Chromosomal localization of P-450 genes is now in progress in several laboratories by in situ hybridization techniques and Southern blot analysis of hybrid cell DNAs. So far several P-450 genes have been determined for their chromosomal localizations in human and mouse genomes (Wolf, 1986; Nebert and Gonzalez, 1987). In view of the evolutionary distance between these various families of P-450, it may not be surprising that their genes are widely dispersed among chromosomes. In mouse, PB-inducible P-450 (equivalent to rat P-450b and P-450e) has been reported to locate on chromosome 7 (Simmons and Kasper, 1983), while MC-inducible P-450 (equivalent to rat P-450c and P-450d) is found on chromosome 9 (Tukey et al., 1984). P-450 pcn and steroidogenic P-450 c21 are on chromosome 6 (Simmons et al., 1984) and 17 (White et al., 1984), respectively.

6. Regulation of expression of cytochrome P-450 genes

Most of P-450 species are known to be regulated for their gene expression by endogenous or exogenous stimuli and show temporal and tissue-specific expressions. Exogenous substances such as PB, 3-MC, 2,3,7,8-tetrachlorodibenzo-*p*-dioxin (TCDD), PCN, ethanol, polychlorinated biphenyl (PCB), isosafrole and others are known to be inducers of their specific forms of cytochrome P-450, while ACTH (adrenocorticotropin), FSH (follicle-stimulating hormone), LH (luteinizing growth hormone), and many steroids regulate the synthesis of other forms of P-450s as endogenous stimuli (Adesnik and Atchison, 1986; Whitlock, 1986; Waterman et al., 1986).

In general, it has been clarified from the studies using translational and transcriptional inhibitors and *in vitro* translational systems directed by the isolated mRNA that the increased levels of various forms of P-450 result from increased rates of de novo synthesis of the P-450 apoproteins which can be accounted for by the increase in the corresponding mRNA within cells treated with inducers. Evidence has been accumulating that the increase in the P-450 mRNA is primarily a consequence of transcriptional activation of the corresponding genes. The availability of hybridization probes to measure precisely the mRNA content and transcription rate makes it possible to clarify the level at which regulation is effected, and facilitate accurate and direct measurement of the changes in levels of gene expression. These observations have established the inducible nature of cytochrome P-450s to open the way toward studies of the induction mechanism at a molecular level and have been extensively documented in the excellent review articles published recently (ADESNIK and ATCHISON, 1986; WHITLOCK, 1986; WATERMAN et al., 1986; EISEN, 1986). Accordingly, we will focus mainly on the molecular biological studies using isolated genes and gene transfer methods to provide deeper insights into the molecular aspects of the induction mechanisms in this section. Several genes of the P-450 superfamily have been isolated and analysed for their structures as described in the previous section, but only a few successful examples have been reported in which the P-450 genes or their derivative genes transfected into the cultured cell lines expressed their products in response to the inducers faithfully mimicking the expression of the endogenous corresponding genes in their specific tissue. These are genes for drug-metabolizing mouse P-450 p1 or the rat equivalent P-450c which is induced in liver by lipophilic xenobiotics such as 3-MC, TCDD and PCB and for steroidogenic P-450 c21 whose expression is positively regulated in adrenal cortex by ACTH.

6.1. Polycyclic aromatic hydrocarbon inducible P-450 gene

Several lines of evidence have suggested that the induction of this group of cytochrome P-450s by TCDD or 3-MC is mediated by the receptor for these chemicals which is believed to be a product of the Ah locus (NEBERT et al., 1972; THOMAS et al., 1972). The inducers taken up in the cells are first recognized by the receptor to form the receptor-inducer complex. It has been proved that the receptor associated with the inducer, in turn, moves to the nucleus to activate the transcription of rat P-450c gene or mouse equivalent (P-450 p1) probably by interacting with a regulatory region of the gene (POLAND and GLOVER, 1973; POLAND et al., 1974). The presence of the *cis*-acting regulatory sequence in response to TCDD or 3-MC has been reported by JONES et al. (1985, 1986a, 1986b), GONZALEZ and NEBERT (1985), and FUJISAWA-SEHARA et al. (1986) in the upstream region of a high-activity variant P-450 p1 gene, normal P-450 p1 gene and rat P-450c gene, respectively. All these groups constructed chimeric genes by ligating the 5' upstream regions of their respective genes to the chloramphenicol acetyltransferase (CAT) structural gene and then introduced them into Hepa-1 cells for investigating the inducibility of the CAT activity. The transfection experiments showed that these fusion genes expressed the CAT activity in response to TCDD or 3-MC. The transient expression system of the CAT activity was shown to mimic faithfully the induction phenomena of P-450c gene in rat livers (FUJISAWA-SEHARA et al., 1986), by all criteria tested such as inducer-specificity, cell-specificity, inducibility (ratio of the induced to the non-induced activity) and correct transcription site determined by S1 nuclease mapping. Essentially similar results have been obtained with transformants of Hepa-1 cells cotransfected with P-450c fusion gene and pSV2-neo. Because of the apparently aberrant construction of the P-450 fusion gene reported by JONES et al., (1985, 1986a), the transcription in their constructs seemed to use a cryptic start site in the middle of the first intron in response to TCDD.

The deletion experiments revealed that the upstream regulatory regions up to −6.3 kb contain at least three functional domains; −44 to −200 bp region immediately upstream of the TATA box functions in the basal level of transcription and the other two which are located in the sequences from −0.8 to −1.1 kb and from −1.1 to −6.3 kb enhance, in combination, the transcription in response to inducers (FUJISAWA-SEHARA et al., 1986; SOGAWA et al., 1986). Besides these functional regions, Jones et al. (1985) have suggested that the sequence from −50 to −350 bp (−700 to −1000 bp in their numbering system) is an inhibitory domain which presumably interacts with the labile repressor.

Further subdivision of the sequence from −0.8 to −1.1 kb which is the most important for the inducibility-conferring activity revealed that two homologous core sequences of 15 base pairs are tandemly arranged in the inverted orientation with each other in the sequences from −1007 to −1021 bp and from −1088 to −1092 bp in rat gene (FUJISAWA-SEHARA et al., 1987). The homologous core sequences are also found in the corresponding positions in the mouse JONES et al., 1986a; GONZALEZ et al., 1985a) and human genes (KAWAJIRI et

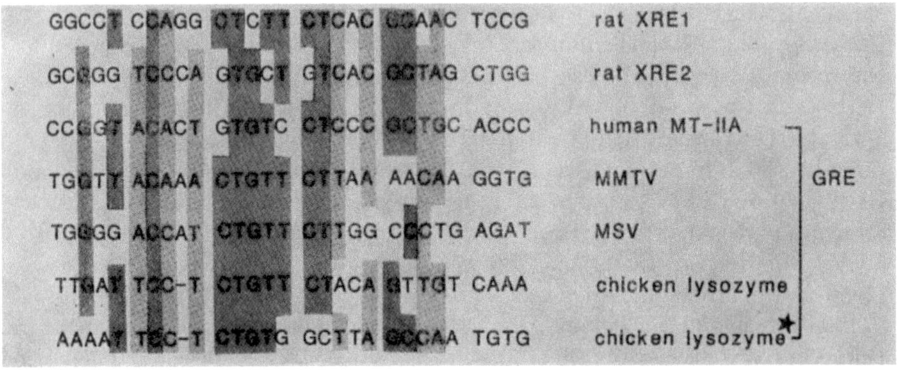

Fig. 9. DNA sequences of XREs and several GREs. DNA sequences of human metallothionein IIA (MT-IIA) gene from −264 to −236 (KARIN et al., 1984), MMTV DNA from −186 to −158 (PAYVAR et al., 1983; SCHEIDEREIT and BEATO, 1984), MSV DNA from −175 to −203 (MIKSICEK et al., 1986; DEFRANCO and YAMAMOTO, 1986), and the sequences of chicken lysozyme gene from −50 to −77 and from −191 to −164 (RENKAWITZ et al., 1984) were compared with XRE1 and XRE2. The distal GRE in chicken lysozyme (marked by an asterisk) is also recognized by progesterone receptor (VON DER AHE et al., 1986). Thickly stippled lanes indicate the positions in which a single nucleotide occupies 5 or more positions out of 7, and lightly stippled lanes indicate the positions in which either of the two alternative nucleotides occupies 6 or more positions.

al., 1986). These two core sequences express the CAT activity in response to 3-MC, even when they are placed separately on the heterologous promoter of SV40 early gene without the enhancer sequence which is fused to the CAT gene (SOGAWA et al., 1986; FUJISAWA-SEHARA et al., 1987). Tandem arrangement of the sequences was found to enhance remarkably the expression of the subordinate gene. FUJISAWA-SEHARA et al. (1987) have found that the core sequence which is designated xenobiotic responsive element or XRE shows a significant homology with the sequence of glucocorticoid responsive elements (GRE) (KARIN et al., 1984) of MMTV, human metallothionein and lysozyme genes as shown in Figure 9. Although XRE1 and 2 have 1 and 2 base mismatches with the consensus sequence for the GRE (CN TGTT/CCT), respectively, the sequence similarity appears to extend to the flanking sequences on the both sides of the consensus. It has been recently reported that a trans-acting factor for 3-MC (or TCDD) receptor which is believed to interact with XRE sequences has many physicochemical and biochemical properties in common with glucocorticoid receptor (OKEY et al., 1980; WILHELMSSON et al., 1986; POLAND et al., 1986; WHITLOCK and GALEAZZI, 1984; CUTHILL et al., 1987). Considered together, sequence homology between XREs and GREs leads us to a notion that the inducible system of XRE and 3-MC (or TCDD) receptor is evolutionarily related to that of GRE and glucocorticoid receptor. This notion will be finally tested by cloning the xenobiotic (3-MC or TCDD) receptor gene.

6.2. Steroidogenic P-450 c21 gene

A unique set of P-450s are localized mainly in adrenal cortex, testis and ovary. The activities of these P-450s are regulated by peptide hormones which are targeted to these organs, that is, ACTH for adrenal cortex, and FSH and LH for ovary and testis (ADESNIK and ATCHISON, 1986; WATERMAN et al., 1986). P-450 c21 is among them. Treatment of primary culture cells of adrenal cortex with ACTH causes approximately 15 fold increase in the synthesis of P-450 c21 within 36 hr (FUNKENSTEIN et al., 1983), together with enhanced synthesis of other P-450s including P-450 scc (DUBOIS et al., 1981), P-450 11β (KRAMER et al., 1983), and P-450 17α (MCCARTHY et al., 1983). The effect of ACTH is known to be mediated by cAMP (WATERMAN et al., 1986). *In vitro* translation assay indicated that the enhanced rates of synthesis of these P-450s correlated with increases in the level of the corresponding mRNA (FUNKENSTEIN et al., 1983; DUBOIS et al., 1981; KRAMER et al., 1983; MCCARTHY et al., 1983). These results are now being confirmed by RNA-DNA hybridization experiments using the cloned cDNAs.

In order to investigate *cis*-acting DNA element responsive to ACTH treatment, PARKER et al. (1985) introduced cosmid DNAs bearing either the mouse P-450 c21A (active gene) or B (pseudogene) into Y1 adrenocortical tumor cells, together with pSV2-neo to isolate transformant cells. Of 29 transfectants of P-450 c21A gene screened, 14 were shown to express P-450 c21 activity, whereas no expression of the activity was observed with transfectants of P-450 c21B gene. Of the 14 positive clones, 8 were strongly responsive to ACTH for the expression of P-450 c21 activity with varying levels of the basal activity. RNA blot analysis and S1 nuclease mapping clearly demonstrated that the elevated P-450 c21 activity by ACTH was mediated by the increase in the corresponding mRNA content. The external deletion of the 5' flanking region of the gene and subsequent transfection experiments revealed that an essential regulatory element for P-450 c21 expression is located between −230 and −180 bp upstream of the transcription start site (PARKER et al., 1986). The equivalent sequence of this element of the murine P-450 c21 gene was also found at the corresponding positions of −241 to −201 bp and −251 to −211 bp of the human (HIGASHI et al., 1986) and bovine (CHUNG et al., 1986) genes, respectively, as shown in Figure 10. The striking conservation of the sequence in the region delineated by the deletion analysis of murine P-450 c21 gene provides further evidence for an important regulatory element in the region from −230 to −180 bp preceding the transcription start site.

Mouse	P450c21	-210	AGGTCAGGGT	CTTGCATCCC	TTCCTTTCTT	CTTGATGGAT	G
Human	P450c21	-241	••••••••••	TGCATT••••	••••••G•••	••••••••G•	•
Bovine	P450c21	-251	••••••••••	TG•TT••••T	••••••••••	•••••A••••	•

Fig. 10. Homologous regions of three P-450 c21 genes. Identical bases are indicated by dots. The positions for the first base pairs of these conserved sequences (−210, −241 and −251 for the murine, human and bovine sequences, respectively) relative to the transcription initiation site are shown (from PARKER et al., 1986).

The nature of this regulatory element has not yet been determined. It is also not known how this element functions in the expression of the P-450 c21 gene or what *trans*-acting factor interacts with this element in the regulatory process. Recent studies have proposed that the consensus sequence, found in the promoter region of several genes that are regulated by cAMP, plays an important role in conferring cAMP response (COMB et al., 1986; SHORT et al., 1986). This consensus does not appear to be included in the conserved sequence as described above. It will be of interest to know whether the conserved sequence of the P-450 c21 gene plays a role in determining an adrenal-specific expression or has something to do with the ACTH-regulatory mechanism.

Structural analysis and transfection experiments of the genes for other steroidogenic P-450s, such as P-450 scc, P-$450_{11\beta}$ and P-$450_{17\alpha}$ in adrenal cortex will help us to elucidate the mechanism of the ACTH-specific and tissue-specific expression of the P-450s.

Acknowledgment

The authors would like to express their sincere thanks to Drs. T. IIZUKA, K. SOGAWA, A. FUJISAWA-SEHARA, T. IYANAGI, K. KAWAJIRI, and Y. TAGASHIRA for their valuable advice and suggestions. The work performed in our laboratories was supported in part by Grants-in-Aid for Scientific Research from the Ministry of Education, Science and Culture of Japan, research grants from the Ministry of Health and Welfare of Japan, and funds obtained under the Life Science Project from the Institute of Physical and Chemical Research, Japan.

7. References

ADESNIK, M. and M. ATCHISON, (1986), CRC Crit. Rev. Biochem. **19**, 247–305.
AFFOLTER, M. and A. ANDERSON, (1984), Biochem. Biophys. Res. Commun. **118**, 655 to 662.
AFFOLTER, M., D. LABBE, A. JEAN, M. RAYMOND, D. NOËL, Y. LABELLE, C. PARENT-VAUGEOIS, M. LAMBERT, R. BOJANOWSKI, and A. ANDERSON, (1986), DNA **5**, 209–218.
AKHREM, A. A., V. I. VASILEVSKY, T. B. ADAMOVICH, A. G. LAPKO, V. M. SHKUMATOV, and V. L. CHASHCHIN, (1980), Dev. Biochem. **13**, 57–64.
ANDERSON, S., A. T. BANKIER, B. G. BARRELL, M. H. L. de BRUIJN, A. R. COULSON, J. DROUIN, I. C. EPERON, D. P. NIERLICH, B. A. ROE, F. SANGER, P. H. SCHREIER, A. J. H. SMITH, R. STADEN, and I. G. YOUNG, (1981), Nature **290**, 457–465.
ATCHISON, M. and M. ADESNIK, (1983), J. Biol. Chem. **258**, 11285–11295.
ATCHISON, M. and M. ADESNIK, (1986), Proc. Natl. Acad. Sci. U. S. A. **83**, 2300–2304.
BEAUNE, P. H., D. R. UMBENHAUER, R. W. BORK, R. S. LLOYD, and F. P. GUENGERICH, (1986), Proc. Natl. Acad. Sci. U. S. A. **83**, 8064–8068.
BERNHARDT, R., A. MAKOWER, G.-R. JÄNIG, and K. RUCKPAUL, (1984), Biochim. Biophys. Acta **785**, 186–190.
BIBB, M. J., R. A. VAN ETTEN, C. T. WRIGHT, M. W. WALBERG, and D. A. CLAYTON, (1981), Cell **26**, 167–180.
BLACK, S. D. and M. J. COON, (1982), J. Biol. Chem. **257**, 5929–5938.
BLACK, S. D. and M. J. COON, (1986), in: Cytochrome P-450. Structure, Mechanism, and Biochemistry, (ORTIZ de MONTELLANO, P. R., ed.), Plenum, New York, 161–216.
BLACK, S. D., G. E. TARR, and M. J. COON, (1982), J. Biol. Chem. **257**, 14616–14619.
BONITZ, S. G., G. CORUZZI, B. E. THALENFELD, A. TZAGOLOFF, and G. MACINO, (1980), J. Biol. Chem. **255**, 11927–11941.
BOTELHO, L. H., D. E. RYAN, and W. LEVIN, (1979), J. Biol. Chem. **254**, 5635–5640.
CAVALIER-SMITH, T., (1985), Nature **315**, 283–284.
CHAPLIN, D. D., L. J. GALBRAITH, J. G. SEIDMAN, P. C. WHITE, and K. L. PARKER, (1986), Proc. Natl. Acad. Sci. U. S. A. **83**, 9601–9605.
CHASHCHIN, V. L., V. N. LAPKO, T. B. ADAMOVICH, A. G. LAPKO, N. S. KUPRINA, and A. A. AKHREM, (1986), Biochim. Biophys. Acta **871**, 217–223.
CHUNG, B.-C., K. J. MATTESON, and W. L. MILLER, (1986a), Proc. Natl. Acad. Sci. U.S.A. **83**, 4243–4247.
CHUNG, B.-C., K. J. MATTESON, R. VOUTILAINEN, T. K. MOHANDAS, and W. L. MILLER, (1986b), Proc. Natl. Acad. Sci. U. S. A. **83**, 8962–8966.
CHUNG, B.-C., J. PICADO-LEONARD, M. HANIU, M. BIENKOWSKI, P. F. HALL, J. E. SHIVELY, and W. L. MILLER, (1987), Proc. Natl. Acad. Sci. U. S. A. **84**, 407–411.
COMB, M., N. C. BIRNBERG, A. SEASHOLZ, E. HERBERT, and H. M. GOODMAN, (1986), Nature **323**, 353–356.
CONNEY, A. H., (1982), Cancer Res. **42**, 4875–4917.
COOPER, D. Y., S. LEVINE, S. NARASIMHULU, O. ROSENTHAL, and R, W. ESTABROOK, (1965), Science, 400–402.
CRAMER, H., (1948), Mathematical Methods of Statistics, Princeton, New Jersey.
CUTHILL, S., L. POELLINGER, and J.-Å. GUSTAFSSON, (1987), J. Biol. Chem. **262**, 3477 to 3481.
DAYHOFF, M. O., ed., (1978), Atlas of Protein Sequence and Structure, Vol. 5, suppl. 3, National Biomedical Research Foundation, Washington.

De Bruijn, M. H. L., (1983), Nature **304**, 234–241.
DeFranco, D. and K. R. Yamamoto, (1986), Mol. Cell. Biol. **6**, 993–1001.
De Lemos-Chiarandini, A. B. Frey, D. D. Sabatini, and G. Kreibich, (1987), J. Cell Biol. **104**, 209–219.
Dickerson, R. E., (1971), J. Mol. Evol. **1**, 26–45.
Dodgson, J. B. and J. D. Engel, (1983), J. Biol. Chem. **258**, 4623–4629.
Doolittle, R. F., (1981), Science **214**, 149–159.
DuBois, R. N., E. R. Simpson, R. E. Kramer, and M. R. Waterman, (1981), J. Biol. Chem. **256**, 7000–7005.
Durica, D. S., J. A. Schloss, and W. R. Crain, Jr., (1980), Proc. Natl. Acad. Sci. U. S. A. **77**, 5683–5687.
Dus, K., (1980), Dev. Biochem. **13**, 129–132.
Dus, K., W. J. Litchfield, and A. G. Miguel, (1974), Biochem. Biophys. Res. Commun. **60**, 15–21.
Eisen, H. J., (1986), in: Cytochrome P-450. Structure, Mechanism, and Biochemistry (Ortiz de Montellano, P. R. ed.), Plenum, New York, 315–344.
Estabrook, R. W., D. Y. Cooper, and O. Rosenthal, Biochem. Z. **338**, 741–755.
Evans, C. T., D. B. Ledesma, T. Z. Schulz, E. R. Simpson, and C. R. Mendelson, (1986), Proc. Natl. Acad. Sci. U. S. A. **83**, 6387–6391.
Firtel, R., R. Timm, A. R. Kimmel, and M. McKeown, (1979), Proc. Natl. Acad. Sci. U. S. A. **76**, 6206–6210.
Frey, A. B., G. Kreibich, A. Wadhera, L. Clarke, and D. J. Waxman, (1986), Biochemistry **25**, 4797–4803.
Friedberg, T., D. J. Waxman, M. Atchison, A. Kumar, T. Haaparanta, C. Raphael, and M. Adesnik, (1986), Biochemistry **25**, 7975–7983.
Fujii-Kuriyama, Y., T. Taniguchi, Y. Mizukami, M. Sakai, Y. Tashiro, and M. Muramatsu, (1981), J. Biochem. (Tokyo) **89**, 1869–1879.
Fujii-Kuriyama, Y., Y. Mizukami, K. Kawajiri, K. Sogawa, and M. Muramatsu, (1982a), Proc. Natl. Acad. Sci. U. S. A. **79**, 2793–2797.
Fujii-Kuriyama, Y., Y. Mizukami, T. Taniguchi, and M. Muramatsu, (1982b), in: Microsomes, Drug Oxidation and Drug Toxicity, (Sato, R. and Kato, R., eds.), Japan Scientific Societies Press, Tokyo 589–596.
Fujisawa-Sehara, A., K. Sogawa, C. Nishi, and Y. Fujii-Kuriyama, (1986), Nucleic Acids Res. **14**, 1465–1477.
Fujisawa-Sehara, A., K. Sogawa, M. Yamane, and Y. Fujii-Kuriyama, (1987), Nucleic Acids Res., **15**, 4179–4191.
Fujita, V. S., S. D. Black, G. E. Tarr, D. R. Koop, and M. J. Coon, (1984), Proc. Natl. Acad. Sci. U. S. A. **81**, 4260–4264.
Funkenstein, B., J. L. McCarthy, K. M. Dus, E. R. Simpson, and M. R. Waterman, (1983), J. Biol. Chem. **258**, 9398–9405.
Fyrberg, E. A., B. J. Bond, N. D. Herskey, K. S. Mixter, and N. Davidson, (1981), Cell **24**, 107–116.
Gallwitz, D. and I. Sures, (1980), Proc. Natl. Acad. Sci. U. S. A. **77**, 2546–2550.
Garnier, J., D. J. Osguthorpe, and B. Robson, (1978), J. Mol. Biol. **120**, 97–120.
Gibson, G. G. and P. P. Tamburini, (1986), Chem.-Biol. Interact. **58**, 185–198.
Gonzalez, F. J. and D. W. Nebert, (1985), Nucleic Acids Res. **13**, 7269–7288.
Gonzalez, F. J., S. Kimura, and D. W. Nebert, (1985a), J. Biol. Chem. **260**, 5040 to 5049.
Gonzalez, F. J., D. W. Nebert, J. P. Hardwick, and C. B. Kasper, (1985b), J. Biol. Chem. **260**, 7435–7441.
Gonzalez, F. J., S. Kimura, B.-J. Song, J. Pastewka, H. V. Gelboin, and J. P. Hardwick, (1986a), J. Biol. Chem. **261**, 10667–10672.
Gonzalez, F. J., B.-J. Song, and J. P. Hardwick, (1986b), Mol. Cell. Biol. **6**, 2969 to 2976.

GOTOH, O., (1982), J. Mol. Biol. **162**, 705–708.
GOTOH, O., (1986), J. Theor. Biol. **121**, 327–337.
GOTOH, O., (1987), CABIOS **3**, 17–20.
GOTOH, O. and Y. TAGASHIRA, (1986), Nucleic Acids Res. **14**, 57–64.
GOTOH, O., Y. TAGASHIRA, T. IIZUKA, and Y. FUJII-KURIYAMA, (1983), J. Biochem. (Tokyo) **93**, 807–817.
GOTOH, O., Y. TAGASHIRA, K. MOROHASHI, and Y. FUJII-KURIYAMA, (1985), FEBS Letters **188**, 8–10.
GOVIND, S., P. A. BELL, and B. KEMPER, (1986), DNA **5**, 371–382.
GUENGERICH, F. P., (1979), Pharmacol. Ther. **6**, 99–121.
GUNSALUS, I. C. and S. G. SLIGAR, (1977), Adv. Enzymol. **47**, 1–44.
GUT, J., C. RICHTER, R. J. CHERRY, K. H. WINTERHALTER, and S. KAWATO, (1983), J. Biol. Chem. **258**, 8588–8594.
HAHN, J. E., K. O. HODGSON, L. A. ANDERSSON, and J. H. DAWSON, (1982), J. Biol. Chem. **257**, 10934–10941.
HANIU, M., L. G. ARMES, M. TANAKA, T. K. YASUNOBU, R. S. SHASTRY, G. C. WAGNER, and I. C. GUNSALUS, (1982a), Biochem. Biophys. Res. Commun. **105**, 889–894.
HANIU, M., L. G. ARMES, K. T. YASUNOBU, B. A. SHASTRY, and I. C. GUNSALUS, (1982b), J. Biol. Chem. **257**, 12664–12671.
HANIU, M., K. T. YASUNOBU, and I. C. GUNSALUS, (1982c), Biochem. Biophys. Res. Commun. **107**, 1075–1081.
HANIU, M., K. T. YASUNOBU, and I. C. GUNSALUS, (1983), Biochem. Biophys. Res. Commun. **116**, 30–38.
HANIU, M., P.-M. YUAN, D. E. RYAN, W. LEVIN, and J. E. SHIVELY, (1984), Biochemistry **23**, 2478–2482.
HANIU, M., D. E. RYAN, W. LEVIN, and J. E. SHIVELY, (1986), Arch. Biochem. Biophys. **244**, 323–337.
HARADA, N. and M. NEGISHI, (1985), Proc. Natl. Acad. Sci. U. S. A. **82**, 2024–2028.
HARDWICK, J. P., F. J. GONZALEZ, and C. B. KASPER, (1983), J. Biol. Chem. **258**, 10182–10186.
HARDWICK, J. P., B.-J. SONG, E. HUBERMAN, and F. J. GONZALEZ, (1987), J. Biol. Chem. **262**, 801–810.
HEINEMANN, F. S. and J. OZOLS, (1982), J. Biol. Chem. **257**, 14988–14999.
HEINEMANN, F. S. and J. OZOLS, (1983), J. Biol. Chem. **258**, 4195–4201.
HIGASHI, Y., H. YOSHIOKA, M. YAMANE, O. GOTOH, and Y. FUJII-KURIYAMA, (1986), Proc. Natl. Acad. Sci. U. S. A. **83**, 2841–2845.
HINES, R. N., J. B. LEVY, R. D. CONRAD, P. L. IVERSEN, M.-L. SHEN, A. M. RENLI, and E. BRESNICK, (1985), Arch. Biochem. Biophys. **237**, 465–476.
HOBBS, A. A., L. A. MATTSCHOSS, B. K. MAY, K. E. WILLIAMS, and W. H. ELLIOTT, (1986), J. Biol. Chem. **261**, 9444–9449.
IMAI, Y., (1987), J. Biochem. (Tokyo), **101**, 1129–1139.
JAISWAL, A. K., F. J. GONZALEZ, and D. W. NEBERT, (1985a), Science **228**, 80–83.
JAISWAL, A. K., F. J. GONZALEZ, and D. W. NEBERT, (1985b), Nucleic Acids Res. **13**, 4503–4520.
JAISWAL, A. K., D. W. NEBERT, and F. J. GONZALEZ, (1986), Nucleic Acids Res. **14**, 6773–6774.
JOHN, M. E., T. OKAMURA, A. DEE, B. ADLER, M. C. JOHN, P. C. WHITE, E. R. SIMPSON, and M. R. WATERMAN, (1986), Biochemistry **25**, 2846–2853.
JOHNSON, E. F., G. E. SCHWAB, J. SINGH, and L. E. VICKERY, (1986), J. Biol. Chem. **261**, 10204–10209.
JONES, P. B. C., D. R. GALEAZZI, J. M. FISHER, and J. P. WHITLOCK, Jr., (1985), Science **227**, 1499–1502.
JONES, P. B. C., L. K. DURRIN, J. M. FISHER, and J. P. WHITLOCK, Jr., (1986a), J. Biol. Chem. **261**, 6647–6650.

Jones, P. B. C., L. K. Durrin, D. R. Galeazzi, and J. P. Whitlock, Jr., (1986b), Proc. Natl. Acad. Sci. U. S. A. **83**, 2802–2806.
Kabsch, W. and C. Sander, (1983), FEBS Letters **155**, 179–182.
Käppeli, O., (1986), Microbiol. Rev. **50**, 244–258.
Karin, M., A. Haslinger, H. Holtgreve, R. I. Richards, P. Krauter, H. M. Westphal, and M. Beato, (1984), Nature **308**, 513–519.
Karn, J., S. Brenner, and L. Barnett, (1983), Proc. Natl. Acad. Sci. U. S. A. **80**, 4253–4257.
Kawajiri, K., K. Sogawa, O. Gotoh, Y. Tagashira, M. Muramatsu, and Y. Fujii-Kuriyama, (1983), J. Biochem. (Tokyo) **94**, 1465–1473.
Kawajiri, K., O. Gotoh, K. Sogawa, Y. Tagashira, M. Muramatsu, and Y. Fujii-Kuriyama, (1984a), Proc. Natl. Acad. Sci. U. S. A. **81**, 1649–1653.
Kawajiri, K., O. Gotoh, Y. Tagashira, K. Sogawa, and Y. Fujii-Kuriyama, (1984b), J. Biol. Chem. **259**, 10145–10149.
Kawajiri, K., J. Watanabe, O. Gotoh, Y. Tagashira, K. Sogawa, and Y. Fujii-Kuriyama, (1986), Eur. J. Biochem. **159**, 219–225.
Khani, S. C., P. G. Zaphiropoulos, V. S. Fujita, T. D. Porter, D. R. Koop, and M. J. Coon, (1987), Proc. Natl. Acad. U. S. A. **84**, 638–642.
Kimura, M., (1969), Proc. Natl. Acad. Sci. U. S. A. **63**, 1181–1188.
Kimura, S., F. J. Gonzalez, and D. W. Nebert, (1984a), Nucleic Acids Res. **12**, 2917–2928.
Kimura, S., F. J. Gonzalez, and D. W. Nebert, (1984b), J. Biol. Chem. **259**, 10705 to 10713.
Kimura, S. and D. W. Nebert, (1986), Nucleic Acids Res. **14**, 6765–6766.
King, C. R. and J. Piatigorsky, (1983), Cell **32**, 707–712.
Kominami, S., Y. Itoh, and S. Takemori, (1986), J. Biol. Chem. **261**, 2077–2083.
Kramer, R. E., E. R. Simpson, and M. R. Waterman, (1983), J. Biol. Chem. **258**, 3000–3005.
Kumar, A., C. Raphael, and M. Adesnik, (1983), J. Biol. Chem. **258**, 11280–11284.
Kyte, J. and R. F. Doolittle, (1982), J. Mol. Biol. **157**, 105–132.
Leicht, M., G. L. Long, T. Chandra, K. Kurachi, V. J. Kidd, M. Mace, Jr., E. W. Davie, and S. L. C. Woo, (1982), Nature **297**, 655–659.
Leighton, J. K., B. A. DeBrunner-Vossbrinck, and B. Kemper, (1984), Biochemistry **23**, 204–210.
Levanon, D., J. Lieman-Hurwitz, N. Dafni, M. Wigderson, L. Sherman, Y. Bernstein, Z. Laver-Rudich, E. Danciger, O. Stein, and Y. Groner, (1985), EMBO J. **4**, 77–84.
Lu, A. Y. H. and S. B. West, (1980), Pharmacol. Rev. **31**, 277–295.
Masters, B. S. S. and R. T. Okita, (1980), Pharmacol. Ther. **9**, 227–244.
Matsubara, S., S. Yamamoto, K. Sogawa, N. Yokotani, Y. Fujii-Kuriyama, M. Haniu, J. E. Shively, O. Gotoh, E. Kusunose, and M. Kusunose, (1987), J. Biol. Chem. **262**, 13366–13371.
Matsuura, S., Y. Fujii-Kuriyama, and Y. Tashiro, (1979), J. Cell. Biol. **78**, 503–519.
Matsuura, S., R. Masuda, K. Omori, M. Negishi, and Y. Tashiro, (1981), J. Cell. Biol. **91**, 212–220.
Matsuura, S., R. Masuda, O. Sakai, and Y. Tashiro, (1983), Cell. Struct. Func. **8**, 1–9.
Matteson, K. J., J. Picado-Leonard, B. Chung, T. K. Mohandas, and W. L. Miller, (1986), J. Clin. Endocrinol. Metab. **63**, 789–791.
McCarthy, J. L., R. E. Kramer, B. Funkenstein, E. R. Simpson, and M. R. Waterman, (1983), Arch. Biochem. Biophys. **222**, 590–598.
McLaughlin, P. J. and M. O. Dayhoff, (1972), in: Atlas of Protein Sequence and Structure, Vol. 5, (Dayhoff, M. O., ed.), National Biomedical Research Foundation, Washington, 47–52.
Mihara, K., N. Kagawa, and R. Sato, (1985), Seikagaku **57**, 1115.

MIKSICEK, R., A. HEBER, W. SCHMID, U. DANESCH, G. POSSECKERT, M. BEATO, and G. SCHÜTZ, (1986), Cell. **46**, 283–290.

MIYATA, T., H. HAYASHIDA, R. KIKUNO, M. HASEGAWA, M. KOBAYASHI, and K. KOIKE, (1982), J. Mol. Evol. **19**, 28–35.

MIZUKAMI, Y., Y. FUJII-KURIYAMA, and M. MURAMATSU, (1983a), Biochemistry **22**, 1223–1229.

MIZUKAMI, Y., K. SOGAWA, Y. SUWA, M. MURAMATSU, and K. FUJII-KURIYAMA, (1983b), Proc. Natl. Acad. Sci. U. S. A. **80**, 3958–3962.

MOLOWA, D. T., E. G. SCHUETZ, S. A. WRIGHTON, P. B. WATKINS, P. KREMERS, G. MENDEZ-PICON, G. A. PARKER, and P. S. GUZELIAN, (1986), Proc. Natl. Acad. Sci. U. S. A. **83**, 5311–5315.

MOROHASHI, K., Y. FUJII-KURIYAMA, Y. OKADA, K. SOGAWA, T. HIROSE, S. INAYAMA, and T. OMURA, (1984), Proc. Natl. Acad. Sci. U. S. A. **81**, 4647–4651.

MOROHASHI, K., K. SOGAWA, T. OMURA, and Y. FUJII-KURIYAMA, (1987a), J. Biochem. (Tokyo) **101**, 879–887.

MOROHASHI, K., H. YOSHIOKA, O. GOTOH, Y. OKADA, K. YAMAMOTO, T. MIYATA, K. SOGAWA, Y. FUJII-KURIYAMA, and T. OMURA, (1987b), J. Biochem. (Tokyo), **102**, 559–568.

MÜLLER, R., W. E. SCHMIDT, and A. STIER, (1985), FEBS Letters **187**, 21–24.

NAGATA, K., T. MATSUNAGA, J. GILLETTE, H. V. GELBOIN, and F. J. GONZALEZ, (1987), J. Biol. Chem. **262**, 2787–2793.

NAKAJIN, S., M. SHINODA, M. HANIU, J. E. SHIVELY, and P. F. HALL, (1984), J. Biol. Chem. **259**, 3971–3976.

NEBERT, D. W., (1979), Mol. Cell. Biochem. **27**, 27–46.

NEBERT, D. W., M. ADESNIK, M. J. COON, R. W. ESTABROOK, F. J. GONZALEZ, F. P. GUENGERICH, I. C. GUNSALUS, E. F. JOHNSON, B. KEMPER, W. LEVIN, I. R. PHILLIPS, R. SATO, and M. R. WATERMAN, (1987), DNA **6**, 1–12.

NEBERT, D. W. and F. J. GONZALEZ, (1987), Ann. Rev. Biochem. **56**, 945–993.

NEBERT, D. W., F. M. GOUJON, and J. E. GIELEN, (1972), Nature New Biol. **236**, 107 to 110.

NEBERT, D. W. and M. NEGISHI, (1982), Biochem. Pharmacol. **31**, 2311–2317.

NEEDLEMAN, S. B. and C. D. WUNSCH, (1970), J. Mol. Biol. **48**, 443–453.

NG, R. and J. ABELSON, (1980), Proc. Natl. Acad. Sci. U. S. A. **77**, 3912–3916.

NISHIKAWA, K., (1983), Biochim. Biophys. Acta **748**, 285–299.

OKAYAMA, H. and P. BERG, (1982), Mol. Cell. Biol. **2**, 161–170.

OKEY, A. B., G. P. BONDY, M. E. MASON, D. W. NEBERT, C. J. FORSTER-GIBSON, J. MUNCAN, and M. J. DUFRESNE, (1980), J. Biol. Chem. **255**, 11415–11422.

OKINO, S. T., L. C. QUATTROCHI, H. J. BARNES, S. OSANTO, K. J. GRIFFIN, E. F. JOHNSON, and R. H. TUKEY, (1985), Proc. Natl. Acad. Sci. U. S. A. **82**, 5310–5314.

OMURA, T. and R. SATO, (1962), J. Biol. Chem. **237**, 1375–1376.

ONODA, M., M. HANIU, K. YANAGIBASHI, F. SWEET, J. E. SHIVELY, and P. F. HALL, (1987), Biochemistry **26**, 657–662.

OZOLS, J., (1986), J. Biol. Chem. **261**, 3965–3979.

OZOLS, J., F. S. HEINEMANN, and E. F. JOHNSON, (1981), J. Biol. Chem. **256**, 11405 to 11408.

OZOLS, J., F. S. HEINEMANN, and E. F. JOHNSON, (1985), J. Biol. Chem. **260**, 5427–5434.

PARKER, K. L., D. D. CHAPLIN, M. WONG, J. G. SEIDMAN, J. A. SMITH, and B. P. SCHIMMER, (1985), Proc. Natl. Acad. Sci. U. S. A. **82**, 7860–7864.

PARKER, K. L., B. P. SCHIMMER, D. D. CHAPLIN, and J. G. SEIDMAN, (1986), J. Biol. Chem. **261**, 15353–15355.

PARKINSON, A., D. E. RYAN, P. E. THOMAS, D. M. JERINA, J. M. SAYER, P. J. van BLADEREN, M. HANIU, J. E. SHIVELY, and W. LEVIN, (1986a), J. Biol. Chem. **261**, 11478–11486.

PARKINSON, A., P. E. THOMAS, D. E. RYAN, L. D. GORSKY, J. E. SHIVELY, J. M. SAYER, D. M. JERINA, and W. LEVIN, (1986b), J. Biol. Chem. **261**, 11487–11495.

PAYVAR, F., D. DEFRANCO, G. L. FIRESTONE, B. EDGAR, Ö. WRANGE, S. OKRET, J.-Å. GUSTAFSSON, and K. R. YAMAMOTO, (1983), Cell. **35**, 381–392.
PHILLIPS, I. R., E. A. SHEPHARD, A. ASHWORTH, and B. R. RABIN, (1983), Gene **26**, 41–52.
PHILLIPS, I. R., E. A. SHEPHARD, A. ASHWORTH, and B. R. RABIN, (1985), Proc. Natl. Acad. Sci. U. S. A. **82**, 983–987.
POLAND, A. and E. GLOVER, (1973), Mol. Pharmacol. **9**, 736–747.
POLAND, A. P., E. GLOVER, J. R. ROBINSON, and D. W. NEBERT, (1974), J. Biol. Chem. **249**, 5599–5606.
POLAND, A., E. GLOVER, F. H. EBETINO, and A. S. KENDE, (1986), J. Biol. Chem. **261**, 6352–6365.
PORTER, T. D., T. W. BECK, and C. B. KASPER, (1986), Arch. Biochem. Biophys. **248**, 121–129.
POULOS, T. L., (1984), Fed. Proc. **43**, 1671.
POULOS, T. L., (1986), in: Cytochrome P-450. Structure, Mechanism, and Biochemistry, (ORTIZ de MONTELLANO, P. R., ed.), Plenum, New York, 505–523.
POULOS, T. L. and J. KRAUT, (1980), J. Biol. Chem. **255**, 10322–10330.
POULOS, T. L., B. C. FINZEL, I. C. GUNSALUS, G. C. WAGNER, and J. KRAUT, (1985), J. Biol. Chem. **260**, 16122–16130.
POULOS, T. L., B. C. FINZEL, and A. J. HOWARD, (1986), Biochemistry **25**, 5314–5322.
PYERIN, W., M. MARX, and H. TANIGUCHI, (1986), Biochem. Biophys. Res. Commun. **134**, 461–468.
QUATTROCHI, L. C., U. R. PENDURTHI, S. T. OKINO, C. POTENZA, and R. H. TUKEY, (1986), Proc. Natl. Acad. Sci. U. S. A. **83**, 6731–6735.
RENKAWITZ, R., G. SCHÜTZ, D. von del AHE, and M. BEATO, (1984), Cell. **37**, 503–510.
ROGERS, J., (1985), Nature **315**, 458–459.
ROISE, D., S. J. HORVATH, J. M. TOMICH, J. H. RICHARDS, and G. SCHATZ, (1986), EMBO J. **5**, 1327–1334.
SABATINI, D. D., G. KREIBICH, T. MORIMOTO, and M. ADESNIK, (1982), J. Cell. Biol. **92**, 1–22.
SAKAGUCHI, M., K. MIRARA, and R. SATO, (1984), Proc. Natl. Acad. Sci. U. S. A. **81**, 3361–3364.
SCHEIDEREIT, C. and M. BEATO, (1984), Proc. Natl. Acad. Sci. U. S. A. **81**, 3029–3033.
SHIBAHARA, S., T. KUBO, H. J. PERSKI, H. TAKAHASHI, M. NODA, and S. NUMA, (1985), Eur. J. Biochem. **146**, 15–22.
SHIFFER, M. and A. B. EDMUNDSON, (1967), Biophys. J. **7**, 121–135.
SHORT, J. M., A. WYNSHAW-BORIS, H. P. SHORT, and R. W. HANSON, (1986), J. Biol. Chem. **261**, 9721–9726.
SIMMONS, D. L. and C. B. KASPER, (1983), J. Biol. Chem. **258**, 9585–9588.
SIMMONS, D. L., P. A. LALLEY, and C. B. KASPER, (1984), J. Biol. Chem. **260**, 515–521.
SLIGAR, S. G. and R. I. MURRAY, (1986), in: Cytochrome P-450. Structure, Mechanism, and Biochemistry, (ORTIZ de MONTELLANO, P. R., ed.), Plenum, New York, 429–503.
SOGAWA, K., O. GOTOH, K. KAWAJIRI, and Y. FUJII-KURIYAMA, (1984), Proc. Natl. Acad. Sci. U. S. A. **81**, 5066–5070.
SOGAWA, K., O. GOTOH, K. KAWAJIRI, T. HARADA, and Y. FUJII-KURIYAMA (1985), J. Biol. Chem. **260**, 5026–5032.
SOGAWA, K., A. FUJISAWA-SEHARA, M. YAMANE, and Y. FUJII-KURIYAMA, (1986), Proc. Natl. Acad. Sci. U. S. A. **83**, 8044–8048.
SOKAL, R. R. and P. H. A. SNEATH, (1963), Principles of Numerical Taxonomy, W. H. FREEMAN and Co., San Francisco.
SONG, B.-J., H. V. GELBOIN, S.-S. PARK, C. S. YANG, and F. J. GONZALEZ, (1986), J. Biol. Chem. **261**, 16689–16697.
SUWA, Y., Y. MIZUKAMI, K. SOGAWA, and Y. FUJII-KURIYAMA, (1985), J. Biol. Chem. **260**, 7980–7984.
TANAKA, T., H. OHKUBO, and S. NAKANISHI, (1984), J. Biol. Chem. **259**, 8063–8065.

TANIGUCHI, H., W. PYERIN, and A. STIER, (1985), Biochem. Pharmacol. **34**, 1835—1837.
TARR, G. E., S. D. BLACK, V. S. FUJITA, and M. J. COON, (1983), Proc. Natl. Acad. Sci. U. S. A. **80**, 6552—6556.
THOMAS, P. E., R. E. KOURI, and J. J. HUTTON, (1972), Biochem. Genet. **6**, 157—168.
TUKEY, R. H., P. A. LALLEY, and D. W. NEBERT, (1984), Proc. Natl. Acad. Sci. U. S. A. **81**, 3163—3166.
TUKEY, R. H., S. OKINO, H. BARNES, K. J. GRIFFIN, and E. F. JOHNSON, (1985), J. Biol. Chem. **260**, 13347—13354.
UNGER, B. P., I. C. GUNSALUS, and S. G. SLIGAR, (1986), J. Biol. Chem. **261**, 1158—1163.
von HEIJNE, G., (1986), EMBO J. **5**, 1335—1432.
von DER AHE, D., J.-M. RENOIR, T. BUCHOU, E.-E. BAULIEU, and M. BEATO, (1986), Proc. Natl. Acad. Sci. U. S. A. **83**, 2817—2821.
WATERMAN, M. R., M. E. JOHN, and E. R. SIMPSON, (1986), in: Cytochrome P-450. Structure, Mechanism and Biochemistry, (ORTIZ de MONTELLANO, P. R., ed.), Plenum, New York, 345—386.
WHITE, R. E. and M. J. COON, (1982), J. Biol. Chem. **257**, 3073—3083.
WHITE, P. C., D. D. CHAPLIN, J. H. WEIS, B. DUPONT, M. I. NEW, and J. G. SEIDMAN, (1984), Nature **312**, 465—467.
WHITE, P. C., M. I. NEW, and B. DUPONT, (1986), Proc. Natl. Acad. Sci. U. S. A. **83**, 5111—5115.
WHITLOCK, J. P., Jr., (1986), Annu. Rev. Pharmacol. Toxicol. **26**, 333—369.
WHITLOCK, J. P., Jr. and D. R. GALEAZZI, (1984), J. Biol. Chem. **259**, 980—985.
WILHELMSSON, A., A.-C. WIKSTROM, and L. POELLINGER, (1986), J. Biol. Chem. **261**, 13456—13463.
WOLF, C. R., (1986), Trends in Genet. **2**, 209—214.
YABUSAKI, Y., M. SHIMIZU, H. MURAKAMI, K. NAKAMURA, K. OEDA, and H. OHKAWA, (1984), Nucleic Acids Res. **12**, 2929—2938.
YOSHIOKA, H., K. MOROHASHI, K. SOGAWA, T. MIYATA, K. KAWAJIRI, T. HIROSE, S. INAYAMA, Y. FUJII-KURIYAMA, and T. OMURA, (1987), J. Biol. Chem. **262**, 1706 to 1711.
YOSHIOKA, H., K. MOROHASHI, K. SOGAWA, M. YAMANE, S. KOMINAMI, S. TAKEMORI, Y. OKADA, T. OMURA, and FUJII-KURIYAMA, Y., (1986), J. Biol. Chem. **261**, 4106 to 4109.
YUAN, P.-M., D. E. RYAN, W. LEVIN, and J. E. SHIVELY, (1983), Proc. Natl. Acad. Sci. U. S. A. **80**, 1169—1173.
ZUBER, M. X., M. E. JOHN, T. OKAMURA, E. R. SIMPSON, and M. R. WATERMAN, (1986), J. Biol. Chem. **261**, 2475—2482.

List of Authors

A. I. ARCHAKOV
Department of Biochemistry
Medico-Biological Faculty
Ostrovitjanova la

Moscow 117437
USSR

P. BATTIONI
Université René Descartes
Laboratoire de Chimie et Biochimie
45 rue des Saints Pères

F-75270 Paris Cedex 06
France

J. BLANCK
Central Institute of Molecular Biology
Academy of Sciences of the GDR
Robert-Rössle-Straße 10

1115 Berlin
DDR

Y. FUJII-KURIYAMA
Department of Biochemistry
Tohoku University

Sendai 980
Japan

O. GOTOH
Saitama Cancer Center
Research Institute
Ina-machi

Saitama 362
Japan

F. P. GUENGERICH
Vanderbilt University
Department of Biochemistry
School of Medicine

Nashville Tennessee 37232
USA

J.-Å. Gustafsson
Department of Medical Nutrition
Karolinska Institute
Huddinge University Hospital

S-141 86 Huddinge
Sweden

D. Mansuy
Université René Descartes
Laboratoire de Chimie et Biochimie
45 rue des Saints Pères

F-75270 Paris Cedex 06
France

E. T. Morgan
Emory University
School of Medicine
Department of Pharmacology

Atlanta Georgia 30322
USA

H. Rein
Central Institute of Molecular Biology
Academy of Sciences of the GDR
Robert-Rössle-Straße 10

1115 Berlin
DDR

K. Ruckpaul
Central Institute of Molecular Biology
Academy of Sciences of the GDR
Robert-Rössle-Straße 10

1115 Berlin
DDR

A. A. Zhukov
Department of Biochemistry
Medico-Biological Faculty
Ostrovitjanova la

Moscow 117437
USSR

Subject Index

Bei Fragen zur Produktsicherheit wenden Sie sich bitte an:
If you have any questions regarding product safety,
please contact:

Walter de Gruyter GmbH
Genthiner Straße 13
10785 Berlin
productsafety@degruyterbrill.com